Struggles for Equal Voice

Struggles for Equal Voice

The History of African American Media Democracy

YUYA KIUCHI

Photo credit: © Rogério Bernardo | Dreamstime.com

Published by State University of New York Press, Albany

© 2012 State University of New York

All rights reserved

Printed in the United States of America

No part of this book may be used or reproduced in any manner whatsoever without written permission. No part of this book may be stored in a retrieval system or transmitted in any form or by any means including electronic, electrostatic, magnetic tape, mechanical, photocopying, recording, or otherwise without the prior permission in writing of the publisher.

For information, contact State University of New York Press, Albany, NY
www.sunypress.edu

Production by Diane Ganeles
Marketing by Kate McDonnell

Library of Congress Cataloging-in-Publication Data

Kiuchi, Yuya.
  Struggles for equal voice : the history of African American media democracy / Yuya Kiuchi.
      p. cm.
  Includes bibliographical references and index.
  ISBN 978-1-4384-4479-6 (hardcover : alk. paper)
  1. African Americans in television broadcasting—History. 2. African Americans on television—History. 3. Television broadcasting—Social aspects—United States. 4. Cable television—Social aspects—United States. I. Title.

PN1992.8.A34K58 2012
791.45'652996—dc23   2012002855

10 9 8 7 6 5 4 3 2 1

To Nichole and Noriko

and

In memory of Tsuneo

Contents

Acknowledgments	xi
Preface	xv
Introduction: Unveiling the Struggles for Equal Voice	1
Chapter 1: The Black Image in the White Pathology	17
African Americans and the Film Industry	18
African Americans and the Television Industry	24
African Americans in Mainstream Visual Culture	32
Chapter 2: Cable Television: Past and Present	35
Prelude: Early Days of Cable Television, 1948–1969	37
Legal, Economic, Professional, and Technological Concerns, 1960–1979	38
Legal Literature and Public and Community Access Channels, 1980–1989	52
Decrease in Cable Awareness, 1990–2010	60
Need for Grassroots Movement for Cable Representation	61
Chapter 3: The Incubation Period of Cable Television	63
Boston's Social and Historical Background in the 1970s	65
Mel King and African American Media Representation	69
Foreseeable Advantage of Cable Television in Boston	72
Detroit's Social and Historical Background in the 1970s	85

Detroit's Twenty-Year Period of Feasibility Discussions and Study	87
Final Draft of the Request for Proposals	99

Chapter 4: Drafting of Democratic Communication Media — 109

Drafting and Issuing the Request for Proposals in Boston	111
Applying to Wire Boston: Submitting Preliminary Application	114
Issuing of Request for Proposals in Boston	129
Submission of Amended Application	133
Drafting and Issuing the Request for Proposals in Detroit	142
Applying to Wire Detroit: Barden Cablevision	148
Emphasis on Public Access and Local Origination in Detroit	176

Chapter 5: Progress and Struggles in the Process of Franchise Decisions for Media Democracy — 177

Boston's Period of Application Review	178
Public Hearings in the Early Summer of 1981	179
Analyzing the Final Applications	189
Choosing Cablevision over Warner Amex	194
Discussion with Cablevision	200
Granting the License to Cablevision	203
Detroit's Period of Application Review	205
Issuing the Final Report	211
Politics Delay Media Democracy	216
Signing the Final Agreement with Barden Cablevision of Detroit	225

Chapter 6: From Agreement to Production: Period of Struggling — 229

Boston and Its Post-Agreement Phase	230
Cablevision's Failure to Meet the Expectations	239
A Beginning of an Alternative Media Form for African American Bostonians	241
Detroit and Its Post-Agreement Phase	243
Delays during the Post-Agreement Phase	246

Conclusion: BET is not the Answer — 249

Historical Lessons from Cable Television in Boston and Detroit	252
African Americans in Cable Television in a National Context	256

Notes	261
Bibliography	303
Index	311

Acknowledgments

This volume attempts to fulfill an immense responsibility of resurfacing a part of rich African American history. This is a task too ambitious for one individual to undertake. I could not have finished this book without the support of my family, friends, and colleagues. I am grateful and humbled by their continuing support, kindness, and generosity. First and foremost, I thank advisors, mentors, and professors who helped me grow as a scholar and as an individual. Of course, all the people whose names appear here generously shared their knowledge and experience with me. What I cherish and appreciate the most, however, is their support and commitment. Dr. Pero Gaglo Dagbovie challenged me intellectually while demonstrating his commitment to African American history and work ethic, which I continue to emulate. He also provided me with useful and empowering advice when needed. Whenever I met with him for a cup of coffee, for lunch, or for just a quick update, I felt strongly empowered and motivated. Dr. Richard Thomas shared with me his personal childhood experiences growing up in Detroit as an African American. He read a very early version of this book in fall 2006. At that time, it focused only on Detroit and it was only twenty pages long. This research paper has expanded to half of this book in a matter of a few years. Without Dr. Thomas' encouragement to pursue the research, half of this work would not exist. Dr. Gary Hoppenstand was always a great supporter of my work and of me. Allowing me to assume a position as an editorial assistant and later as a member of the Editorial Advisory Board for the *Journal of Popular Culture* for which he serves as editor, Professor Hoppenstand provided me with precious opportunities to expose myself to the latest academic discourse in television studies, issues concerning representation, and many other pertinent fields in Popular Culture Studies. Whenever he saw me in the journal office, he shared his enthusiasm and support for my work with me. Dr. Joe Darden always offered different approaches and perspectives that I had not thought about on my own. After spending twenty to thirty minutes discussing various issues from my dissertation to urban culture, to African American identity and representation, and even to trans-Pacific Black migration, I always left Professor Darden's office with new perspectives on my research and writing. Dr. Darden asked me one of the toughest questions during my dissertation defense. At the time, I was nervous. But now I know that the experience made me even sharper as a scholar. I must also thank especially three professors from the University of Massachusetts at Boston. Drs. Shirley Tang, Rachel Rubin, and Peter Kiang gave me advice and encouragement while I was working on my master's degree final project, which later became a basis for part of this book.

A bulk of my research took place at Boston City Archive, Boston Public Library, Burton Historical Collection at Detroit Public Library, and Wayne State University Arthur Neef Law Library. I would like to thank especially Zachary Enright and Kristen Swett from Boston City Archive, Ashley Koebel and Romie Minor from Burton Historical Collection, and Doug Card from the Arthur Neef Law Library.

I also thank Larin McLaughlin, Andrew Kenyon, and Michael Rinella for their professional support. We all know that after attending a few conferences, we begin to see familiar faces not just in conference rooms but also in our favorite part of the conference: the book exhibit. I would see Larin sitting at the booth for the State University of New York Press at different conferences. I think she began to recognize me because I would swing by the exhibit hall looking for one more book to purchase between panels, a few times a day. I was, at the time, still a master's degree candidate. Larin, nevertheless, expressed her interest in my project. For several years to come, I would update her on the progress of my work, in hopes that it would be considered for publication one day. She was excited for me when I finally submitted my proposal and manuscript for review. After Larin left the Press, Andrew guided me through the manuscript review phase. Michael assisted me through the rest of the process. I cannot thank them enough for their help and support.

I thank my close friends for their encouragement and support. Mathew Bartkowiak, Dwight Branch, Adam Capitanio, Kelly MacDonald, Gregory Stoller, and Shawn Young read my drafts and manuscripts numerous times. They corrected my grammar, gave me honest feedback and suggestions, asked some of the most difficult questions, and expressed their support for my academic endeavor. I also thank Francisco "Chico" Villarruel for his encouragement. Although we did not initially meet as academics, but as soccer referees, he became my mentor and my role model balancing academic and personal responsibilities. From our early morning meetings at a local coffee shop to hour-long drives to and from our soccer matches together, he was always filled with words of wisdom.

My family was always with me. I owe so much to my wife, Nichole. Despite the vast intellectual disciplinary gap between us, she has been the most avid supporter of my work, even if it sometimes meant that I sat in front of my computer until late at night. I could not have finished this book without her support, encouragement, and understanding. Last but not least, I want to thank my parents. I know it was not an easy choice for them to let their only child live abroad for years. My mother, Noriko Kiuchi, always supported my decisions, even when it meant she would see her son for only a few weeks a year and she would be alone in our house for the rest of the time. Her daily emails have meant so much more than she can even imagine. Hearing her say "welcome back" with a smile at the airport always told me that I had a place to return to. My father, Tsuneo Kiuchi, is not here with me to share the excitement of completing a significant stage of my life. But when I touch five small scars on my back from the bone marrow transplant from 2004, I can easily remember all the advice that I received since my birth. If he were here with me, I know he would give me a small nod or a pat on my back, which always meant so much to me. May you rest in peace.

I started my scholarly career with a small three-shelf bookcase. Fifteen five-shelf bookcases and eight plastic tubs later, I am happy and proud to complete this book. Any errors that remain in the text are solely my responsibility.

Preface

As a high school student in Tokyo, Japan, I nearly failed my history class. Approximately ten years later, I was in the final process of finishing up my doctoral dissertation—the base of this book—writing about history. A few years later, I was standing in front of more than one-hundred students at Michigan State University, giving lectures twice a week in my African American history class. A lot changed over those ten years, both personally and professionally. But a change that affected me on both those levels was my deepened understanding and appreciation of history. As a high school student, history simply meant rote learning of different names, facts, and years. It is true that this field requires much reading and remembering. No matter how much the students in my history class may complain, they still have to spend hours reading hundreds of pages, writing essays, and memorizing names and years in order to pass my class. It is a foundational part of the profession of historians. Facts and years create the sense of objectivity, an important responsibility that we have to fulfill. Without them, we would not have the basis to discuss, analyze, or explore history. Of course, this is not to say that there is singular objectivity or that an amalgamation of proper nouns makes up history. However, our day-to-day work as historians often is far from the kind of glory that other fields may have, at least in my eyes more than ten years ago.

What I learned as I pursued my career as a historian, however, is that this field can be filled with personal interests and passion about topics of study. We spend days and weeks, if not months, deciphering handwritten primary documents from decades or centuries ago. In an archive, we go through dozens of storage boxes until we finally find the one with all the information we need. We spend years typing our manuscripts in a converted home office surrounded by stacks of photocopied documents. Probably the only colors we see around us are from the covers of the books we have accumulated over time that now seem to serve as a fort. But as we immerse ourselves in our rich history, we become increasingly filled with passion. This book is a fruit of my enthusiasm in African American History and Study, a field that is filled with African American individual and collective memories—often times erased from the mainstream academic discourse—of struggles, achievements, disappointments, rejoice, displacement, and empowerment. As historians, we can give voice to those who never had a chance to have their voice heard while alive. We also can unearth the voice of the silenced. This is where my very personal, not just professional or intellectual, passion exists.

The origin of my fascination in this field exists in my experience of awakening in 2002. During my three-year experience in Boston as an international student from Japan where racial homogeneity is so strong and pervasive that seeing a non-Japanese or non-Asian–looking person at a local supermarket could be a topic at my grandparents' dinner table, I was shocked by the intricacy of American society. In the United States, there not only were different skin colors but also different color tones. The more I familiarized myself with Boston, the more aware I became of what Washington Street signified in the mind of many Bostonians. "Things could be rough on the south side of Washington Street. You don't want to go there after dark," one of the administration workers at my undergraduate institution once told me. This is when my rosy idea about American diversity and pluralism began to subside and take on a different meaning in my life. Except for one year I spent in France, I had spent twenty-one years of my life in Japan. I believed that the United States was a true melting pot where racism was a thing of the past and skin color no longer mattered. As much as it sounds like a cliché, that was what I believed before my arrival in the United States. I was only ten years old when Rodney King was beaten by Los Angeles police officers. I was eleven when the controversy around Anita Hill and Clarence Thomas took place. I was fifteen during the O. J. Simpson murder trial. More importantly, I was on the other side of the Pacific Ocean. For a high school student, I was relatively familiar with current affairs. But I knew little about what was happening more than six thousand miles away. So when I arrived in the United States and was confronted with reality, I had many questions. What does diversity mean? What is pluralism? What does it mean to coexist? What is race? Without much knowledge about or firsthand experience with the other side of Washington Street, my curiosity grew. Out of that curiosity came a reason to conduct an academic investigation through which I learned the role of cable television as inter- and intracommunity building and as a tool of empowerment in Boston's African American communities.

Pursuing my academic career in Michigan was an optimal opportunity to continue my project. About an hour away from Detroit, which once was known as "arsenal of democracy" and "Detroit the Dynamite," but is now as an epitome of perceived urban decay, I also witnessed the reflection, or perhaps the essence of American social intricacy about the popular myth. As I finished my doctoral work and started my career as an assistant professor at Michigan State University, the U.S. economy plummeted. I would hear from my family members and friends from Japan talking about how Japanese news media showed the General Motor's headquarters building in Detroit, as well as abandoned neighborhoods not even a mile away from it. The powerful message that Chrysler's "Imported from Detroit" television commercial possessed when it premiered in February 2011 is seldom shared in Japan. It is clear, however, that no matter on which side of the Pacific Ocean people might live, they are well aware, probably only subconsciously, of America's perpetuating dilemma. On the one hand is a country whose social progress is represented by the election of President Obama. On the other hand is continuing struggles of African Americans. The longer I lived and experienced American society, the more questions I had.

Although we must acknowledge achievements of past community leaders, we should not forget that the United States is not colorblind. As I discussed with colleagues and panelists at the National Council for Black Studies meeting in 2010, the myth of colorblindness is pervasive. Just because a Harvard-educated African American made it to the White House or some African American popular culture icons can afford to live an upper-class lifestyle does not mean the majority of African Americans live in the same conditions. The National Urban League's annual publication eloquently reveals the truth of American racial politics. The reality that many Americans choose to ignore or deny continues to afflict many African Americans. Their ignorance and denial in many cases make social conditions less ideal for many Blacks.

As any writer of a scholarly book wishes, I hope that this volume will be widely read in classrooms. My hope is particularly strong because I have talked with many undergraduate students from within and outside my classes and realized that many of them blindly believed in the myth of colorblindness and of a post-racial society. "I'm not a racist, because my best friend is Black," "My high school was very diverse. So I'm not a racist," "I can make racial jokes because I don't really mean it," and many other statements are innocently uttered. Teaching freshmen writing classes with a particular focus on race and ethnicity was a moment of revelation. Students who willingly admit they have lacked personal experiences with African

Americans or other racial or nonracial minorities claim very confidently that the minorities are in much better situations than the nonminorities. The ideas of institutional racism, tacit discrimination, systemic and structural racism, and so on, slip their mind.

But I do not intend to blame these young students only in their late teens and early twenties. Their parents, family members, teachers, neighbors, and many others have repeatedly told them that the United States has become colorblind. In a country where national pride and patriotism are so cherished, erasure of historical memories and denial are convenient tools. This is why this academic study should be treated not only as a descriptive and corrective work, but also as a prescriptive work. As past African American leaders have helped American racial awareness move from racism and animosity to racial ambivalence, the effort must be continued toward racial comity. My hope is that this work will contribute to this African American intellectual tradition.

Introduction

Unveiling the Struggles for Equal Voice

The history of the relationship between African Americans and the cable television industry is complex and intriguing. Although Black Entertainment Television empowered many African Americans through its establishment in January 1980 and its listing on the New York Stock Exchange in 1991 as the first Black-owned corporation, many felt betrayed when its founder, Robert Johnson, sold the company to a white-owned major media conglomerate, Viacom, in 2003. Exploring the experience of African Americans with cable television systems unearths the development and under-examined history of their struggles for equal voice, or of the Black image in the Black mind.[1] In other words, the question is how to publicly share the images of African Americans as seen by themselves, not seen or imagined by whites, who have historically owned the means of image production in the mainstream and broadcasting media. In the idea of public access and cable television as the property of local communities and residents, a significant group of African Americans believed that communications technology would bring social and community justice and equality across color lines. Although such optimism has yet to materialize even to this day, many African Americans hoped that cable television would empower their neighbors. This new kind of televisual image production and consumption provided African Americans of the post-Civil Rights era, who had been disproportionately left unnoticed, erased, silenced, or marginalized in the broadcasting media, with the locus and opportunity to produce their own collective and communal memory, to realize self-representation, and to create a sense of community membership. Cable television was a vehicle of community justice for many African Americans in the second half of the twentieth century. Employing a comparative analysis using Boston, Massachusetts, and Detroit, Michigan as case studies, this book argues that since the early 1970s African Americans in both cities possessed and exercised political and social agency and influenced the decision-making processes in their respective municipalities. These cities experienced many of the demographic and political changes many other urban areas underwent: the rise of ethnic communities, socioeconomic domination by the white population while their Black counterparts tried to obtain their agency, economic instability, and so on. By using cable television, both as a concept and as a technical means, they raised their self-esteem through empowerment, strengthened community ties, and reversed the negative images of African Americans that the media had disseminated in visual media-dependent American society.

The scholarly trend of examining network television and major movies and not focusing on narrowcasting visual culture such as cable television makes this study particularly important. In mainstream media industries, African Americans had little presence as producers. Production for Hollywood and network television that increasingly became popular in the mid-twentieth century required financial capital, technical and specialized education and training, equipment, and other resources to which very few African Americans had access. As a result, even as African Americans appeared more on movie and television screens in the mid-century, the images of African Americans on television often were simply reflections of the preexisting Black stereotypes in white minds.

In the film industry, especially during the 1970s, blaxploitation films such as *Sweet Sweetback's Baadasssss Song* (1971), *Super Fly* (1972), and *Blazing Saddles* (1974) embodied very subtle but enduring racism after the Civil Rights era. The 1980s and 1990s continued to witness numerous film representations of distorted Blackness. *The Cosby Show* (1984–1992) and other situation comedies, as well as the monumental series *Roots* (1977) and other more documentary-like shows, continued to portray African American images that were considered safe for white audiences. Although both film and television industries often depended on the disposable income of African Americans as a major source of their revenue, whites dominated the production part of the content and aimed to satisfy the white audience. J. Fred MacDonald explains that although television promised its African American audience to "[overcome] hatred, fear, suspicion, and hostility," the reality was "the tale of persistent stereotyping, reluctance to develop or star Black talent, and exclusion of minorities from the production side of the industry."[2] Similarly, Donald Bogle reminisces that during his childhood "[he] was aware, as was most of Black America, of a fundamental racism or a misinterpretation of African American life that underlay much of what appeared on the tube."[3] Although Martin Luther King Jr. and a few other activists used network television as a political tool, such attempts were rather limited in the 1960s.

The development of cable television systems took place in this historical and social context. The Civil Rights era did not revolutionize the media industry. The relationship between media and African Americans was one way. In other words, serving as the masters' tool, media did not inherently serve for the interests for African Americans and other minority groups. Those who had been underrepresented and underserved, if not unrepresented and unserved, had to reconfigure the medium of their interest so that it could work for their benefit. The increasing concerns about distorted representations of African Americans that African American leaders and intellectuals shared coincided with the augmenting interests in and attention to cable television as a separate communication medium from television. The new technology seemed to have its own strengths and potential to materialize the promise "that television [was] free of racial barriers."[4] Unlike the industries with which African Americans experienced difficulty accessing the production side of the media, the new industry had the appeal of allowing African Americans to be a part of content production once the cable system was introduced to their community. Such an effort was their strategy to use cable television as a tool for social change during the time when many African Americans continued to be disappointed with the distorted images of themselves on television and struggled against geographic, psychological, occupational, and emotional segregation.

Theories from African American Studies serve as useful platforms for examining the history of African American involvement in the cable industry. The idea of "the Black image in the Black mind" embodies the continuum that connects the existing historiography in African American Studies and modern mass communication technology. In 1971, George M. Frederickson published his seminal study titled *The Black Image in the White Mind: The Debate on Afro-American Character and Destiny, 1817–1914*. In 2000, Mia Bay wrote *The White Image in the Black Mind: African-American Ideas about White People, 1830–1925*. In the same year, Robert M. Entman and Andrew Rojecki published *The Black Image in the White Mind: Media and Race in America*.[5] These scholars analyze dimensions of Black life through the lens of what W.E.B. Du Bois called "double consciousness."[6] As these scholars untangled the binary between Black identity and "American identity," this study also attempts to understand this intricate history of the African American experience. The dualistic idea promoted by Du Bois and others still endures and serves as a fundamental framework in African

American Studies scholarship. In one of his most representative works, Ralph Ellison wrote, "[y]ou ache with the need to convince yourself that you do exist in the real world, that you're a part of all the sound and aguish, and you strike not with your fists, you curse and you swear to make them recognize you. And, alas, it's seldom successful."[7] In Franz Fanon's words, "not only must the Black man be Black; he must be Black in relation to the white man."[8] Although it is true that African Americans have often negotiated their identity vis-à-vis white American society, there also has been a series of attempts to separate the African American consciousness from it.

This study is, in a certain way, Afrocentric in orientation. It does not, however, fully follow Molefi Kete Asante's theory. It rather is influenced by Alice Tait and Todd Burroughs' idea of Afrocentric media, or "media created by and reflecting the worldview of people of African descent." For them, Afrocentricity in the African American context "involves a systematic exploration of relationships, social codes, cultural and commercial customs, mythoforms, oral traditions, and proverbs of the peoples."[9] Especially because what is arguably the largest television station catered to African Americans, Black Entertainment Television, is actually owned by a white media conglomerate, Viacom, and the distinction between what is Black and what is white is no longer straightforward in media, it is important to explore Black media history with such Afrocentric perspectives.

Because this project focuses on "local" African American experiences, culture, and values, it not only looks at the conventional political discourse but also at more community-based movements. Attention to "bottom–up" politics and "infrapolitics" deserve as much, if not more, attention than how city officials, including mayors and chairs of relevant committees, influenced citizens in a traditional top–down manner. Bottom–up politics refers to the active involvement of African Americans in the municipal decision-making processes. This is particularly important because Boston was led by a white mayor and a superficial analysis of the city politics risks duplicating existing researches examining how white officials affected Black lives. I argue that Black citizens influenced opinions and the decisions made by the white mayor. Simply picturing the cable-related politics as white-led efforts is to dismiss African Americans' rich history of community building and infrapolitics.

James Scott and Robin D. G. Kelly make convincing arguments concerning infrapolitics. They use the term to emphasize the influence of what often is invisible to the public eye on what is more visible. As I demonstrate in this work, African Americans were never mere consumers of cable television. The fact that we seldom read about their contribution to the development of this new form of media does not suggest the lack of Black involvement. Kelley argues, "the political history of oppressed people cannot be understood *without* reference to infrapolitics, for these daily acts have a cumulative effect on power relations."[10] The magnitude of such activism by African Americans merits scholarly attention.

Previous scholarship has neglected African American infrapolitics in cable and network television history.[11] Not only is the scholarship on Blackness from African American perspectives limited, it also has often overlooked the relationship between African American empowerment and the media. Even recent scholarship that has successfully examined how white perceptions of Blacks changed after the emergence of African Americans in movies, comedy shows, talk shows, music entertainment, news programs, and other television programs, insufficiently study these social movements in relation to African Americans history.

Starting with Du Bois, Carter G. Woodson, and other early twentieth-century African American thinkers, many scholars have published historical analyses on the various economic, professional, and legal implications for African Americans. Much of this scholarship shared the concerns about the emotional and psychological impact on African Americans. Once more African American actors and actresses began to appear on the screen in the mid-twentieth century, television stations hired more African American producers. More Black characters with whom the general African American public could identify emerged. The involvement of national Black organizations such as the National Association for the Advancement of the Colored People (NAACP) increased. Despite the body of scholarship on those topics, few have realized the fact that grassroots African American citizens, leaders, and organizers made proactive commitments to make such shifts happen. This study is one of the first to examine two actual cases of such grassroots movements.

How have African Americans tried to obtain access to cable television in order to improve their self-image in the future? This work seeks to answer this multilevel question. This inquiry is important because African Americans have not only been exposed to the statistical evidence of their disadvantaged social condition in the United States, as introduced in Tavis Smiley's radio and television shows or the annual publications of the National Urban League, but also have been surrounded by negative messages distributed by mainstream media, as was made even more apparent in the aftermath of Hurricane Katrina, as Michael Eric Dyson presented.[12] Additionally, from the Jena Six controversy in 2007 to Henry Louis Gates Jr.'s "break-in" incident in 2009, media have portrayed African Americans in ways that forces critical thinkers to reassess how far American society has left to go in its racial politics. It is for this reason that it is particularly important to put emphasis not on how media treat African Americans, but rather on how African Americans designed and reconfigured media. Analytical emphases must be placed on how African Americans in Boston and Detroit affected and interacted with cable television, rather than on how cable television affected the lives of African Americans. This is not just a matter of semantics.

My study analyzes the proactive engagement of African Americans in media, whereas many previous scholars have focused on their reactions. As is discussed more extensively in Chapter 1, there was an obvious color line in the media industry. On the one hand, many studies have examined the images of Blacks on television. Since whites have dominated the broadcasting industry, such scholarship has studied how whites viewed Blacks and how the perspectives of whites affected African American self-images.[13] When a scholar criticized a racist portrayal of an African American on television, he or she conventionally focused on the cause of racism in the white mind and society, rather than on its impact on the Black mind. Black psychologists such as Na'im Akbar, Wade W. Nobles, Claude Steel, and others, have analyzed these dimensions. They, however, rarely recorded how African Americans actively responded to these expressions of racism and attempted to establish a new media form that portrayed them in a fairer manner. On the other hand, this work explores the history of African American agency, the act of both tangible and intangible resistance based on their personal, communal, ethnic, and racial experiences, and its application in the municipal to community-level politics, via their use of cable television.

When exploring African American history in such social contexts, historians must ask numerous questions to fulfill the aforementioned goals. Some of these questions are basic. How has white-dominated American society made it difficult for many African Americans to achieve high self-worth? What are some of the ways in which African American leaders have sought to reverse this trend to empower their peers? What kinds of roles did African Americans expect media to play in relation to African American self-esteem and empowerment? What has been the common relationship between African American leaders and community organizers, and media technology? How have Black communities and community members felt they would benefit from and be hurt by cable television?

More geography-specific questions will provide an avenue to the understanding of the history of cable television in the two cities. What did African American leaders and community organizers in Boston and

Detroit know about cable television once such technology became available? How did they look at the potential of such media? What were the implications of cable television for African American communities? How influential were African Americans in the political arenas of the two cities? What were community leaders' reactions to their city's decision to study and adopt cable television? How did they influence the municipal decision-making process for its cable television franchise agreement? What was the relationship among city officials, African American leaders, and African American community members like as they discussed the introduction of cable television? How did franchise candidates attempt to appeal to local African American communities and citizens? How much did the cities listen to the needs and requests of local African Americans? What kind of service did the franchisee provide African Americans? How did community members obtain public access and produce content to air? How did the introduction of cable television empower African Americans and affect their self-esteem? How did Black communities change after the arrival of cable television?

Chapter 1, "The Black Image in the White Pathology," provides the historical overview about the relationship between African Americans and their presence in American media. This chapter challenges the persistent view that Blacks suffer from a cultural pathology that prevents them from social advancement. This chapter shows that if such a thing as cultural pathology exists, it was in the white mind, which systematically denied and neglected the value of African American cultural contribution. Indeed, I show that broadcasting visual media frequently deprived African Americans of their agency on one level. Aspiring actors, for example, had to face a conundrum of becoming successful in a national arena or remaining loyal to their racial background and roots. It is, of course, important to remember that being successful meant two different things for whites and Blacks. To appear on a screen was often more than a success for many African American actors, whereas it was less so for their white counterparts. Although this is not to say that African American artists were fully deprived of agency, their efforts were just not as impactful as they wished.

Chapter 2, "Cable Television in the Past and Present," discusses African American agency in the media industry. Across the color line, scholars and researchers have examined African American representation in television and media (or popular culture) studies. Despite the systematic, technological, economic, and operational differences between broadcast television and narrowcast television systems, this chapter benefits from the historiography pertaining to both types of media as both share the apparatus in the family room and later in bedrooms, interact with their audience aurally and visually, and ascribe to the common technological root. It shows how historically researchers, professionals, and policymakers have discussed cable television and minority interests. The relationship between broadcasting and narrowcasting is especially important because the latter emerged and developed as the remedy to the problems encountered by the former, not only in its technological inadequacy but also in its social shortcomings.

Chapter 3, "The Incubation Period of Cable Television," examines the interests of city officials, community leaders, and residents in cable television. They viewed cable television as possibly something beneficial for African Americans. This section covers the period between the initial discussions about the future introduction of cable television in the Black neighborhood and the issuing of the request for proposals (RFPs). The RFP was the official document issued by the local municipal office or committee in charge of arranging the franchise agreement. Calling for applications, the RFP outlined all the requirements applicants were expected to meet. The RFP serves as an optimal source to understanding what each municipality expected from its cable system, what it tried to achieve through it, and how the cable system should benefit its residents. A turning point in the relationship between African Americans and televisual images came in 1963 when news programs and documentaries featuring African American civil rights struggles politicized the nation. More specifically to Boston and Detroit and their use of narrowcasting media as a political tool, the history goes back to the late 1960s and the early 1970s. This incubation period ended in August 1980 in Boston and in August 1982 in Detroit when the RFPs were issued. At the time, African Americans were not only aware of the potential of cable television, but also expressed their interests and participated in discussions regarding whether or not cable television would be beneficial to the city and local communities. Their interests existed more than ten years prior to the release of the RFP. There are numerous examples of the involvement by African Americans in casual and unofficial discussion meetings on cable television with other citizens and city officials, in study groups to examine its feasibility in their locality, in the formation of the RFP, and other major moments.

The years covered in this chapter were dynamic. Once the cities determined to officially examine the feasibility of cable television, they established cable communications commissions. African Americans both in Boston and Detroit were heavily involved in their respective commissions. When the cities finalized their decisions to introduce the cable television system in the cities, each city established a formal committee to work toward the introduction of cable systems. Public involvement increased. The commission selection process was more serious. The discussions became more intense. Commission members strove to secure independence from the mayor or other parties that could assert their power over the commission. African Americans were not mere witnesses to such a dynamic period in history, but a major part of it. To explore the lively engagements by African Americans, this chapter particularly benefits from publications by aforementioned study groups, letters written by residents to city officials, and memos distributed in the city office and feasibility study groups.

Chapter 4, "Drafting of Democratic Communication Media," focuses on the short period of time between the release of the RFP and the submission of final applications by franchise candidates that took place in February 1981 in Boston and in December 1982 in Detroit. Although African American citizens had little time to intervene with potential service operators during this period, a review of corporate efforts to meet the needs and wants of local African American residents reveals the actual impact of African American grassroots movements on corporations and municipal politics. A close examination of the proposals shows that different operators specifically attempted to serve local African Americans and suggests that without the presence of African American infrapolitics, the cable television systems in Boston and Detroit would have been different. Proposals were specific about production training, equipment, and facilities that would be available to community members once the cities and their bid winner signed the franchise agreement. They elaborated on public access channels, financial assistance, and other intangible resources for the communities. While both the Federal Communications Commission (FCC) and the RFPs required that bidders include information about aforementioned issues in their application, content analyses of the proposal reveal that potential operators attempted to appeal to the cable communications commissions with their interests in public and communal engagement in content production more than they were required to. The applications are the most informative documents in this chapter. Relevant documents such as the addendum to the application, revisions to the original document, and newspaper articles evaluating the candidates will supplement the proposals.

Chapter 5, "Progress and Struggles in the Process of Franchise Decisions for Media Democracy," explores the municipal decision-making processes after the proposal submission. I examine how city officials came to their final decision to award their franchise agreement to Cablevision in Boston and Barden Cable in Detroit. Both in Boston and Detroit, cable communications commissions and cable operator candidates exchanged many letters and sat together for discussions after the submission of the final application. Questions and answers were sent back and forth between the two parties. I critically analyze correspondences and face-to-face meeting records that took place after the submission of the application. This chapter covers the time between February 1981 and December 1982 in Boston and December 1982 and 1985 in Detroit. This period has ample examples of the African American involvement in decision making, reviewing of the applications, and drafting of the final decision. The decision makers made efforts to evaluate the feasibility of the public involvement expressed in applications, willingness of the companies to realize their promise, and the appropriateness and sufficiency of the proposed facility and equipment, as well as other forms of resources. The commission also emphasized the importance of minimizing the subscription fee for the basic service so that the financial burden would not be too great and thus made it more accessible to all. This chapter demonstrates how the involvement of African American community leaders encouraged decision makers some of whom were Black themselves, to balance the profitability of the system that would eventually benefit the city as the franchising fee, and the low entry barriers for local residents. Commissions' minutes, internal memos, and other commission member documents help reconstruct this part of the urban history.

Chapter 6, "From Agreement to Production: Period of Struggling," probes into the post-franchise agreement phase. For Boston, the period from December 1982 to the end of the 1980s is covered. For Detroit, the covered period is from 1985 to the end of the decade. A shift in the history of the cable television development occurred in both cities in 1989. Debated issues were no longer the way to get access to the technology or how to control the medium. Citizens were more concerned about making sure that promised services would be provided. During several years in the 1980s, the cities built production facilities. African Americans had initial contact with the technology, and were offered free training. Many other tactile changes took place in the cities. Even after the agreement was signed between the city and the cable operator, both parties continued the detail finalization process in which they discussed the exact service, facility, equipment and others provided to the local community. They also discussed how they could encourage community members to join training sessions, produce programs, and attract subscribers. Starting the system did not mean that people would know about the program or want to automatically join. Community organizers had significant responsibilities to maximize the positive effect that cable television could bring to the city. It was also during this time when community members expressed their frustration about the slow process. Some wrote letters of complaint to the municipal office or cable commissions. Such letters vividly show, even twenty years later, the sense of excitement and expectation these citizens had for cable television programs.

"Conclusion: BET is not the Answer" amalgamates the historic lessons from the African American community experience within Boston and Detroit to establish an intellectual framework for the history of the post-Civil Rights movement period. It underscores the importance of these small-scale media outlets such as public access and local programming as a part of the larger discourse on media representation and African Americans. This section also suggests and outlines how African American communities will be able to continue to benefit from technology. I argue that African Americans influenced the development of cable television, determined what they could get out of the franchise, and remained intact as a community to create positive images about and for themselves. I suggest various potential ways to use media technology for community-building purposes. The prescriptive role of this chapter not only allows its readers to better understand about the history of urban Black communities but also gives a model of social change that can further be applied to future community formation efforts particularly in the era of communication technology development.

These chapters show what happened to Boston and Detroit after their mid-century "crises" when manufacturing industries moved to their suburbs. For the last decades, both have witnessed an increase in ethnic communities. As for Boston, the minority population ratio increased to the point where the city's population is more than 50 percent non-white, or minority majority. Comprising only 4.9 percent of the city's population in 1950, the African American population more than doubled by 1970, from 40,057 to 104,707, making up for 16.3 percent of the population. The numbers continuously increased. In 1980, there were 126,229 Blacks in Boston, or 22.4 percent of the city's population. In 1990, 25.8 percent of the population, or 146,945 residents, were African Americans.[14] It is also noteworthy that the increase partially took place while the city underwent the busing controversy in the early 1970s.

As for Detroit, white flight left the city predominantly Black. Although there were only 4,700 Blacks in the city at the turn of the century, the number had gone up to almost a half of the 1.5 million residents by the early 1970s.[15] In 1950, 300,506 Blacks made up for 16.2 percent of the central city's population. In 1960, the percentage went up to 28.8 percent, or 482,223 Blacks. As the number of whites decreased, the number of Blacks increased. In 1970, 43.6 percent of the population was Black with 660,428 residents. The 1980 census for the first time showed the majority Black condition of Detroit with 758,939 Blacks making up for 63 percent of the population. In 1990, 75.6 percent of the Detroiters were Black. Between the forty years from 1950 to 1990, the number of whites decreased from 1,269,377 to 222,316, while that of Black more than doubled from 300,506 to 666,916.[16] In many ways, the memory of riots from 1943 and 1967 affected the cognitive map of many Americans about Detroit. These industrial and demographic shifts, however, were not unique to these two cities. Many other American urban cities have undergone similar changes. Chapter 3 briefly introduces some of the relevant urban Black history of the two cities that created an environment in which discussions on cable television implementation took place.

As the size of Black populations in each city increased, African Americans in Boston and Detroit continued their long tradition of organizing themselves for municipal political representation. In Boston, for example, local organizations helped strengthen community ties. African Americans being the largest minority group, Boston's Dudley area was a place of "low income, high crime, poor schools, burned-out buildings, acres of vacant lots used as dumping grounds, abandoned cars, and the night lit by fires."[17] The successful implementation of the Dudley Street Neighborhood Initiative in 1984, which enabled the "Don't Dump on Us" meetings, the "Development without Displacement" campaign, establishment of the Vine Street Community Center, and other programs to revitalize the area, is just one of the many examples.[17]

Although the racial tension in the city was high as the controversial busing case attests, African American residents reinforced their community strength through all-Black weekends and summer activities led by local churches.[18] In Detroit, religious institutions often provided political support to Black leaders. When Richard Austin became the first African American to win a mayoral primary election in the United States in 1969, his support base existed in Detroit's religious community, which reflected the rise in African American political representation in the city.[19] Churches not only provided political support, but also served as the "keepers of the fire" of African American self-help on the community levels.[20]

Despite these examples of municipal-level political and social participation by African Americans in Boston and Detroit, few studies on these cities focus on visual media as a tool to advance such objectives among Blacks. Many study psychological, economic, and political impact of media on African Americans. Scholarship on cable television with particular emphasis on local origination and community programming, however, has been missing. This study adds a new layer of understanding in the history of African American self-help, empowerment, and visual culture.

Boston and Detroit have their own particularities that make this study indicative. Boston resurged as a large service industry city. It is also a major hub of intellectualism with numerous universities and hospitals. In the history of cable television in Boston's African American communities, these academic institutions played significant roles. For example, Mel King, one of the major African American community organizers in Boston, taught at the Massachusetts Institute of Technology (MIT). His successors, including Ceasar McDowell, Richard O'Bryant, and others, now teach at MIT, and Northeastern University and there have been strong cooperation efforts between the academy and the local community. By examining Boston, where scholars served as community organizers and where community leaders have taught in higher education classrooms, we are able to learn about different roles that intellectuals must play outside of academia.

Detroit's case, on the other hand, is indicative of the power of African American self-help particularly due to its heavy African American concentration. For example, the city has similar experiences to other urban areas such as its experience with rebellions, white flight, and industrial decline. Unlike Boston where most of the city officials were whites—a fact that makes the Boston study meaningful to other cities with whites dominating the municipal office—Detroit had African American decision makers, including Mayor Coleman Young. Although the number of African American mayors and high officials is still limited nationally, Detroit's case provides us with the actual case in which African Americans were fully responsible for the outcome of the cable television introduction. This is a representative case of African American self-sufficiency.

Boston and Detroit, respectively, have unique histories of their relationships to media. Boston, as a center of American intellectual community, witnessed nationally popular visual culture production. One such example is *Say Brother,* produced by the Public Broadcasting Service and WGBH, Boston's public television station. Started in 1968, the program is a "public affairs television program by, for and about African Americans" in which Maya Angelou, Amiri Baraka, Stokely Carmichael, Louis Farrakhan, Jesse Jackson, and many other prominent Blacks appeared. It was also a program that provided "an opportunity to expose the true facts about Black history through music, song, discussions, art, drama, fashion, and educational scholarship."[21] Similarly, Henry Hampton established Blackside Film & Video Production, the largest African American–owned film company at the time, while in Boston in 1968. Blackside's production portfolio includes the fourteen-part series *Eyes on the Prize,* which was on air in early 1987.[22]

In the early second half of the century, Detroit also experienced both positives and negatives in its relationship to media. On the one hand, radio and television worked against the city when it underwent the 1967 rebellion. Unlike the efforts made by Martha Jean, a popular disc jockey on radio station WJLB whose primary audience was African American to appease the rioter, most radio stations followed Damon Keith, an African American co-chairperson of the Michigan Civil Rights Commission to "observe the moratorium on riot news." Consequently, many African American residents unknowingly found themselves in a riot zone, although they would have stayed away if they had been informed. Additionally, ABC radio, a broadcasting radio station whose office was located in Detroit, transmitted the breaking news about the riot nationally and partially initiated a set of negative images the city suffers from even today.[23]

On the other hand, despite being the site for Director Arthur Marks' blaxploitation movie *Detroit 9000* in 1975, in September of the same year, WGPR, or the first full-time FM station in Michigan and the pioneer of a minority-owned radio station, successfully bought out Channel 62 to become the first minority-owned television station in the United States. This change allowed African Americans to obtain skills and exposure in television production regardless of their past experience. The station produced local programs such as *Access Hollywood* and *Strictly Speaking* with African American crews.[24] Such developments served as precursors to the rise of cable television in Detroit during the following decade.

As is the case with any regional or local historical study, the arguments, discoveries, and achievements that surface from this study should not be overly generalized as a national trend. Although both cities have a history of municipal-level political activism involving African Americans and interactions with media, there are more differences than similarities between the two. Both Boston and Detroit have their own particularities that enabled, facilitated, and encouraged the materialization of certain objectives as well as those that hindered, slowed, or complicated others. The significance of this study exists in the similarity between the two cities in terms of the successful outcome of using cable television as a community empowerment tool, despite each city's differences and uniqueness in racial composition, municipal leadership, industrial background, social history, and other issues that affected Black communities in Boston and Detroit. Identifying what exactly allowed African Americans in two very different cities to use the same technology for their own benefit is a major objective of this study.

It is also important to remember that the value of this research that focuses on these two cities does not rise because they were pioneers in the introduction of cable television or in African American engagement in its development. By the time Boston was first wired, the state of Massachusetts had twenty cable television systems serving 53,324 subscribers. William A. Lucas' report published by the Rand Corporation in 1973 makes an extensive analysis of the cable system in Boston's neighboring city, Somerville, Massachusetts. Additionally, out of fifteen population-dense cities around Boston, seven municipalities had a cable system or a franchise agreement by 1973.[25] As for Michigan, long before Detroit was wired, the state had 126,379 cable subscribers served by forty-nine systems. One of them was the system in Kalamazoo, the forty-second largest in the nation with approximately 13,000 customers.[26] The system in Kalamazoo was even considered the "best access center in the U.S."[27] Therefore, Boston and Detroit had years to learn from other neighboring cities' experiences while they planned for the system introduction. This is to say that Boston and Detroit are fit for this study not because they were pioneers in achieving fair minority representations through cable television or wiring their local areas so effectively that African American residents took full advantage of it without any troubles. In reality, both cities experienced many delays, controversies, disagreements, and other problems. All of these were common in other urban areas planning to wire the city with coaxial cables. Concurrently, when compared, the stories that these cities tell reveal how effective African Americans were in voicing their needs and wants and securing proper access to the technology.

This work benefits from diverse sources, including municipal study reports, economic analyses, legal documents, and technical reports that do not mention African Americans or even any ethnic or racial issues. Many documents, especially the ones that focus on financial or technical issues do not make direct reference to African Americans. It is, however, important to pay attention to the fact that in many cases, more encompassing words like so-called "minorities" or "economically disadvantaged" often included many African American residents. Although many African Americans in Boston and Detroit belonged to the upper middle class, there were many more African American residents that struggled economically. The lack of direct reference to African American residents, therefore, does not mean the process of cable introduction ignored African Americans.

In Boston, the non-Black minority population is rapidly increasing. There have been interracial community-building efforts since the mid-century. African American leaders, who are mentioned in this study, led some of these efforts. Although their struggles are significant and should never have been neglected, it is also an undeniable truth that no other ethnic or racial group has been subject to collective racism and violence with a history as long as African Americans.

As Robin Kelley argues, scholars have so far paid more attention to *how* people are involved in a movement than to *why*.[28] By centering African Americans and closely examining their day-to-day struggles and activism, I attempt to unearth their discourse of resistance against white-dominated society and media. This approach will identify both why and how African Americans successfully reversed discrimination on television. My study seeks to be applicable to African American communities in other cities, as well as to other minority communities in the United States. One of the major contributions of this work is that it helps us understand not only the African American experience with media, local political entities, the cable industry, and others, but also the larger American experience through the contemporary history of Black resistance. Lila Abu-Lughod claims that resistance is "*diagnostic* of power."[29] This is because resistance does not exist exterior to the power.[30] By understanding how African Americans in Detroit and Boston reacted to whitestream society, we are able to obtain a macro-view about American racial experiences and history.

Chapter 1

The Black Image in the White Pathology

Visual culture has been a locus of both reflection and production. It is reflective because it is formed within its social, political, economic, cultural, and other environments. Visual culture showcases both what is and what has been. It also is productive because it has the power to influence our consciousness and even push us to a new direction. Cable television, a major source of visual culture in the 1970s and 1980s, also reflected its social milieu while offering agency to its users. Analyzing cable television, therefore, calls for a multifaceted approach, at minimum, studying the images produced prior to the popularization of cable television, examining the expected contribution cable television was going to make, and understanding how consumers and producers introduced cable television to their locality.

In relation to African Americans whose image had long been erased and distorted by white-dominated mainstream media, the promise of African American–oriented cable television was significant. How were African Americans represented in mainstream visual culture, especially during the few decades immediately leading up to the 1970s? What else did they do to produce their own media content? What were their objectives? Answering these questions reveals that African American participation in the development of cable television since the 1970s was a response to the century-old history of the erasure and distortion of Black images in the white-dominated society and media. It will also attest that despite the progress African Americans had made in the United States since slavery, the nadir, or even the civil rights era, there still was much to be done to achieve a fair representation. Boston and Detroit were no exceptions. African American Bostonians and Detroiters needed more than a decade to fully deploy cable television as their tool.

This chapter also reminds readers that the struggles for media affirmative action in the 1960s and 1970s by African Americans were not unprecedented. Similar struggles have existed, and many African Americans have won those battles. There is a century-long history of Black endeavor to generate fair televisual representations. African American involvement and their significant roles in cable television, therefore, should not be considered independently from a broader racial discourse on their identity politics, particularly through the use of visual popular culture. Furthermore, African American experiences that may seem to be relatively distanced from the experience of white Bostonians and Detroiters in the 1970s, such as the broken promise of the second Reconstruction, minstrel-like stereotypes, and other phenomena, are connected to their experiences in the post-Civil Rights era. The issues concerning access and representation in cable television are integral parts of a larger racial, political, sociological, and historical discourse in African American studies and history.

African Americans and the Film Industry

The history of Hollywood and the film industry in general is one of visual prejudice against African Americans. A founding father of motion pictures, Thomas Alva Edison, is responsible for the racist image of African Americans eating watermelon for his *The Watermelon Patch*. Donald Bogle also showed in his study on Black Hollywood during the first five decades of its history, that the film industry capitalized on racial stereotypes and grew at the expense of African Americans. Despite some African Americans who worked in the industry, early films repeatedly distributed racist images in theaters. In particular, *The Birth of a Nation* (1915) by D. W. Griffith pioneered in providing the nation with "the shocking and degrading stereotypes that were to plague African American movie images throughout the twentieth century."[1] Linda Williams explained that the film essentially "turned the nation to southern sympathy."[2] It was no coincidence that less than a year after the premier of Griffith's movie, the second Ku Klux Klan was founded. Additionally, ten years after Edison's infamous rendering of African Americans, the cinematic production of Thomas F. Dixon Jr.'s literary work from 1905, *The Clansman,* epitomized and anticipated Hollywood's upcoming trend to perpetuate negative images of Blacks.[3]

The first several decades of American film industry were filled with racist images. Although some exceptions, commonly known as "race films," existed in the 1930s and 1940s, African American characters on the movie screen played subservient roles that entertained white audiences but did not threaten their dominance in society. Gertude Howard, Libby Taylor, and

Louise Beavers were just a few of the African American actors who faced a limitation in the kind of roles they could play. By the time Hattie McDaniel and Butterfly McQueen acted in *Gone with the Wind* (1939), this trend had been ingrained in the industry.

Prejudiced images of African Americans, however, are not only the creation of the early American film industry, but also of the Civil Rights and post-Civil Rights eras that gave birth to cable technology. Although the first half of the twentieth century offers an endless list of movies with prejudice against African Americans and the War Department, during World War II, financed movies that portrayed African American soldiers heroically to gain support among Black Americans for U.S. involvement in the war, the film industry continued to produce tacit, but in essence very similar, images of African Americans. Although Oscar Micheaux and other African American directors strove to make films for African American audiences that portrayed only positive images, the overall trend in Hollywood changed little. Reflecting the socially volatile days during the 1960s, African Americans were turned into either militants or "model Negroes."[4] Even when Hollywood portrayed them heroically, it often was nothing but "more subtle and masked forms of devaluing African Americans on the screen."[5] These cinematic manifestations of racial politics reflect a series of negotiations between the rapid social change in the mid-century and American popular culture and Hollywood's old values.

Characterizing the 1960s film period as a time when "problem people [turned] into militants," Bogle focused on African American actor Jim Brown as an exemplary case for understanding how Hollywood treated Blackness on the movie screen. Brown was a prototypical African American actor who frequently appeared in movies during the 1960s. Although moviegoers in the 1960s saw a rise in artistic, and often independent, films that showed "untyped" African Americans, Hollywood offered them "dangerous" Black characters.[6] There is a significant parallel between their images to those of the Black Panther Party members that often raised fear of social instability and concern about safety among whites. Of course, historically, it was not a rarity for whites to distort African American images in order to generate fear. Brown fit well in this historical trend, at least on the surface level. Bogle explained, "[Brown] was big. He was Black. He was outspoken. He was baaaddddd."[7] His characters reflected Black separatist idealism. With such a portrayal of his African American identity, on the surface, he seemed to satisfy some African American audiences' need for a viable Black power figure and actually refute the idea that Hollywood only showed Black characters who were safe to white society.

In most cases, socially dangerous Black movie characters such as Brown did not actually infuse fear among whites or encourage African Americans to pursue their political goals. Bogle said that "their strength was always used to work with the dominant white culture rather than against it."[8] This means that although Brown's characters seemed, on the surface, to reflect the African American independence and self-determination that the civil rights activists strove to attain, they were, in essence, turned into accommodationists whose power was safe for whites. His militant attitude was only by appearance. His contained and guided force became benign because white audiences were not ready to see "a politically militant Black man" at their neighborhood movie theater.[9] Despite the militancy that Brown's characters seemed to reveal, it was gentrified enough to be safe for whites but dangerous enough for Hollywood to casually problematize and criticize African Americans' political movements.

Hollywood also stripped some African American characters and actors of even a hint of militancy or power that could be a threat to white American society. Sydney Poitier was one of them. The movie industry portrayed him as an "ideal" African American man. Although *Guess Who's Coming to Dinner* (1967) featured a potentially controversial interracial relationship with a white actress, Katharine Hepburn, it was a "safe" relationship because Poitier's personality was highly intelligent and kind, personal traits reserved for white actors.[10] Poitier's skin was Black, but the character he played was what white actors would usually assume in other movies. Poitier rarely suggested a sense of rebelliousness or dissatisfaction, if at all. He was not alone. Such African American characters had existed since the 1920s as in *The Scar of the Shame* produced in 1927 by the Colored Players Company, which actually was a white-owned company in Philadelphia.[11] These African American characters were "diligent and morally upright if not also refined and prosperous."[12] In these movies, prejudice against African Americans as explicit as the ones in *The Birth of a Nation* did not manifest. Poitier was even content to play "a likeable guy." He explains, "I played a good, decent, useful human being—and mind you, much of what was being made in Hollywood at that time, with very rare exceptions, wasn't complimentary from the Black perspective."[13]

As he reminisced his own career, Poitier seems to realize that decades ago, he had been unaware of the significance of roles he played. The movie industry of the 1960s recognized the agency that African Americans and their culture possessed. But it also pushed African Americans to "the narrow confines determined by a hostile white world." It still lived with its old racist values. This pessimistic analysis should not be read as evidence of failure among African American activists, actors, and producers. However, in the mind of many Hollywood movie producers, the report by the President's National Advisory Commission explaining the social reality that the nation was moving toward two separate societies of Black and white meant very little. Hollywood continued with "its brotherly-love-everything's-going-to-be-dandy escapist movies" through Poitier, and neutralized African American militancy that Brown's characters realized.[14]

The characters that Poitier was satisfied to play were precursors to those in the blaxploitation era, which contained even more latent racism. Produced roughly between 1969 and 1974, blaxploitation films attracted large Black audiences. It is significant that this period coincided with the rise of discussions on the introduction of cable television to Boston and Detroit, as well as many other cities. By the end of the 1960s, the African American public was politically and socially mobilized enough to want to "see their full humanity depicted on the commercial cinema screen."[15] They wanted to see not only just "nice guys," but also heroes. Hollywood met their demand by featuring numerous African American heroes. After *Shaft* came out in 1971, both Black and white critics celebrated the "breakthrough production in terms of expanding Black representation in commercial cinema."[16] The movies seemed to suggest that African Americans could finally "turn the table on an unjust racist society."[17]

The irony of this new trend was that these films did not exist to promote positive images of African Americans or to help them advance their political agenda. They simply tried to exploit African Americans as a new source of revenue. By the end of the 1960s, the movie industry was at a near economic demise. As popular television played a more central role as a source of entertainment, it became increasingly difficult for film to maintain its dominance as a visual entertainment source. One solution was to capitalize on the African American economic power that was on the rise in the mid-century.[18] It did not take long until it became clear that blaxploitation films were just Hollywood's reproduction of "ebony saint[s]" while capitalizing on their prejudice.[19] The film industry was not interested in advancing African American social status. Guerrero explained:

> As the popularity and influence of Blaxploitation began to explode, Black intellectuals, community leaders, and film critics became increasingly uneasy and came to see the cinematic production and celebration of these unsavory inner-city heroes and narratives as, at least in part, the responsibility of Black film artists, actors, and directors in collaboration with Hollywood.[20]

The blaxploitation formula used in *Shaft* (1971), *Super Fly* (1972), *Blazing Saddles* (1974), and numerous other films often consisted of "a pimp, gangster, or their baleful female counterparts, violently acting out a revenge or retribution motif against corrupt whites in the romanticized confines of the ghetto or inner city."[21] Produced during this period and featuring African American heroes, these movies sent contradictory messages to African American audiences. On the one hand, Hollywood seemed to have shifted toward visual equality by the mid-1970s. At least, there was an increase in the number of African American characters on the movie screen. On the other hand, the reality was that such movies only "played on the needs of Black audiences for heroic figures without answering those needs in realistic terms."[22]

Critical examinations of white-produced cinematic images of African Americans must coexist with celebrations and critical analyses of the African American images from their own perspectives put on a movie screen. In other words, it was the images of African Americans as representations of "us," and not as "them," that carried significance. Spike Lee was an internationally famous movie director who was able to do so and to receive extensive scholarly attention. Lee's movies often are identified as the ultimate case of African American popular culture. Not only is he from the African American community, but he has been engaged heavily in Black issues throughout his cinematic career. Lee's movies have dealt with urban poverty, interracial tension within inner-city communities, a controversial Black activist hero, and other relevant issues. His works place African American experience in the center of the story.

Furthermore, Lee's production precisely meets Stuart Hall's definition of African American popular culture. Defining popular culture in general, Hall argued:

> popular culture always has its base in the experiences, the pleasures, the memories, the traditions of the people. It has connections with local hopes and local aspirations, local tragedies and local scenarios that are the everyday practices and the everyday experiences of ordinary folks. . . . The role of the "popular" in popular culture is to fix the authenticity of popular forms, rooting them in the experiences of popular communities from which they draw their strength, allowing us to see them as expressive of a particular subordinate social life that resists its being constantly made over as low and outside.[23]

In terms of "Black," he claimed that the word referred to:

the Black community, where [African American] traditions were kept, and whose struggles survive in the persistence of the Black experience (the historical experiences of Black people in the diaspora), of the Black aesthetic (the distinctive cultural repertoires out of which popular representations were made), and of the Black counternarratives we have struggled to voice.[24]

Although some critics such as Wahneema Lubiano and Ed Guerrero have questioned whether Lee dealt with Black problems adequately or how effective his movies had been to ameliorate the social struggles of American Blacks, he nonetheless dealt with racial, gender, and class politics as well as issues of sexuality in *Do the Right Thing* (1989), *School Daze* (1988), and other topics that few other movies had focused on. He also practices what Hall defines as Black popular culture through his focus on African American activism in *Malcolm X* (1992) and others.[25]

Spike Lee's intent to produce Black films has received positive responses. Along with two other Black independent filmmakers, Mary Neema Barnette and Reginald Hudlin, Lee has been described as a producer who had "a fluent command of Black and white cultural languages, and an insistence on counterposing them in an aesthetic dialectic." James Snead argued:

> Instead of seeing Blacks purely in terms of white norms and practices, these films show Blacks securely positioned in their own environments, discussing and dealing with their own problems, ignoring or at best belittling the toys and games of the dominant white culture.[26]

Do the Right Thing well embodied this assessment. It featured actors and actresses who were recognizable for African Americans but who had rarely been seen on large movie screens during the pre-Lee era. Its language and humor seemed to achieve an authentic voice of the community it was representing. Amidst humor, the movie asked important questions, such as one asked by Mookie (a character played by Lee himself) "Hey, Sal, how come you ain't got no brothers up here?"[27] Examining the messages that African American independent film producers communicated through their work, Snead's analysis is highly celebratory about the fact that young Black filmmakers attained unprecedented prominence in the film industry in the 1980s.

Lee's contribution to African American communities, however, did not go unquestioned. Even celebratory Snead wondered if African American independent film production would be able to compete economically and culturally with the other mass-produced movies that had already witnessed enormous box office success. Additionally, Snead was not naïve enough to believe that white audiences would be ready to accept African Americans' view of American society.[28] Bogle answered Snead's question with pessimism. He explained that although critics frequently talked about Lee's films, most of them were commercially unsuccessful.[29] Lee's films continued to bear the important political messages that attracted vast attention to *Do the Right Thing*. But such movies did not necessarily translate into large box office hits.

Other critics problematized not only Lee's relatively small commercial success but also more fundamental messages that his movies carried. Lubiano, for example, questions what Lee had in mind when he was "trying to make it real" while filming. She claimed that media's "deification" of Lee resulted in the marginalization of other independent Black filmmakers as being not real enough, especially with regard to female filmmakers. Additionally, she argued that Lee's films tend to connect African American masculinity to employment and African American femininity to a subject of male gaze. In Lubiano's analysis, this gender divide went so far as to associate Blackness to manhood, creating a homophobic subtext.[30]

Lubiano was not alone. Guerrero also was critical of Lee's portrayal of Black society. He does not believe that Lee's films were a particularly powerful political tool. Just like Lubiano who questioned how Lee could detach himself from "the constraints of Euro-American film discourse,"[31] Guerrero pointed out that Lee downplayed the social conflict in his movies. He wrote, "Lee inadvertently gives way to dominant cinema's reflex strategy of containment that of depicting complex social conflict as disputes between individuals, where deliberated collective action is either impossible or necessary."[32] When Lee depicted interracial conflicts in Mookie's community, the animosity often was personal, rather than communal or collective, minimizing the degree of tensions. Even after the scene of the mob, "relations of domination remain exactly the same, while neighborhood consciousness remains unelevated, unchanged by the night's events." As a result, "any possibility of interrogating the *causes* of the racism and violence that vex the community is safely contained and reduced to merely depicting the *spectacle* of racism and violence."[33]

African Americans and the Television Industry

For African Americans, contested terrains did not exist only in the film industry. Black images on television also have been controversial. Just like the film industry, the television industry had been dominated by whites. African American presence was very low both in front of and behind cameras. Even when they appeared on a screen for a rare occasion, their image was based on mid-twentieth–century stereotypes. The 1960s and 1970s witnessed an increasing number of discussions about the roles and responsibilities of the television industry in American society, including its effect on racial politics. By the 1960s, television had become an integral part of American lifestyle. It became "an environment." In other words, it was "a symbolic environment into which one is now born and raised."[34] What people saw on television was as influential as what they encountered in their own lives off the television screen. It was in this social and industrial milieu that stereotyped and "safe" African American characters became popular in the white-owned television industry. *The Cosby Show* was one of such programs that showed African Americans but also implied that they were only acceptable in the American living room after they no longer posed a threat to whites.

After the introduction of television to American middle-class living rooms, it continued to develop socially as African American struggles during and after the Civil Rights movement persisted. During the 1960s, Martin Luther King Jr. and other activists strategically used television to advance their own agendas. Images of struggles distributed by television helped them gather a sympathetic audience for their cause. Even after the Civil Rights era was over, Black-centered programs such as *Roots* (1977) continued to politicize African Americans. It is also important to remember that television has been particularly important for African Americans. Unlike non-Black households that watched forty-seven hours of television a week, Black households spent almost seventy hours a week watching television. Blacks between the ages of two and seventeen watched 68 percent more television than non-Blacks of the same age group. When African Americans made up 12 percent of the nation's population, 20 percent of the prime-time audience was African American.[35] This is why in analyzing African American engagement in popular culture, understanding the relationship between Blacks and television, through which "the value-structure of society is symbolically represented," is especially important.[36]

Television was a segregated space at least until 1969. The history of televisual representation of African Americans in Jackson, Mississippi, eloquently shows their absence on television screens. Jackson's local television stations were dominated and controlled by whites. Both behind and in front of the television camera, few African Americans existed. As audiences, they were able to receive more news information about their own predominantly

African American neighborhood from a national news program that was aired on local programs run by and for whites. On rare occasions when African American characters appeared on television, stations suddenly had "technical problems," Blackening out the screen.[37]

Exclusion of Blacks from the television industry and Blackening out did not erase African Americans from the medium. African Americans sometimes still appeared on television. But they often reinforced existing racial thoughts. Throughout the 1960s, *Star Trek* (1966–1969), *Mission Impossible* (1966), *The Outcasts* (1968), and other shows included African American supporting actors. The social impact of their images was more significant when they began to play the protagonist. *The Nat "King" Cole Show* (1956–1957), *I Spy* (1965), *Julia* (1977), and other programs in the mid-century featured African American characters. To consider their presence as integrating in the television industry is premature because such shows only "attempted to make Blacks acceptable to whites by containing them or rendering them, if not culturally white, invisible." Nat Cole, for example, was an entertainer who NBC carefully distanced from the jazz culture that often was associated with drug, sex, rebellion, and so on.[38] Even with this whitening effort, the program failed to attract sufficient sponsorship in the South and NBC ended up canceling the show after only one season.[39] Social sentiments in the mid-century still made African American presence a risky business decision.

What whites considered urban social malaise was also on television, in shows such as *Good Times* (1974–1979), *Stanford and Son* (1972–1977), and *What's Happening!!* (1976–1979). These shows made it seem that only African Americans suffered from poverty, lack of housing, and unemployment. Just as television news did, these programs made the connection seem so strong that "many whites' perceptions of poverty are difficult to disentangle from their thinking about African Americans."[40] This further reinforced existing prejudice against them.

It also was assumed that despite their difficulties in life, African Americans were "good-humored and united in racial solidarity" without seemingly causing significant threats to the white society. Such images continued to idealize and legitimize "a normative white middle-class construction of family, love, and happiness."[41] If not involved in illicit or immoral activities, or suffering from poverty, African Americans were pictured as someone similar to Poitier. The presence of African American characters on the television screen, therefore, did not mean racial prejudice was gone.

Of all the similar programs in the mid-twentieth century, *The Cosby Show* attracted most scholarly investigation. The show revealed that the key to appear on the white-dominated television screen was to hide Blackness. The Huxtable family was only acceptable because it was not a typical African American family. This show took the strategy of what Michael Eric Dyson calls "accidental Blackness." This attitude that "we are human beings who by accident of birth *happen* to be Black" is opposite of "incidental Blackness," which believes that "we are proud to be Black, but it is only one strand of our identity."[42] Actor Bill Cosby did not speak with a Black accent or in broken English. He did not mug or make a scene out of a minor incident.[43] As a result, Cosby portrayed the Huxtables as a replica of the white family, which just *happened* to have Black skin.

Cosby's occupation was also a symptom of American mainstream popular culture's resistance to total acceptance of African American popular culture. Cosby, acting as Heathcliff Huxtable, was a gynecologist, and later an obstetrician. This calculated choice of job should not be ignored because of the historical portrayal of African American males as hypersexual. Numerous accounts by Black slaves state that they were frequently accused of uncontrollable sexual desire that resulted in violence and intimidation. During the nadir of American history, it was also a reason for lynching. Even in the mid-century, African Americans knew, especially in the South, not to make eye contact with white females.[44] Allowing an African American male to be on a screen, as stated earlier was a risky business decision, but as Michel Foucault assessed, medicalization of sexuality was an effective way to control and contain the idea of sex. It was the exact strategy that *The Cosby Show* employed without desexing the Huxtables.[45] Fiske speculated whether or not "making Cliff into a gynecologist is a way of rendering Black sexuality safe by medicalizing it and thus of defusing its threat to the white imagination."[46]

The hidden message was that African American families were welcome in white society only after they erased their African American characteristics and became upper middle class.[47] Privileged African Americans who had been robbed of their Blackness appeared on prime-time fictional television programs like *The Cosby Show,* whereas lower-class Blacks remained confined in news programs as troublemakers.[48] The sitcom was, in essence, a sign of intraracial class divisions.[49] On the one hand, a privileged few from the African American community who were economically fortunate and who were willing to associate themselves with Cosby could enjoy the fictional world that American popular culture offered. On the other, a vast majority continued to suffer from the limiting images that white-owned network shows and news programs put forth. As Alan Nadel argued, television created "a nation of fragments."[50]

The Cosby Show was also a representative case of the counterproductive relationship between African Americans and mainstream American popular culture. American popular culture was only willing to incorporate Black popular culture on a very selective basis.[51] On the one hand, African Americans were no longer excluded from popular culture. Similar to the heroic roles that blaxploitation film actors played, "respectable" African American characters on television also presented a positive image of the race. By doing so, they were allowed to cross the color line. On the other hand, such programs did not truly show the struggles of African Americans as they were experiencing them. What was more detrimental to African Americans was not just exclusive television company policies or negative images. It was the fact that television had the power to legitimize what it showed.[52] Because of television shows like *The Cosby Show*, whites' investment in the idea that the United States had become colorblind increasingly seemed valid.[53] *The Cosby Show* allowed its white audience to believe that race was no longer a problem.[54] The problem of the twentieth century was no longer that of the color line. As a result, such programs deprived African Americans of the tools to realize social advancement. Fiske argues:

> Members of disempowered races . . . cannot afford the luxury of color blindness, for their social survival depends upon racial awareness. They need to see racial difference constantly in order to defend themselves against the power inscibed [sic] in it, and consequently they are at times able to make tactical uses of the technology whose strategy is normally hostile to them.[55]

Mass-produced images of African Americans from the perspective of the white industry, which served to satisfy white audience's stereotypes, hurt African Americans. *The Cosby Show* was only one such television program. Television excluded African Americans in some cases. It included them if they could be contained. But it also legitimated the false ideas about American racial politics and made the concept of colorblindness realistic.

African Americans did not remain silent. They also used televisual images about themselves to advance their political agenda. Television during the Civil Rights era could "cover the grievances and abuses of the Black masses and send them nationwide, perhaps worldwide."[56] Similarly, "television has a unique ability to take [its audience] directly to the scene of a racial conflict—and then to take [them], in a documentary, to the story behind the scene. . . . The Negro revolution of the 1960's [sic] could not have occurred without the television coverage that brought it to almost every home in the land." Television was "a medium with great emotional impact" that civil rights activists could mobilize to affect the psychology of both Blacks and whites.[57] Emerging approximately around the same time, both television and the Civil Rights movement followed similar paths to attract wider audiences.[58]

Martin Luther King Jr. was fully aware of the power of television. When he attended or organized a protest, he made sure that television cameras were present. He believed in the effectiveness of television so much that when he learned that a camera crew would not show up to an event, he would sometimes cancel his marches.[59] His belief in television was twofold. Molefi Kete Asante speculates that "if the demonstrators were attacked by police and hecklers the nation would be repulsed by the cruelty received by the peaceful marchers. Secondly, he thought that policemen and hecklers who observed their violent behavior toward peaceful demonstrators would, in a moment of reflection, be ashamed."[60] King's assumption was correct. After the Little Rock incident was televised, there was a "new seriousness" in the relationship between television and the Civil Rights movement.[61] Stokely Carmichael remembers thinking, "Now there is a man who's desperately needed in this country" every time he saw King on television.[62] Television's portrayals of Birmingham and Selma "helped turn the tide of world opinion in favor of civil rights." Media began to recognize civil rights as "a moral good."[63]

During his civil rights activism, King maintained a close relationship with network television stations. He not only expected television programs to show racial struggles in the South, but also tried to appear on television programs himself. His interest in television worked particularly well because the television industry also wished King to appear on television. Sasha Torres elaborates on this reciprocal relationship stating "from 1955 to 1965, both the Civil Rights movement and the television industry shared the urgent desire to forge a new, and newly *national*, consensus on the meanings and functions of racial difference." As much as King wanted to advance his agenda to the national audience across racial boundaries, network television also wanted to attract wider audience across geographic boundary.[64] Images of Bull Connor in Birmingham and Jim Clark in Selma that television stations were willing to air for their own good also allowed King and other Black activists to advance their agenda.

On the national level, commercial networks heavily reported on the Civil Rights movement. On May 26, 1963, James Meredith, the first African American student to be enrolled at the University of Mississippi appeared on *Meet the Press* to talk about the Civil Rights movement. An episode of

Issues and Answers on ABC, on June 16, 1963, discussed the merits of the civil rights legislation proposed by President Kennedy. About two months later, *Meet the Press* invited Sen. Richard Russell of Georgia who challenged the legislation. ABC broadcast a five-part special series *Crucial Summer* and NBC aired *The American Revolutions of '63* to offer documentary presentations. The former analyzed the nature and history of racism in the United States and showed African Americans trying to overcome this struggle. The latter analyzed what civil rights activists were demanding, from voting rights to housing, employment, and so forth. These images successfully stirred up the American public. Television was reflecting the ongoing change in society while at the same time reinforcing new social consciousness.[65]

Pro-Civil Rights African American images also existed outside of documentary and news programs. One of the most influential productions from the post-Civil Rights era was *Roots*. Based on Alex Haley's seminal work, the show was on television for eight evenings from January 23 to January 30, 1977. Its sequel, *Roots: The Next Generations* was put on air between February 18 and February 25, 1979. The show was "the epic story of the Black American odyssey from Africa through slavery to the twentieth century." When the show was aired, approximately 130 million people, or almost half the nation's population, tuned in.

> [The program] brought to millions of Americans, for the first time, the story of the horrors of slavery and the noble struggles of Black Americans. This television representation of Blacks remained anchored by familiar commitments to economic mobility, family cohesion, private property, and the notion of America as a land of immigrants held together by shared struggles of hardships and ultimate triumph.[66]

The show was a big success, although before *Roots*, "dramatic portrayals of nonwhites held little appeal for most viewers." Once the show was over, hundreds of colleges started offering courses on *Roots*, and many African Americans asked the National Archives in Washington, DC, how they could trace their family genealogies.[67] This show proved television possessed the power to mobilize the masses.

The show impacted African Americans in two ways. First, it emphasized and taught them the importance of knowing and appreciating African American history and heritage. *Roots* allowed "the popular media discourse about slavery [to move] from one of almost complete invisibility (never mind structured racial subordination, human degradation, and economic exploration) to one of ethnicity, immigrants, and human triumph."[68] It also boosted the self-esteem of African American audience. Many African Americans felt that their heritage was worth a television program and public attention. The history of Black exclusion from the television screen had prevented them from feeling such a way. Although the program did not deal in depth with human degradation, economic exploitation of Black labor, racial subordination, structural racism, and other key concepts, *Roots* effectively dealt with what had been a taboo subject in American visual popular culture—slavery.

The story also was that of "African American cultural struggles over the sign on Blackness." It opened "a discursive space in mass media and popular culture within which contemporary discourses of Blackness developed and circulated." Young African American audiences had a chance to consider the history of Black media representation, as well as Black history as represented in *Roots*. Furthermore, it encouraged them to think about what could be done to improve the way in which African Americans were portrayed on television. The show also "helped to alter slightly, even momentarily interrupt, the gaze of television's idealized white middle-class viewers and subjects" about African Americans.[69] Not only did the program affect the psychology of African Americans, but also that of whites.

Although the show attracted a large audience, *Roots* was accused of creating numerous misunderstandings. The show was misleading and it omitted some of the significant issues about Black history. Critics such as Herman Gray argued, however, that *despite* its weaknesses, the show was successful in that it encouraged people to talk and think about issues concerning African Americans. MacDonald, on the other hand, was more critical. He argues that *because of* such misleading information, the show actually had negative impact on African Americans. After watching the show, white viewers could "leave the series blaming impoverished contemporary Blacks for their own social deprivation." They also could feel as though they had learned sufficiently about Black struggles and that "contemporary Black poverty was the product of individual weakness and lack of appreciation." Much of the white audience detached its own racial ideologies from the program. Even when they took what the show presented more personally, MacDonald argued that "in measuring their prejudices against those in the *Roots* programs, most whites could feel better about themselves, [because they] did not physically maim or kill Blacks in contemporary America."[70]

As later chapters show, production was a major part of the effort to attain fair representation on television. *Roots* generated debates on Black experiences and much was discussed regarding its production process.

Positive evaluations of *Roots* emphasized content so much that it paid little attention to its production crew. The writers for *Roots* were all white. The production process was no different from typical melodrama program. As a result, Bogle concludes that the show became not necessarily a "recounting of Black history," but rather "twelve hours of glorious, high-flung soap opera."[71] This was not a criticism limited to *Roots*, however. Classen, Fisher, and other historians of African American visual popular culture had stated that despite the name "African American," popular culture consumed by African Americans had often been produced by whites. The Black Entertainment Television's (BET's) contemporary white ownership continues to be debated years after Robert Johnson sold the company.[72] Because African Americans historically have been excluded from the production side of popular culture, a repeating question is: Who has the legitimacy to represent their voice? Is it whites who have access to production resources? Or is it privileged Blacks like Spike Lee? Or should it be someone else? African American broadcasting visual popular culture in the 1960s and 1970s did not have the answers to these questions.

African Americans in Mainstream Visual Culture

African American visual popular culture in the 1960s and 1970s was a site of struggle and negotiation. Blacks in movies and on television screens during this period were not as stereotyped as they were earlier in the century. Both radio and film in the early twentieth century generated, popularized, and promulgated racism images. They were not the first to do so. The history of American popular culture shows that theatrical performances, magazines, and other forms of cultural productions were precursors to this trend even during early slavery. American popular culture in mid-century continued to live with racist thoughts. Racial prejudices existed only less explicitly but more systematically. Because of its latent nature, it was more difficult to detect. Even those who participated in popular culture, such as actors like Sidney Poitier, did not seem to understand the intricacy of the subtext embedded in their own films and other movies and programs. Such cultural productions reflected and reinforced American mainstream culture's process of accepting African Americans only if they seemed safe to whites. More importantly, Black presence in visual culture functioned as legitimate evidence that race no longer mattered. Just as the popularity and wealth that African American hip-hop artists achieved, or the success story of Barack Obama are wrongfully used as evidence of a post-racial society today, these actors were considered as the proof that the United States was no longer racist. Not only could African American actors appear in movies and television, they were portrayed in a positive manner. Many white critics argued that it was a sign of progress. A hidden message, however, was the fact that only African Americans who fit into whitewashed images of Blackness could be a part of American popular culture.

African American artists, activists, and leaders tried to reverse the negative impact of white-produced African American images by producing their own visual cultural products. In movies, Spike Lee was a prominent director who attracted much attention from media and critics. Martin Luther King Jr. was another figure who proactively put Black experiences and struggles in front of the television camera to reach national audiences and to gain support for the Civil Rights movement. Although produced by whites, *Roots* stimulated discussions and made Americans aware. To different degrees, Lee and *Roots* successfully mobilized African American political consciousness. However, these images contained questions about authenticity and appropriateness. Critics argued that Lee's movies were homophobic and downgraded women. *Roots* failed to introduce some of the most tragic experiences of African Americans, so as not to alienate white audiences. African American visual popular culture could not find the answer to this problem by the end of the 1970s.

As African American film and television culture developed during the two decades in the mid-twentieth century, it constantly battled with the hegemonic presence of American popular culture that historically capitalized on Blacks but that was unwilling to accept them as equal participants. It is important to recognize that the mid-century history of African American popular culture is not just a history of misrepresentation and a lack of representation. It also is a history of African American activism that used popular culture to gain support for the Civil Rights movement, to improve Black self-esteem, and to achieve social equality. As popular culture was a site where the popular resisted the elite, African American popular culture was a site where Blacks resisted the American popular, which was relatively more powerful than the former.

Chapter 2

Cable Television

Past and Present

The trend of televisual resistance by whites against Black representation took a more concrete shape with cable television. Decades of erasure, distortion, exclusion, and discrimination in visual culture culminated in the debate on fair representation of African Americans and other minorities on the new medium.[1] There is approximately a sixty-year history of cable television and African American involvement in it. Since the mid-century when the first body of scholarship on cable television began to appear, studies tended to focus more substantially on technical, economic, and legislative analyses than on social and political issues. Most referred to minorities only in passing and few conducted full-scale studies on public access, minority production, fair representation, and other pertinent concerns. Additionally, despite the slight increase in Afrocentric literature toward the end of the twentieth century and early twenty-first century, a majority of the literature was written from Euro-, techno-, or bureau-centric perspectives with paternalistic, tokenist, or essentialist preconceptions about African Americans. It was the same within the academy. The rise of interdisciplinary fields such as American Studies, African American Studies, Media Studies, and others, did not result in scholarships focusing on cable television and African Americans.

A dearth of literature, however, does not mean that African Americans have no history or place in the development and history of cable television. As academics in African American Studies have attested for the past four decades, historians in this field have the responsibility to unearth the buried past of Blacks. To find such hidden, erased, and distorted experiences in history necessitates a study focusing on the pertinent environment that contributed to the creation of the environment in which Blackness was seldom discussed. This is why we must understand how federal and state policies were shaped and interpreted, the Federal Communications

Commission's (FCC) restrictions were evaluated and criticized, how the African American presence on television was appreciated and rejected, and how the industry's focus on profit generation succeeded and failed. It also is important to remember that without the cable applicants' success in satisfying the FCC's legal, economic, and technical requirements, local African American communities would not have been able to mobilize technology as their tool for empowerment.[2] Unfortunately, there has been a lack of comprehensive review of scholarship on African Americans and cable television, or more broadly minorities and media, including issues concerning legislation, economy, and technology, especially with a focus on African Americans. To understand and share the perspectives of those involved in historical events adds more objectivity and validity to the scholarly and community knowledge that a research project generates. Such an approach helps us fulfill both descriptive and corrective duties of African American Studies scholarship.[3]

Early studies focused on how cable television worked, how it differed from the conventional broadcasting television service, what kind of training people should undergo to obtain a job in the new industry, and other technical and occupational information. From the 1960s through the 1980s, there was an augmentation in legal publications. As the FCC increased its control over the cable television market, both industries and consumers were concerned about the implication. As a result, legal journals and other academic writings examined issues such as copyrights, federal control over the industry, and acts concerning the media. As an increasing number of cities adopted cable television systems, studies began to focus more on the social aspect of the technology. Some of the topics included how minority groups could benefit from cable television, how public access channels could enrich urban life, and how educational institutions could take advantage of the cable network. The 1970s was the richest decade in terms of the amount of literature produced on cable television.

The general trend around the 1980s was a large number of publications in the field of law focused on legislative amendments that affected the cable industry. Although some sociological and political publications exist from the decade, most focus on the status quo of the industry as a business investment opportunity. Few make extensive references to minorities or African Americans and their involvement in cable television franchising or development.

The trend continued in the 1990s and still continues today. However, not only did the amount of publications decline in general, race almost disappeared as a framework through which to study cable television. Additionally discussions, concerning public and community access also began to disappear. Most publications focus on cable channels for broader audiences. When African Americans are discussed in such a context, it is often because of BET. But an important question to ask is if BET and other commercial cable media giants including MTV truly offered the ultimate answer and solution to the century-long history of misrepresentation. The answer, of course, is no. It had been a series of smaller-scale grassroots attempts that made the difference in African American communities.

Prelude: Early Days of Cable Television, 1948–1969

The history of cable television dates back to 1948. John Walson, a lineman for Pennsylvania Power & Light, and the owner of a local electronic appliance store in Mahanoy City, Pennsylvania, realized that the low-quality television signal reception in his neighborhood accounted for his poor sales in television sets. The quality of the off-the-air signal reception was easily affected by the distance from the television stations and the topographical impediments such as mountains and buildings. To many of the audience in Mahanoy City, the signals from Philadelphia stations were too subtle to receive off-the-air. The bad signal reception resulted in the poor image quality on the screen. Consequently, the sales of television at Walson's store were stagnant.[4]

Walson's solution was to build a reception tower on top of the local hill and lay a wire between the new structure and his home television set. This solution brought a high-quality signal reception to the Mahoney neighborhood. Once Walson successfully established his personal cable television system, many of his customers who had purchased television sets expressed their interests in being wired to his antenna. Walson's business expanded far more rapidly after the launch of the system. Charging $2 a month as the service fee, he had 727 customers by the middle of 1948 and encouraged other entrepreneurs to start similar services in rural areas where television signal reception was inadequate. By 1955, there were approximately four-hundred cable television systems serving 150,000 subscribers.[5]

As cable television expanded, academics in different fields such as media studies, sociology, science and technology studies and others, as well as those in the relevant industries began publishing studies. Although few studies were done initially, scholarship increased by the end of the 1960s and peaked in the 1970s and 1980s. By the new millennium, discussions on cable television became a part of a larger discourse of media and technology studies that encompassed commercial cable television services and the Internet.

Legal, Economic, Professional, and Technological Concerns, 1960–1979

The vast majority of studies on cable television published in the late 1960s and 1970s focused on state control, profitability, occupational opportunities, and technical features. Although cable television had more than twenty years of history by then, its popularity was much smaller than that of network television services. It was natural that capitalist society strove to find a way to turn the new media tool into a profit-generating and job-creating tool. Simultaneously, after twenty years, cable television had become too large to avoid federal regulation. With high interests in legal issues, only a handful scholars made any kind of extensive reference to cable's impact on African Americans and other minority groups. Even though television served as a powerful community-building tool for African Americans during their civil rights efforts, as examined earlier in this volume, American media society and its surroundings were yet to be proactive in exploring the potential of new media. As a result, issues concerning community channels attracted less attention than those concerning the fee structure, governance, or marketing.

The lack of scholarship on race in cable television does not mean that racial minorities including African Americans remained unaffected by the technology. Cable television was "still unknown to vast majority of Americans, for it serve[d] only about 6 or 7 percent of the nation's population."[6] Although as a communication medium it was small compared with other media forms, what was discussed and agreed on would indeed affect Blacks during the decades to come. Many of the studies investigated FCC policies, federal decisions, and the political environment surrounding cable television. It was in this decade when both Bostonians and Detroiters were most active in studying the feasibility to introduce cable television to their municipalities and what they learned affected the outcome in the 1980s.

The 1970s began with Stephen R. Barnett's extensive legal research, which signaled the decade's interest in legislative policies. Having served as a consultant to the President's Task Force on Communications Policy, Barnett originally prepared the document to submit to the FCC as his comments to its rulemaking procedure. He argued that "[a]s part of [the] new media-mindedness, there is a growing struggle over control of and access to the media."[7] In this lengthy study, Barnett explored the impact of the antitrust laws, common control between various media systems serving the same geographical market, and other questions that came up in the 1960s and what he considered to be consequential in the industry's development in the coming decade. Borrowing Marshall McLuhan's language, he claimed that "[t]he medium these days is not only the message; it is also the topic."[8] Although Barnett successfully brought readers' attention to the fact that cable television still had to undergo a series of discussions to capitalize on its potential, he made no substantial reference to the impact of control and medial concentration on racial and ethnic minorities.

The FCC's restrictions over the cable television industry imposed in 1966 and 1968 were common topics of discussion. Reflecting both commercial and legal interests of the cable industry, Williams S. Comanor and Bridger M. Mitchell explained that the industry was free from the FCC's control until 1966. The commission's publication of its *Second Report and Order* signaled the first federal attempt to restrict the market. Two years later, the FCC announced additional rules. The most controversial of these regulations was the distinction made by the FCC between top one-hundred markets in the United States and the rest. If the market was located in one of the largest one-hundred cities designated by the FCC, the system was not allowed to import distant signals. This meant that these cities were not allowed to bring programs from outside of the market, unless the system had been established before the onset of the regulation to meet the "grandfather provision." The authors argued that because of this restriction, there were a limited number of cable television systems in urban areas. Although Comanor and Mitchell acknowledged the positive impact of the FCC's effort to relax its regulation regarding the top one-hundred markets proposed in July 2, 1970, they remained skeptical about the profitability of cable television systems in such markets even after the reregulation.[9]

The myopic commercial interests in cable television prevented the development of the medium for the benefit of minorities.[10] Being obsessed with generating profit, interest groups had paid little attention to program production for African Americans and other minority populations. Minority audiences, who already constituted a large portion of media audience, would further increase in number. Stating that "[o]ne of the great advantages of cable television is that it can cater to some degree to minorities," Richard Roud in 1971 saw a potential in this technology for serving minorities that other broadcasting media could not attain.[11] But few were aware of the potential.

Judy Crichton similarly argued that residents in low-income areas could benefit from cable television systems. Although she cautioned that "no matter how hobble our intentions, this system would also be competing with the great power and seduction of commercial television," she also claimed that "A community oriented cable television system could provide the inner city consumer with direct, personal information."[12] She identified the likelihood of the custom-made format of cable television access channels attracting particular interest groups as core audience.

Harald Mendelsohn's work, "The Neglected Majority: Mass Communications and the Working Person," focused on the socioeconomic status of the audience to argue the importance of media, including television for working-class viewers. Although his study prioritized class over race and ethnicity, it was indicative of the fact that many of the discussions about bringing cable television to urban areas had to deal with African Americans' disadvantage due not only to their Blackness but also to their economic status. He explains that television was a major, and often the only, source of entertainment for working-class people because of their lack of financial resources to spend on other forms of entertainment.[13] He argued that such a reliance further necessitated a form of provision so that people with lower socioeconomic status must be given opportunities to voice the "needs, grievances, and problems" on air that they encountered in their daily lives. This was expected to allow them not only to consume television but also to participate in the production. It also would add authenticity and a new flow of information that would positively impact their lives.[14]

Mendelsohn continued his argument by enumerating seven objectives and benefits of cable television for working-class people. He claimed that through cable television, this population could be better aware of the "outside world"; reach out of their "psychologically comfortable but socially stalemated peer-centered tradition-bound narrow social environments"; better understand social institutions that many of them feel are threatening; realize their lower socioeconomic status is not their personal fault but is due to "remote institutional breakdowns, long-term social processes, misguided public policies, and inept social decision-making"; learn to adapt to social changes around them; prepare themselves against political exploitation; and produce entertainment to match their own taste.[15]

By 1971, it was clear that cable television could contribute to the surveillance of public areas to protect citizens from crime, violence, and other social concerns as well as to make immediate moves against fire, flood, and other disasters. A cable system proposed in Florida was slated to warn its subscribers about severe weather, the one in Kansas for emergency alerts, and the one in Virginia against fire.[16] Using media in such a way to monitor and possibly to discipline could raise concerns among African

American communities especially with their history of disproportionate control by the white-dominated authorities. Gilber Cranberg agreed with the President's Commission on Law Enforcement and Administration of Justice by stating "serious problems of Negro hostility to the police [exist] in virtually all medium and large cities." He optimistically predicted that cable television enhanced the mutual understanding between the police and the local community by encouraging dialogues between the two, explaining police procedures, inviting minority engagements, and forming committees with minority members.[17]

Ithiel de Sola Pool and Herbert E. Alexander argued that "CATV [cable television] is not one thing, but a family of things sharing only the physical fact of a wire in common."[18] Focusing on the political implication of cable television, they explained that political candidates were able to select this over-the-television campaign audience and avoid more expensive multiconstituency media such as broadcasting television.[19] Based on their case study in Waianae, Hawaii, Herbert S. Dordick and Jack Lyle also claimed that local political candidates were able to benefit from sending messages to their voters via cable television.[20] Although they tended to be more reserved in terms of their estimate of cable television's future expansion, Pool and Alexander claimed that cable television "would permit various communities to be reached selectively, giving ingress into homes with selected socioeconomic or demographic characteristics."[21] All of this was possible because of narrowcasting. Community members obtained the political information that mattered most to them. The first two authors explained the value of cablecasting. They argued that most of the reclusive communities in American neighborhoods are ethnically defined, including Black inner cities. Because of the abundance of channels, however, such neighborhoods *could* have their own channel to establish "an outlet for their own distinctive cultural products."[22]

Lionel Kestenbaum presented three ways to realize the original expectations about cable television. He argued that "[c]able communication is thought to open vast opportunities for diversity and creativity in informational, entertainment and instructional programming, and to establish a basis for an array of information, data and communications services" that was not available with network television or the other forms of medium. As solutions, he suggested that an increase in "the channel availability can eliminate technical obstacles to broad access." By having an increased number of channels, without major technical advancement, the broader public would be able to benefit from media content. Second, he argued, "the availability of cable channels on an existing system reduces financial barriers to access, and distinguishes such channels from unused UHF allocations, which require substantial investment." Because the start-up cost was often a large obstacle for many network stations, cable system's low financial entry barrier was appealing. Last, "the cable system operator should have an economic incentive to welcome use of any unused channel for an additional service, so long as the cost to him of using the channel is less than the incremental revenue received, either from an increase in the number of subscribers, from payments for uses of the channels, or both."[23] The channel abundance that the cable system would bring to the local community had the potential to solve some of the shortcomings of network television.

Referring to the reregulation proposal by the FCC, Don R. Le Duc and Richard A. Posner agreed with Comanor and Mitchell's statement from the previous year and warned that the commission's restriction had hindered the development of cable television systems in major urban areas. They are particularly critical about the FCC's restriction. Le Duc emphasized the significance of the matter especially because cable television had been considered "a medium destined to revolutionlize the existing impersonal program delivery process through its broad promise of public access and message diversity."[24] Despite this initial hope, the cable coverage had been kept as low as 60 percent in saturation in the market that was not affected by the FCC's regulation for the top one-hundred markets. Because approximately 90 percent of the potential cable television audience lived in those one-hundred markets, he suggested that only 60 percent of one-tenth of the overall market had actually been served. Rolla Edward Park also agreed by stating, "FCC rules have effectively prohibited the importation of any outside signals, while cable operators have maintained that they need outside signals to attract enough customers to operate profitably in most large markets." No matter how willing cable operators might have been, without a profitable prospect, they could not take the risk involved with establishing a new cable system that would not be able to import distant signals.[25]

The restriction by the FCC was particularly consequential both to the industry and to its audience because by the 1960s, half of the customers still could not receive all three networks' signals and could have potentially benefited from the cable service. Within the FCC's historical context, its regulations appeared problematic because Congress had never passed any legislation that gave the FCC the authority to assert its control over the cable television industry.[26]

Monroe E. Price's work examined the implication of the rules set by the FCC in 1972. Explaining the background leading up to the regulation, he portrayed the optimism in the early 1970 by arguing:

In 1970, dreams of a new communications system were upon us, and vast implications were being forecast. Cable television, it was said, would have dramatic consequences for the quality of life in the cities, the delivery of health and educational services, the dissemination of news and public affairs, the methods of purchase and sale, and the exercise of the franchise.[27]

The benefits that the new technology was going to bring did not entail racial or ethnic representation. Although Price mentioned the "desirability of pluralism" on cable television as a solution to social malaise, he was not specific as to how to bring about the "participatory democracy."[28] An important question regarding "participation by whom?" seemed to have slipped his mind.

Richard C. Kletter claimed that cable television could democratize the media because it could focus on local issues. He argued:

> The importance of localism to local viewers should not be underestimated. Look at the treatment of civil issues. Broadcasting portrays such issues in abstractions: unemployment, housing, health care problems far removed from their local context. Issues such as "civil liberties versus civic order," or "freedom versus authority" are not analyzed in terms of frightened neighbors or threatened individuals; instead, people's faces turn into Arabic [sic] numerals in statistical tables as some high-ranking official discusses "a national problem." This kind of approach breeds frustration.[29]

Kletter was not the only one to emphasize the importance of research on locality. Walter Baer understood that by the mid-1970s, cable television was no longer a technology for better signal reception. Baer explained that it is "turning into a genuine urban communication system, with profound implications for our entire society." He also suggested that the decisions made in the 1970s would "reverberate through the 1980s."[30] This was particularly true because, as Robert K. Yin claimed in the previous year, cable television was still in the development phase in the 1970s. New capabilities would arrive over the next few years. In order to best use such potentials for local services, decisions made on control, access, site selection, costs, and other issues in the 1970s would be consequential.[31]

The research publication by the Community Research Incorporated (CRI) made specific recommendations about how Price's democratic idealism could be brought to cable television. This nonprofit organization conducted the study to fill in some of the issues that the nine-month–long analysis by the Rand Corporation, a major research company in cable television, had failed to address concerning the role of cable television in Dayton, Ohio. The Rand project focused on technical and economic issues, as was the often the case with other research projects in this decade. The CRI, on the other hand, examined social and political issues in depth. As a result, it generated a model of what municipalities could expect from cable television. This is one of the few resources that made an extensive reference to cable's impact on minority groups.[32]

The CRI's report emphasized minority concerns by addressing an alternative hiring policy and promoting the idea of access programming. In terms of employment opportunities, its recommendation based on the American Civil Liberties Union (ACLU) Model Code read:

> No grantee shall discriminate on the basis of sex, race, national origin, religion, creed, or arrest or conviction records in hiring and promoting employees. Each grantee shall seek out and train employees so that minority groups are represented in its employee work force in the same relative proportion as they are represented in the population of the franchise area. Each grantee shall file an affirmative action plan to this end annually with the franchisor. This plan shall include a report of persons employed, together with their positions and salaries, by categories listed in the first sentence of this section.

Very similar hiring principles based on affirmative action appeared in the franchise agreement in Boston and Detroit approximately ten years later. The CRI claimed that the minority-based organizations such as the Urban League and the National Association for the Advancement of Colored People (NAACP) would benefit from access channels reserved for public use. Making such recommendations and analyses, the CRI research foresaw positive changes that cable television with concrete visions could bring.[33]

Dell Keehn positioned the issue of African American involvement in cable television in a larger community-planning context. He argued that because the urbanization and concentrations of African Americans rapidly became more apparent, community control had been a major topic of interest among citizens. African Americans owned less than 1 percent of the total businesses in the United States. Only 12 of 350 radio stations with predominantly Black listeners were owned by African Americans. When it came to commercial television stations, none of the 800 stations were owned by African Americans. Having the potential to generate $4.4 billion and millions of jobs by the end of the decade, cable television could become a major impetus for Black and other minority residents to secure a sense of control over their community and reverse the lack of African American presence in the media industry. The author claimed that African American residents were not likely to subscribe to cable television to get better reception as was originally designed in the late 1940s. Especially in the twenty-five cities where there were more than 100,000 African American residents, cable television would serve as the drive for the African American community advancement.[34]

Leonard Ross and Walter Baer's studies in the mid-1970s maintain a relatively optimistic hope about the potential of cable television particularly because of the termination of the top one-hundred market freeze on March 31, 1972, by the FCC. It had been ongoing since 1966.[35] Cable television was described as "potentially a major corrective to other media, especially over-the-air TV . . . which are governed largely by three networks. . . . [T]his could be the turning point leading toward a less centralized, more diversified, communication media."[36] Ross continued that despite "years of hesitation," Congress and the FCC were ready to deregulate. Even though the reregulation would not be the total elimination of restriction, due to the large number of channels available through the new system and lowered entry barrier to many of the top one-hundred markets, programs produced particularly for the minority audience would be possible. It also was feasible theoretically for such channels to compete against football games and other programs that tend to attract large audience.[37]

Similarly, Baer predicted that "[i]f all goes well, some experts believe that cable television will influence the way we live as radically as the automobile and telephone have done."[38] The emphasis by both Ross and Baer on the positive aspect of cable television, which had not yet been realized, proves that the development of cable television had been stagnant since 1948 primarily because of regulation by the FCC and operators' hesitations due to economic reasons, despite the numeric increase in subscribers. Baer's study showed that by June 1973, 7.2 million, or 10 percent of, American households subscribed to cable television. Approximately 3,000 operating systems served about 5,700 communities. In-depth study of his statistics, however, presented the consequences of FCC rules that had complicated the entry to large markets. Only 105 systems had more than 10,000 subscribers. Two-thirds of the systems had fewer than 2,000 subscribers.[39] These figures suggested that more than twenty years after the initial introduction of the technology, it had not fully served inner-city communities.

Gilbert Gillespie also painted an optimistic picture of cable television. He outright agreed with Frank Korman from the Center for Communications Research at University of Texas at Austin who had claimed more ambitiously than Baer that "[t]he educational and social impact of cable technology is likely to be greater than that of any other foreseeable advance in telecommunications technology" and that "[t]he social impact of cable will alter our basic economic and political institutions in unforeseen ways."[40] Although Gillespie did not refer to specific examples of cable television implementation in African American communities, he based his belief on the historical case studies in urban areas such as Kansas City, New York City, and others that "new and extensive developments that were afoot in the field of telecommunications can work considerable influence upon the quality of human life."[41]

Charles Tate's analysis of cable television as "the last communications frontier for the oppressed" provided both risks and potentials that he saw with the technology. Although he understood that community leaders and organizers were interested in mobilizing new technologies for their own benefit, not many were knowledgeable of sophisticated communication systems. As a result, he cautioned that "Blacks and other minorities [may become] *more* powerless and dependent" if access to technology was based on the existing power structure. If it happened, $40 billion worth of African Americans' annual disposal income would be left untapped.[42]

On the other hand, if African Americans could find a way to partake in the communication system, they would make major contributors to the cable economy with their economy. Tate suggested that it is possible to reverse the history of Black erasure and subjugation in media. Compared with traditional communication technologies, "[c]able television is a better vehicle for achieving sizable gains in community organization, unification, control, and development." Even inner-city communities would be able to benefit from it. He listed six reasons for his belief. Tate argued:

> First, cable television systems are not presently installed in Black communities and central cities. Therefore, no entrenched interest group or power bloc can claim public protection for its investments. Second, franchises are issued by local, municipal governments, and the FCC has recommended the continuation of this process. Third, installation requires the actual stringing of cable on poles or the laying of cable underground along the streets of the ghetto. Individual hookups must then be made from these trunk lines to homes and apartments, and outlets must be

installed within these living units. Fourth, Black communities are a substantial segment of the urban subscriber market. Fifth, the great potential of cable in technology, economics, and the power of mass influence is ultimately tied to cablecasting or local programming origination. Sixth, cable will be used in a wide variety of applications, apart from entertainment programming. Education, health, welfare, safety, crime prevention, and police operations are a few of the likely uses.[43]

He also claimed that the successful history of soul radio in Black "ghettos" attested that communication technology could serve minority communities and their plan for community organizations and participation.[44]

Marrion Hayes Hull's analysis was one of the most extensive works on minorities and cable television published in the 1970s. Referring to the importance of television as the most popular news source for Americans "in the ghetto," and the strengthening sense of disillusionment about broadcast television among minorities, the author explained that major minority organization including the National Urban League and the National Black Media Coalition had "established study groups to investigate the potential use of cable technology for the economic, social, and political betterment of their constituents."[45] Hull predicted that although there were only two African American–owned cable systems in operation—Gary, Indiana and Oshtemo, Michigan—there would be at least one African American–controlled system in all major cities in the country within ten years. She listed Atlanta, Georgia, Seattle, Washington, Dayton, Ohio, Los Angeles, California, Tuskegee, Alabama, and other cities where Black systems were on the way. She did not mention Detroit where another Black-owned and -operated cable television system appeared within ten years. Her prediction about the rise of African American systems, it seems, was correct.[46]

Hull suggested that the reason why African Americans had difficulty developing telecommunication technologies was due to "the exclusion of these groups from national policy-making committees, research projects and design activities for the system." Little was different from radio, television, or film. She continued:

> Minorities are not generally in a position to assess those needs that can be satisfied by cable and other technology, to test the technology's effectiveness in meeting these needs and to evaluate the impact telecommunications systems will have on the people they are intended to serve. For these reasons minorities

tend to adapt, belatedly, systems for which experimentation was performed outside minority communities or by persons who can only speculate as to its potential.[47]

Despite this problem, Hull was not pessimistic.

> Recently, however, minority institutions such as Black colleges, have begun to take advantage of the unique opportunity cable offers for participation in the development of new education and learning systems, the production of a wide range of new educational services, and the distribution and delivery of new services to Blacks, whites, business, government and other consumers.[48]

In reality, Hull introduced his readers to *The Black Corner* and *Gospel Time*, locally produced programs that served African American communities. The former was a weekly program in Beloit, Wisconsin which "explores Black thoughts concerning the past and present." The latter featured gospel groups and encouraged both Blacks and whites to engage with the program.[49]

In order to realize the type of community engagement Hull encouraged, the Cabinet Committee on Cable Communications, headed by Clay T. Whitehead, recommended that "[p]articipation by minority groups in cable system ownership, operation and programming should be facilitated."[50] The committee continued:

> The development of cable represents a unique opportunity for minority racial and ethnic groups to become actively involved in a new communications medium. Minority groups should have not only employment opportunities, but also full opportunity to participate in all aspects of cable ownership, operation, and programming.
>
> The general policy for the structure and regulation of the cable industry that we recommend would facilitate participation by all segments of society in cable ownership or control of channel use. Moreover, the local franchising authority should ensure opportunities for minority ownership and control in cable systems and programming.
>
> At the Federal level, the Equal Employment Opportunity Commission should devote special attention to the development of the cable industry to assure ample employment opportuni-

ties for minority group members. We also recommend that the Office of Minority Business Enterprise and the Small Business Administration of the Department of Commerce be directed to give high priority to cable and to propose any necessary special provisions, such as goal guarantees, to foster significant minority ownership or control of cable operations.[51]

The committee made practical suggestions about how on different levels cable television could serve local minority communities.

Practicality of cable television assumed the mobilizing power of the technology. The new style of television had the potential to cater to local needs. Public access channels were very different from traditional mass media. Through public access television, anybody without formal training in television content production would be able to share their work through cable systems. The same was true for narrowcasting, which promised to appeal to narrowly defined targeted audiences. In order to realize such a new service, the mobilization of the audience was vital. Furthermore, cable television might function "as a tool for the *internal communication purposes* of a community organization." In other words, "the motivation for viewing will be different and a small audience figure may be far more significant."[52]

Building a cable television system in a metropolitan market was a risky business for cable operators that could cost as much as $75,000 per mile just to lay cables. Because such operators did not believe that the business would be profitable without distant signal import in urban markets, growth in the top one-hundred markets remained slow. Although the cable penetration rate was 23.3 percent in rural areas, it was only 1.6 percent in major metropolitan areas by the end of the 1960s. Even though the average penetration rate from 1.7 percent to 9.7 percent between 1962 and 1972 seems to imply general popularization of the system, the growth was disproportionate and was not beneficial to inner-city minorities.[53]

There also existed more positive views of the development of the cable television industry during the post-1972 reregulation phase. George R. Townsend and J. Orrin Marlowe argued that no other industries except for cable television had been subject to as many and as strict rules by the FCC in the history of communication media, and that the restrictions imposed upon the cable television industry reflected FCC's interests in nurturing the UHF television stations and its "overly protective attitude toward television broadcasters" which had gained political clout by the mid-1960s.[54] They claimed that reregulation would have a salutary impact on the "information-hungry society" by enabling minority groups to find extra television channels available on the cable.[55] Media had always been restricted to the privileged few. Even locally produced news programs on radio could not avoid stereotypes of a limited number of anchors and other production staff. Additionally, no matter how conscious radio stations were, they only had a channel for twenty-four hours. It was only after the arrival of cable television that the daily amount of available hours increased thanks to channel abundance. Townsend and Marlowe continued, "An individual citizen can gain access by means of television to make a speech, present a play, sing a song, tap dance or to do anything not obscene or otherwise illegal." Such a system did not guarantee an audience. What mattered most was the simple fact that they could try and do it.[56] Robert E. Jacobson further summarized this aspect of cable television as a community tool by saying that "[o]nly when cable is seen as a public good, rather than as a luxury, will resources become available for its development. When cable is ruled a public good, it can be operated as a public enterprise, even as a 'yardstick' to measure the quality of current service provided by private corporations."[57]

The Rand Corporation was a major organization that carried out an extensive research project under a grant from the National Science Foundation. Starting its study in 1969, the Rand Corporation released more than ten reports covering from the FCC regulation to the social implication of the cable television in the 1970s. These reports, in essence, outlined the issues that Walter Baer introduced in *Cable Television: A Handbook for Decisionmaking* in 1973. Baer's work served as the scaffolding of the Rand reports in order to understand what his colleagues later called the "heavily forested area" of new media technology. Although some were critical of these works because, as the CRI explained in its study, they tended to focus more on economic and regulatory analyses than on social and political ones, the Rand Corporation remained in the forefront of the cable business during the decade and influenced other works at the time.[58]

Walter Baer, in the first Rand report that followed his own extensive work in 1973, reemphasized the importance of public access channels for minorities. He claimed:

> [We need to make] [p]ublic access channels available to individual citizens and community groups. Often using portable cameras, all sorts of groups—churches, Boy Scouts, minority groups, high school classes, crusaders for causes—can create and show their own programs. With public access, cable can become a

medium for local action instead of a distributor of prepackaged mass-consumption programs to a passive audience.[59]

He foresaw that between 1972 and 1980 there would be more local origination and public access programs catering to minority groups. Additionally, he encouraged local government officials and citizen groups to pay attention to and abide by not only the economic or technological requirements of the FCC but also the agency's emphasis on local involvement. Officials hoped to use the new technology for community use by providing training, delegating the public access responsibility to community entities, support access programming financially, and other means.[60]

Leland L. Johnson and Michael Botein made an extensive reference to the hiring and training processes and its implication amongst minorities. Although they acknowledged that "there are few clear-cut 'right' answers to the many questions" of cable television, the report suggested that "[a]n assessment of community needs, objectives, and alternatives" would facilitate community engagement.[61] After introducing a potential provisional framework for equal hiring processes, the authors argued:

> no matter how well intentioned this type of provision, it creates problems of which the franchising authority may not be initially aware. First, hiring requirements must match with existing bands on discrimination; these may include local or state laws and definitely will include the FCC's antidiscrimination rules, the Civil Rights Act of 1964, and the equal protection clause of the federal Constitution. . . . [A]ttempts to define that action which is affirmative often end up in either tokenism or quotas, the former of which is inadequate and the latter of which promotes backlash. . . . The franchising authority should move carefully in drafting these provisions in a manner that is both practical in enforcement and consistent with applicable laws.[62]

This concern shows that not only adequate provisions but also actual implementation of such policies were vital in ascertaining the nondiscriminatory introduction of cable television to municipalities. If the promise of unbiased training and hiring opportunities was realized, as Richard C. Kletter hoped, "[cable television] can become part of a genuine urban communication system, an instrument for social change, and an outlet for a realistic image of life that ordinary TV cannot offer."[63]

Ensuring the equal participation in public programs, rather than just establishing a policy outline, was particularly important for African Americans.[64] American popular culture and media have a long history of erasing, ignoring, and distorting the images of African Americans. Although community organizers in the 1960s began to see the potential of cable television to correct, if not reverse, such a negative trend, many African Americans remained suspicious and "partially cynical" about using media for their good. The suspicion was particularly significant because they felt that financial benefits for the cable operator would be prioritized over what could be good for the benefit of African American communities. A report stated:

> Suspicions may be especially acute, in both inner city and suburban areas, because of the huge profits people imagine to be realizable from cable systems. Residents may assume that every cable-related action, no matter how beneficent in appearance, really stems from profit-maximizing motives. . . . Suspicion may also be aroused because cable is still so dominantly an issue of technological hardware, and residents may wonder whether the new brand of technologists, like the traditional planners before them, will have the sensitivity and compassion to understand that, in the final analysis, the quality of human social life is at stake.[65]

Although it recognized that "a pluralistic process" entails certain risks such as the difficulty reaching an agreement, in order to serve the public interest, municipal members were expected to take these risks.

Legal Literature and Public and Community Access Channels, 1980–1989

There were two loci for cable-related research to take place in the 1980s. The first were conventional venues such as academic journals and books in social science. The other included those in law. The latter characterized the 1980s which witnessed a rise in the amount of attention paid to cable television in the law community. Numerous law review journals published analyses of the recent decisions made by the FCC, assessments of their impact on the business and industry, and forecasts for future development and potential of cable television.

By the end of the 1970s, the financial burden of maintaining access channels for public, educational, and government use as per the FCC regulation in 1972, had become conspicuous to many cable carriers.[66] Many corporations had developed their resistance against citizens' demands for access. It was, therefore, only natural that the U.S. Supreme Court ruled in *FCC v. Midwest Video Corporation*, on April 2, 1979, that the FCC had no authority to require cable operators to carry public access channels. The irony was that by the time the ruling was made, both the national economy and the cable industry had recovered, making the access programming feasible. Additionally, the decision confused both the cable television industry and the communities they were going to serve. Operators were no longer sure what was required of them, what was possible, what was recommended, and what was banned. Similarly, local communities were not certain what they could demand and expect, either. These legal struggles resulted in the big rise in law scholarship.[67]

In this context, the Cable Television Information Center and United Cable Television Corporation published their first comprehensive documents to highlight the importance of cooperation between the operator and the community in order to achieve a successful cable television experience. The two organizations argued that the operator must provide the studio facilities, the video equipment, and public training. On the other hand, the community and its members must maintain their enthusiasm to generate further awareness of the potential of cable television for community building. The reports also provided a detailed list of equipment for community access, syllabi for different workshops for local community members, examples for workshop brochures, and even an example of a certificate given at the end of the training session.[68]

Some law reviews paid closer attention to the social implication of legal decisions than others. For example, Stanley M. Besen and Robert W. Crandall's analysis effectively reconstructed the legal landscape of cable television by tracing the history of federal communications regulations starting its discussion with the Communications Act of 1934. The authors, however, did not take into account how regulations affected audiences differently depending on the social and legal context of the time.[69] Jules F. Simon's work was similar in nature by focusing on legal negotiations.[70] Some had more interests in social environment. But on average, most emphasized the implications of legal decision on society as a whole.

Daniel Brenner's work reflected the concern of how much the FCC could regulate and how much flexibility the audience and the operator could have in programming and expression. He correctly voiced his concern that any type of regulation that ignored the uniqueness of cable television would prevent the technology from serving community members. He argued:

> [g]eneralized analogies do not offer much help. Analogizing cable to a newspaper overstates the editorial nature of cable

enterprises as they actually have developed. Analogizing cable to a common carrier or broadcaster, with their more regulated regimes, ignores the considerable editorial activity in which cable operators engage. And it sells short what cable might accomplish as a medium of communication, given the encouragement of first amendment status.[71]

Although Brenner did not posit what exactly should be done to maintain a good balance between regulation and flexibility, his work showcased decade-long negotiations among the federal regulatory body, cable operators, and consumers.

Alison Melnick positively evaluated the FCC's access channel proposal. Despite her lack of reference to racial minorities, she correctly pointed out that once free access channels were established, the cable provider had to make sure that the programs on air were attractive. She contended that:

[t]he public might be bored by a host of nonprofessional talking to a camera in a studio, without skillful production, direction, and editing. Furthermore, the small budgets and the lower quality of equipment used by cable operators as compared to television broadcasters contribute to an overall lower quality of programming, making it even less likely that noncommercial access programs will be popular.[72]

Investigation of cable television from the legal perspectives in the 1970s did not have such a sociological focus.

Although these legal documents help us understand the nature of a major concern in the cable industry, they do not directly address issues concerning African Americans. It did not mean African Americans could stay indifferent. The FCC's increased or decreased control, or more or less flexibility for operators, affected how African Americans experienced cable television. The 1980s, in this way, witnessed a new development in the scholarly environment which looked at cable television from legal perspectives.

Similar to Don Schiller's work in 1979, which closely examined the needs of socioeconomic interests of local communities by cable operators,[73] United Cable Television Corporation argued that the operator must conduct socioeconomic studies of the local area to identify the unique needs to the community members.[74] Roland E. Frank and Marshall G. Greensberg also argued that public television must make sure that its programs could attract Black viewers because African Americans had increasingly been subject to political and commercial programming for their buying power during the 1980s. African Americans' disposable income was an attractive revenue source, just as it was the case with the film industry in the mid-century. The authors also claimed that the responsibility of content producers included the production of programs that were not necessarily Black programming per se but that could appeal to the general interests and the needs of a large portion of the African American population.[75]

The study also identified some of the topics that African Americans were particularly interested in, including elderly concerns, arts and culture, home and community-centered information, indoor games, and social activities. Additionally, Frank and Greensberg found that African Americans tended to show more interest in informational and educational content than other minority groups.[76] The authors explained:

> Blacks appear to look toward PTV [public television] as a source of information and education. The types of shows they would most welcome in greater frequency on PTV are: educational programs, programs about local events and issues, news programs, children's programs, and "programs appealing to certain kinds of people (i.e., women, Blacks, Spanish-speaking, older people, etc.)." More than half the Black respondents said they would be "somewhat" to "very" interested in seeing each of these types of shows more often on PTV.[77]

Identifying such wants and needs of local African Americans allowed cable television systems to be beneficial to the community.

Gil Noble, former producer and host of an African American public access series program *Like It Is*, as well as a seven-time Emmy Award winner, also emphasized the importance of African American–produced television content. He explained that network stations had hired Blacks as tokenism. When he applied for a position, he was told to his dismay, "You're fine, but we already have a Negro on staff."[78] Even when he was hired, many whites resented the fact that their boss was an African American.[79] The working environment for Noble was far from friendly. African Americans slowly entered the television industry as producers and anchors. Their presence increased. But behind the television camera, Blacks and whites were still separate.

With his experience and expertise, Noble was the first African American television producer to problematize the shortage of Black presence on television. African Americans were almost nonexistent on television screens.

Noble argued in 1981 that although African Americans made up about 12 percent of the American population, they accounted for less than 1 percent of those on television.[80] Additionally, when they were seen on television, "[a] normal, intact Black family is rarely shown in the media. Blacks are depicted as being at peace with America and at war with each other."[81] Considering the significant roles that African Americans played in American society and the amount of television African Americans were exposed to, Noble believed that it was only fair for Blacks to be given opportunities to talk about what mattered to them on television, instead of relying on whites to do so for them.[82]

Despite the status quo, Noble believed in the power of television to change the world. Particularly, from his experience as the producer and the host of documentaries, he believed in the power of social change that documentaries could possess. He said, "[t]elevision can change things. It is a powerful mind-molding machine, and it could make people concerned about their very real problems, instead of preoccupying them with the problems of fictitious families of the soap operas and situation comedies."[83] Noble was well aware that television could formulate an environment in which change could happen.

Noble was correct. Television was a major force for social change. Cable television had even more potential because of its channel abundance. Focusing primarily on art on cable television, Kristen Beck made suggestions that were helpful to advance the understanding of the power of public access for specific interest groups. Because cable television was not "a medium of scarcity" and did not use the public airwaves, it was free from restrictions that existed with broadcast television. She argued that "[a]ccess gives a television voice to the dissenter, the unpopular, and the minority as well as to organizations working in the public sector and in the public interest. Access is a video voice for the people."[84]

Just as Beck focused on the arts, Lawrence N. Redd emphasized the importance of cable television to solve educational problems. He suggested that African American students have struggled from inequality even after the *Brown v. Board of Education of Topeka* decision. Although busing and other efforts to mend problems were implemented, African Americans suffered from a lack of equal education. Redd suggested that through the use of cable television, along with satellite and other communication systems, new opportunities for African Americans students would arise by "creating human interaction." One example is the teleconference. He argued:

> [s]chools can hook their own studios into the cable systems to deliver a wide range of instructional and administrative services,

parental communication, and teacher in-service programs. For people scattered throughout a school district, time, energy, and frustrations associated with travel can be reduced greatly. Many services would be one-way television communication. However, telephones provide an excellent feedback loop.[85]

Such a sentiment, which argued that students could learn by watching television, was not a unique idea. Edward V. Dolan was another strong proponent of the educational use of cable television.[86] He explained:

[o]nce the schools and the homes of students in the district have been wired for cable, many scheduling options are possible: a week at home taking instruction, the next week in school; or two days at home and three days in school one week, three days at home and two days in school the following week.[87]

Their assessments were correct. Many urban cities, including Boston and Detroit, witnessed their educational institutions adopting cable systems for their use. Schools turned out to be a major users of cable television.

Susan Wallace recognized that there were many ways to program access and community channels. She surmised:

[i]f there is a strong city, company, and community support for these channels, the workshops, equipment, and staff necessary to assist the community producers in producing their programming can result in strong, successful access channels.[88]

In many systems, there were one to four access channels that aired content programmed by "organizations, individuals, church groups, schools, self-help groups, libraries, and government agencies from all levels, ages, occupations, and socio-economic status." Citing studies focusing on Iowa City, Iowa, and Kettering, Ohio, she claimed that at least 5 percent to 8 percent of cable television audiences were only attracted to such access channels. Although many operators supported access and community channels because they were required to do so, Wallace showed that customers did hope to have their own channels on which they could share their ideas and see and hear what their neighbors had to offer.[89]

Claiming that "[t]he future of public access seems very bright," Joshua Sapan elaborated on the value of public access. He defined public access as "special channels reserved for members of the local community for transmitting programs they have produced." Even after the change in federal law, almost all cable systems had at least one public access channel. He explained that "[p]ublic access is often viewed by community leaders as an important service to the community." Most franchise agreements included a clause on public access. Even when the content and style of public access differed from system to system, the basic idealism of community-oriented media remained intact. Not only community individuals, but also religious groups, high schools and colleges, local political organizations, and other entities were able to send their messages to the local community and share their knowledge, experience, and information. Sapan argued that due to the rapid increase in the urban cable penetration in the early 1980s, communities were able to expect more ambitious public access schemes from the service provider. He foresaw that such an environment would increase the sense of community toward the end of the 1980s.[90]

Contrary to some optimism on cable television in the 1980s, Edward Dolan was highly critical about its past and suggested more work in the future in order to better serve minorities. He argued that "it would be naïve to expect fundamental changes in broadcasting."[91] It was because operators had pursued their business interests, rather than what was good for their audience.[92] The federal power was not designed to serve the community, either. The government had been protective of broadcasting corporations.[93] In addition to the FCC's refusal to disclose information about the profitability of television licensees, its prima facie appropriate restrictions had never met its original promise to be independent from the federal government. The president appointed all commission members. The Senate then approved them. The Office of Management and Budget approved the FCC budget. All these facts strongly indicated that the FCC had never been an independent regulatory agency. It was controlled by the government that was more inclined to protect the industry. Dolan saw this as one of the major impediments against the service for the public interest.[94]

Dolan also examined cable television as a power structure. He argued that "[p]ublishing is a right; broadcasting is a privilege. Newspapers use private property; broadcasters use public property." Based on this analysis, he continued to claim that "because of its special character, television has a duty to serve the public interest. Newspapers, as instruments of private power, have no such obligation." Ultimately, newspapers "are free to be biased or objective, fair or unfair, accurate or inaccurate." On the other hand, television has to be objective and fair.[95]

Despite television's promise to give its audience what it wanted, those without sufficient disposable income and many minorities did not truly appreciate what media had offered. That minority groups watched television programs did not mean they were in favor of the status quo. It simply meant that they preferred them over turning off the television. Even in a society where there is no incentive to produce quality diverse programs, when content is produced, producers simply need to make sure that it is something that the minority audience would consider to be better than turning off their televisions off.[96] This was not a productive environment.

To materialize public and community access, Dolan encouraged interested community members to continue pursuing their work. He argued that cable television had the promise to deliver "a wide range of diversified programming, doing what broadcast television can not or will not do. Broadcast television seeks the largest possible audience, and profits, largely ignoring specialized and less profitable programming." Due to its available interactive features via audio, video, and digital technology, Dolan was convinced that if used appropriately, cable television could become a major tool for minority groups.[97]

Jan Bone analyzed the arrival of cable television systems as opportunities for employment. Not only did he enumerate basic skills required in the industry and explain different roles and positions in the business, but he also introduced different academic institutions where students were able to obtain necessary training. Howard University in Washington DC, for example, had a School of Communications and a Department of Radio–Television–Film. Instead of focusing on the technology of cable television, the school taught political management and prepared students for positions as producers or directors. Bone, however, also saw problems with the employment style in cable television. The industry's starting salary was low. There were more degree-granting programs that emphasized on broadcast television than on narrowcast television where students were more likely to be hired. Those in technical positions could expect quicker advancement in the office than those in managerial positions. Although Bone did not provide solutions, his work shed light on some important issues concerning African Americans and their employment in the industry.[98]

James W. Roman saw the potential of cable television in its heterogeneous programming and its diverse ownership. Although network television had "homogenized programming that makes it difficult to distinguish one network from another," cable could provide diverse programs.[99] Additionally, media ownership could "provide a valuable service to a segment of the population that feels disenfranchised by existing media." He acknowledged that there were lobbying associations such as the American Association of Cable Television Owners who paid particular attention to minority ownership. In reality, led by the largest Black-owned system in Gary, Indiana, more than seventy-five minority-owned cable corporations, eight operating systems, and thirty franchises were awarded to minority-owned companies.[100]

Decrease in Cable Awareness, 1990–2010

By 1990, approximately 60 percent of American households subscribed to cable television.[101] That number continued to increase during the decade. Despite the growth, fewer literatures discussed cable television, and especially its relationship to race and access channels. Many continued to discuss business and legal issues, whereas on the rare occasion when race was discussed, it was often in the context of commercial channels. The rise of MTV, BET, and other channels, resulted in more focus on these channels while access channels and local origination channels began to lose hold of analysts' interests. Local origination programs, or the programs produced locally, tended to interest those who lived in the neighborhood where the program originated. But the rise of big cable channels undermined this type of production that frequently lacked resources, quality, and sufficient number of new programs. This lack of focus on race and other issues relevant to minority audiences in urban areas using cable television interestingly coincides with the rising colorblind assumption of the booming media of the decade, the Internet.

Linda K. Fuller, despite little reference to racial politics over cable television, wrote a comprehensive review of the history and status quo of the cable television industry as of the mid-1990s. She correctly identified the four basic ground rules of cable television as "freedom of expression," "media education," "localism," and "public service." Through freedom of expression, the new media technology promised to ensure that "diverse ideas can be expressed on television." Although the industry and the federal agency had a difficult time balancing regulation and freedom of expression as previously mentioned, it was one of the major parts of the service. By "media education," Fuller meant that "[a]ccess provides an opportunity for individuals and small organizations to learn to use video and television." As later chapters demonstrate, both African American Bostonians and Detroiters indeed benefited from such media education before and while they produced their content. Localism also was important to "[strengthen] the local infrastructure by increasing and enhancing local communications." Unlike broadcast media, cable television aimed to improve the ties within communities. The last tenet, "public service" meant that "[a]ccess provides informative, educational, and cultural programming which is not otherwise available on television."[102]

Donald R. Browne, Charles M. Firestone, and Ellen Mickeiwicz advanced Fuller's arguments and connected them to minority concerns. They argued that even in the future, access channels were "another form of potential involvement for ethnic minorities." They claimed that "[a]ccess channels are a useful way for ethnic minority groups to gain experience with TV, and many have taken advantage of the opportunity, which usually costs them nothing but their own labors." Despite the potential that access channels possessed, they identified two problems. The first was that "rarely is there any guarantee of regularly schedule airtime, provided that groups would want it." The second was that "many present-day cable operations feature 30, 40, or more channels. With so many choices, and with limited publicity for any one group's access channel program, most viewers probably will remain unaware of programming." Although the authors believed that the best situation was one where minorities would not have to worry about not being covered or wrongfully covered, they understood that it was highly unlikely. Hence, they argued that the government should actively make financial assistance available to minority channels.[103]

Crandall and Furchtgott-Roth's work reflected the shift of cable television's role "from community access television to cable casting." They argued that as deregulations took place in the 1970s, the amount of programming increased. Additionally, better distribution systems and equipment became available. Consequently, cable television became more of a business tool. During the Reagan administration, for example, Congress passed the Cable Communications Policy Act of 1984 that deregulated the subscription rating policy. As a result, the rate increased on average by 39 percent to 43 percent between 1986 and 1989. Unlike in the past, deregulation also allowed telephone companies to partake in the industry.[104]

David Waterman and Andrew A. Weisse were interested in the changing structure of the cable television industry. Due to regulations, deregulation, and reregulation, the industry had undergone various restructuring processes. Despite their claim that their objective was to assess the effect of such changes, they particularly focused on the influence on business. Audiences were rarely a topic of their concern. Discussions on public access or community access were nonexistent. Although the work was intended to be a comprehensive analysis of the industry, their lack of reference to viewers reflected the general trend of cable television literatures of the decade. As Crandall and Furchtgtt-Roth demonstrated, what cable television meant to society changed dramatically in the 1990s. Waterman and Weisse's approach was not a surprising or uncommon one.[105]

Need for Grassroots Movement for Cable Representation

African Americans have experienced under- and mis-representation in visual media over the century since the popularization of visual culture. Radio, a precursor to the twentieth-century style mass-targeting media, had, of course, generated many stereotypes and prejudices against African Americans. In the 1920s, radio allowed African American jazz musicians to disseminate and share their music outside of live concerts. In the 1930s, Duke Ellington, Fats Waller, Art Tatum, and other African American musicians regularly appeared on radio programs. But the radio and audio entertainment industry as a whole continued to marginalize African Americans. The *Amos 'n' Andy Show* is a prototypical example. Theatrical performances, magazines, newspapers, and others have done the same. The development of film and television, however, was more impactful because of their wider reach and visual appeal. In many cases, these highly exclusive industries did not allow African Americans to participate. Even when more African Americans were seen on television, they played stereotypical roles, were turned into "safe" characters, or were only given limited presence in specific types of programs. From news programs to fiction shows, examples are abundant. Past scholarship has correctly identified many of these problems over the past several decades.

When it comes to scholarship on cable television, scholars have ignored or under-discussed African American and other minority interests since the mid-1960s at the national level. Despite the power as a social agent that cable television possessed, researchers did not seem to be aware enough to examine its implications on race-based groups. Despite the hope that cable television would finally bring a colorblind television to African American living rooms, it was mostly controlled by companies, municipalities, and governing bodies that were dominated by whites. Very few scholarly works investigated the issue during the last few decades of the twentieth century.

The dearth of scholarly literatures, however, does not mean that African Americans did not strive to use cable television as their community tool. African Americans have a long history of grassroots and community-based activities that have contributed to the progress they have made in a highly racist society. The number may be small, but there have been research projects that demonstrate a continuation of the discussion as to how best to serve Blacks and minorities in local communities. They were not *completely* ignored in the post-Civil Rights development of cable television. The rest of this book further demonstrates how proactively African Americans involved themselves in the introduction and development of cable television in their own communities in Boston and Detroit.

Chapter 3

The Incubation Period of Cable Television

Although the history of cable television dates back to 1948, the technology did not attract the attention of Boston and Detroit's city planners and community organizers until the early 1970s. It was only in the late 1960s that early users of cable television even began to explore its potential for social services.[1] Both in Boston and Detroit, residents did not benefit from cable television until the late 1980s. City planners, however, did not remain idle during the 1970s and 1980s waiting for the technology to develop on its own. Both cities witnessed various extensive and city-specific studies to investigate whether cable television would be beneficial to their communities. Such analyses examined how residents could benefit from the new technology, what the best introduction method was, and other detailed questions to prepare residents for developments the cities foresaw. Even though this incubation period did not reach a consensus as to how Boston and Detroit residents would benefit from a cable system, what planners and organizers discussed during this time impacted how the new technology was introduced, implemented, and developed in both cities at a later time.

As much as Boston and Detroit witnessed various kinds of discussions and studies during this period, there also were significant changes taking place in cable and other media industries on the national level. The *TV 9* case under the District of Columbia U.S. Court of Justice ruled that the FCC could no longer "ignore the effects of past racial discrimination in choosing applicants for broadcast licenses." After this decision, race became one of the criterions in deciding a potential ownership of a broadcasting system. Although this ruling immediately affected the broadcasting media in 1974, it also influenced the decision-making process of a cable television franchisee selection during the 1980s.[2]

The incubation period of cable television covers slightly different periods for Boston and Detroit. This was the time when planners and organizers prepared for more concrete and detailed efforts to bring cable television to their respective cities. It covers the years between the launching of initial investigations and the issuing of requests for proposals (RFPs). RFPs were public and official announcements that the issuing city was interested in introducing a cable system. Following the instructions given in an RFP, service providers who were interested in wiring the cities applied for licensing. Therefore, the incubation period refers to the period when studies and discussions took place among city officials and residents, before cable providers made any official offers. As for Boston, the official RFP was not issued until February 1981. Corporations that were interested in wiring Boston, however, began to submit their offers and study outcomes in 1980. Therefore, the incubation period for Boston ended in 1979. Detroit's RFP was issued in summer 1982. Its incubation period ended in August of that year.

Encompassing more time than any other period examined in this study, the incubation period is vital in understanding what drove city officials and community organizers to start considering cable television as a tool for community empowerment. Furthermore, this period is dynamic in two ways. On one level, it encompasses both the Civil Rights and Black Power eras. It also includes a very early period in the 1980s that would eventually be known for Reaganomics, increasing concerns about drug use, and other social issues. African American activists involved in the studies of cable television in Boston and Detroit were impacted by many of the social transformations that took place on the national level. In other words, African Americans' efforts for community justice through the use of cable television did not take place in a void.

On another level, African Americans in both cities clearly saw benefits of cable technology. As Chapter 1 demonstrated, cable television's narrowcasting capability was appealing to racial and ethnic minorities. It was going to offer opportunities for self-representation without relying on major media corporations. African Americans could possibly produce their own media content without using or relying on white-dominated resources. By the early 1970s, African American communities had used visual media to advance their political and social agendas, despite the continuing presence of racially demeaning images on mainstream television. Cable television was thought to reverse the trend of assumptions about African Americans. Kristal Brent Zook explained that "[w]hat some programmers think Black people might like to watch is often a far cry from what Black people do, in fact, spend their time watching."[3] Whereas the previous chapter showed how cable television became appealing to African Americans in the post-Civil Rights era, this chapter demonstrates specific discourse on the shift to the introduction of cable television in Boston and Detroit.

Boston's Social and Historical Background in the 1970s

In the middle of the twentieth century, Boston suffered an economic downturn with traditional manufacturing industries and a large portion of the population moving from the city to suburbs. Concurrently, several different city neighborhoods witnessed an increase in their African American populations. The popular expression, "chocolate city, vanilla suburb" was true for Boston. Although some American cities continued to benefit from the lasting effect of the "post-war glory days" in the early 1970s, that was not the case with Boston. The decade was one of "distress and decline."[4] Ranked as one of the 154 largest cities in the United States, Boston was ranked higher than Detroit, Gary, Indiana, or Newark, New Jersey, in terms of city decline.[5] In such an urban environment, Boston's African American residents continued to suffer from financial, occupational, and educational difficulties.

During the second half of the twentieth century, the city of Boston underwent uneven levels of social change. On the one hand, Boston grew from "a veritable urban basket case," to a hub of not only the state of Massachusetts and New England, but also of the United States. Despite the fear among city officials that it was just a matter of time until Boston would become a hollow city due to the absence of mills and other manufacturing industries that had once supported the city's prosperity, it made a quick recovery within a few decades during the latter half of the century.[6] On the other hand, Boston in the 1960s and 1970s continued to show little, or no, evidence of approaching economic recovery. It suffered from lingering racial problems such as school segregation and boycotting. The busing controversy, which began with Judge Arthur Garrity Jr.'s ruling in 1974 affecting Boston's West Roxbury, Roslindale, Hyde Park, North End, and other neighborhoods, reflected the racial tension at the time.[7] Additionally, as Hubie Jones, a Boston activist remembers, African Americans in Boston faced latent racism and tacit discrimination.[8] Although the positive recovery of the city often made the severity and the seriousness of the matters difficult to observe, African Americans in Boston were victims of what George Lipsitz called "the possessive investment in whiteness," which he defined as "a delusion, a scientific and cultural fiction" that also serves as a "social fact, an identity created and continued with all-too-real consequences for the distribution of wealth, prestige, and opportunity."[9] Whites and African Americans were in different situations while waiting for the city's urban recovery. A large number of middle-class whites in Boston were ready to benefit from the city's revival and the reconfiguration of urban spaces by subscribing to and partaking in the possessive investment in whiteness that tacitly made the lives of African Americans more restricted. They were disadvantaged through residential zoning or limited access to well-paying occupations. For example, the redevelopment of the Back Bay showcased by the John Hancock and Prudential towers made African Americans in Boston feel that their needs were being ignored. What they needed in the neighborhood was not business districts that would push them away from the downtown area. Many minorities tried to stay in their familiar neighborhoods. Even after the urban renewal programs of downtown Boston ended in the 1960s with the completion of the aforementioned towers, the Government Center, the Hynes Convention Center, and other buildings, the area was still filled with mostly working-class people of color, including many African Americans.[10] In reality, the number of African Americans continued to rise through the twentieth century. However, as Lipsitz argued, "the structural weakness" of African American communities resulted in continuing discrimination and "psychological domination" by the middle- to upper-class white population that managed Boston's economy.[11]

As the number of African Americans increased in Boston, dualistic racial tensions between whites and Blacks continued to intensify.[12] Even though Boston had the reputation of being a progressive city, it was, for many people of color, a racist city, and was referred to as "up South."[13] Many of the same racial and ethnic problems still remain in current times. African Americans and Latinos tend to be less successful academically. Whereas only 7 percent of white Bostonians have not completed high school, the number was 24 percent and 58 percent for African American and Latino populations, respectively. There are a disproportionate number of African American Bostonians in service occupations, particularly as security guards or night watchmen (men) and as aides at hospitals and nursing homes (women). Their hourly wages are often lower than that of the white population. Occupational opportunities often are restricted from them because of statistical discrimination, or a type of discrimination that is based on general beliefs—often stereotypical—about the characteristics of average group members of the same category.[14] In other areas, from home ownership to geographical distribution of residents, Boston's racial minorities continue to be disadvantaged.[15]

Demographic, industrial, and spatial changes explain how the city reinforced the "possessive investment in whiteness" as a result of which many African American residents strove to use cable technology as an alternative tool to reverse the trend. Boston's demographic shift had two major components: a general decrease in the total population and an increase in the population of people of color, including African Americans. The city's overall population decreased by approximately 150,000 people between 1960 and 1980. During the same period, the white population almost decreased by half, whereas the number of African Americans increased from 67,873 in 1960 to 125,983 in 2000. Through the 1970s alone, the African American population increased approximately 20 percent.[16] Amidst these changes, African Americans continued to be highly segregated. In the 1990s, for instance, more than 80 percent of the residents of Roxbury and Mattapan were African American. On the other hand, African Americans composed less than 2 percent of the population in South Boston, Charlestown, and other areas.[17]

The rise in African American populations and the decline in white populations naturally resulted in the increased presence of African Americans in the city. Consequently, it affected the "cognitive maps" of city residents. Employers continued to avoid hiring Black Bostonians or moving their business to the city. "I'm not unaware that this is considered to be a dangerous area. Its reputation far exceeds the reality, but it's certainly been a longstanding concern," a director of a medical facility once explained.[18] A hotel manager in Boston explained that there was no inner-city Boston resident employed at the reception desk of his hotel because he did not want to have "somebody standing behind the front desk with a long face on . . . looking at their watch every five minutes."[19] Regardless of the credentials of a potential employee, the very fact that he or she resides in a discriminated community often places him or her as an unsuccessful candidate.

Such an aversion to communities of color was partially because of denied access to a quality and adequate educational system in the city. Since the middle twentieth century, Boston's public educational system had been a contested arena due to its racial composition. For example, public schools in Roxbury had a high concentration of Black students, whereas schools in South Boston were predominantly white. As the white population moved into the suburbs and many private schools admitted wealthy white students, more and more schools were considered racially unbalanced. Minority students faced discrimination. School committeeman, William O'Connor from South Boston said in 1964, "We have no inferior education in our schools. What we have is an inferior type of student." The disproportionately minority educational system became questionable in quality. Simultaneously, as well-performing students who were often from middle-class families moved away from local schools, schools in African American neighborhoods increasingly had more African American students and less academically advanced students. Based on this prejudice that for decades viewed African Americans as low-performers, employers often avoided inner-city candidates when making their employment decisions.[20]

Although certain businesses were actively recruiting employees from communities of color, the jobs were mostly blue-collar and the business plans of these companies were usually based on the corporate economic strategy of hiring African Americans who earned only 83 percent of the hourly wage of white employees for the same positions, or Latinos who earned 71 percent on average.[21] In other words, the lack of extrinsic political capital deprived minorities of the possibility to nurture intrinsic political capital. Due to *de facto* segregation, potential employers showed limited interested in African Americans, although these individuals might have had more chances of employment if they lived in a less-segregated area where employers were more serious about finding an employee in the region.[22] The demographic change somehow allowed the city of Boston to grow with little participation from people of color during the last thirty years. This partially explains why such uneven urban growth happened in Boston.

The second tenet of the triple shift—the economic and industrial change—also contributed to limiting the potential of professional participation by African Americans.[23] After the industrial revolution in the nineteenth century, many immigrants and minorities, including African Americans, filled the large demand for low-skilled jobs.[24] The economic change in the second half of the twentieth century, however, altered the industrial focus from "mill-based" to "mind-based."[25] Boston was no exception to this change. In reality, the change took place more in Boston than in many other urban areas. Boston's industries started to require hard skills, as well as soft skills.[26] The businesses that supported Boston's growth in the second half of the century, including those in high technology, medicine, higher education, and other nonmanufacturing industries such as trade and finance, are mind-based. The new industries required more skilled workers than the old ones. Many of the employers in the new economy looked for employees with basic technical skills in addition to good personal skills instead of simple manual skills. They later required rudimentary computer literacy even for starter jobs, as well. By the end of the 1980s, 69 percent of employment in Boston was white-collar, a 22 percent increase from 1950. Only 19 percent was blue-collar, a 23 percent decline from forty years prior. The shift toward a white-collar job market in Boston was much faster than that seen at the national level.[27]

As mentioned earlier, Boston's education system was imbalanced in its quality in the 1960s and 1970s, failing to prepare many African American students for the changing economy despite the increasing availability of white-collar jobs. With the shrinkage of the job market where African Americans would find occupational opportunities, and the fleeing of manufacturing jobs from Boston, this population experienced difficulty finding jobs where they were more likely to be hired. Additionally, race-based cognitive maps kept some industries available in the city out of African American population's reach.

As these shifts that the city experienced in the post-Civil Rights era attest, Boston's paradoxical development hindered the successful implementation of the second reconstruction. One of the remedies for this situation seemed to exist in the narrowcasting capability of cable television. Boston's city officials and particularly community organizers began to examine such a possibility in the early 1970s. Their strategy was to identify how exactly the public could take advantage of the community-based media format and invert the relationship between producers and consumers. Narrowcasting, local origination, community access, and other features, as well as some successful cases in other municipalities suggested that minority residents in Boston might be able to achieve fair access and representation in the city that once claimed to be "a city upon a hill."

Mel King and African American Media Representation

Mel King's life is a history of contemporary African American representation. As a community organizer and activist, and later as a faculty member at the Massachusetts Institute of Technology (MIT), King has dedicated his life to improving the quality and quantity of Black representation and involvement in various forms of media. Although he was not alone in the African American community to organize its members or to use media for minority representation, his lifelong commitment to date truly reflects the seriousness and importance of African American grassroots movement in image production and representation. Examining his early life until the 1970s when he began to be engaged in the introductory phase of cable television in Boston allows us to better understand the street- and community-level background in which new media technology started to be appealing.

Born in 1927, Mel King spent his childhood in the South End of Boston. The city's population was about 3 percent Black at the time. Most were concentrated in the South End and Lower Roxbury neighborhoods. Many memories of his youth are of his African American friends and family members. He remembers being proud to be "a West Indian, and Black." However, he simultaneously was conscious, as a school child, of prevailing negative images of African Americans as clowns or apes. Although he was proud of his heritage, he knew that people outside of his community did not recognize the richness of it. King still recalls that whenever movies were released that mocked African Americans, he and his friends would end up in fights because other students made fun of them. This childhood experience was one of the earliest in which he faced the negative power that visual media possessed.[28]

Once he moved to the all-Black school Claflin College in Orangeburg, South Carolina, in the mid-1940s, King became aware of what positives visual culture might bring to African American communities. He witnessed something that he had never seen before. Unlike any of the theaters he had visited before in Boston, the college theater was owned and operated by African Americans. It was a moment of "awakening." African American students produced their own entertainment. They decided what movies to show and what plays to run. Under segregation, African American students successfully used similar media that portrayed stereotypical images of African Americans in Boston to organize, produce, and distribute image content for their own benefit. "Cowboys were Blacks," he remembers. His fellow students were empowered and experienced a higher sense of self-worth through such film and theatric productions. This was King's first experience feeling the empowerment that could come from image and media content production.[29]

The power of visual imagery that King experienced in his youth was a precursor to what later African Americans would witness as African American visual popular culture developed in the mid-century. When *Jet* magazine was established in 1951, for example, it "provided a common frame of reference for Blacks who were separated by class and religion." As the magazine published a picture of Emmett Till's body in 1955, it politicized African Americans and made them even more aware of the social injustice in America's racist society.[30] The power that just one picture possessed was immense. Appealing to the visual sense was nothing new for King by the Civil Rights era.

The school theater in South Carolina was just one embodiment of African American self-sufficiency. African American students organized and empowered each other through the various types of activities they planned and implemented. King also witnessed the impact of similar organized efforts of the Youth Progressives and NAACP, in which he participated during his college years. Through these experiences, the future urban activist learned what could be done for African American communities through image and content production.[31]

Especially after his awakening experiences in South Carolina, King became acutely aware of the "real delusion" that permeated throughout the city of Boston. Boston's racial dynamic was different from that in Orangeburg. He felt that a large majority of the white population in Boston believed that things such as racism and discrimination could not exist in Boston, simply because "it was Boston and not Mississippi." They seemed to ignore Boston's racist attitudes, which excluded a number of African Americans and other minorities from public schools, eventually erupting in the 1970s with the busing controversy.[32] He was determined to change the status quo. King's strategy to rectify the environment was founded on his negative memories of fighting in school in Boston and the positive experiences in college that happened for the same reason: the power of image production. If image content could be produced by African Americans, King believed that it would help them boost their self-image. This idea was especially effective because television and other mass media were becoming increasingly popular. He felt he could positively impact the social condition around him and his community, because, as he wrote in his book, "I knew there was nothing wrong with me."[33]

Although various Civil Rights movements aiming to ameliorate the living conditions of the African American population were represented on both national and local levels between the early 1950s and the 1970s, King noted that majority of messages from various media such as radio, newspaper, magazines, and television largely reinforced the racist social values of the time. He believed that as a consequence of the disproportionate representation of African Americans in the media, they suffered from a lack of "believability" even when they saw a positive portrayal of their race, which inevitably lowered their self-esteem. In order to rectify this condition, King saw one of his goals as being the ability to gain access to such media and creating an environment in which African Americans were able to produce their own media content.[34]

If his experience at the college theater was his first awakening experience, his second came in 1963. King had by then obtained his master's degree in education from the Boston Teachers College in 1952. He was a social worker and an activist in Boston, particularly in his local communities in the South End and Roxbury. Just as Malcolm X in Spike Lee's movie does, in a hotel room, on May 3, King and his daughter witnessed police officers led by Bull Connor, the public safety commissioner of Birmingham, Alabama, use fire hoses and police dogs against nonviolent marchers on television.[35] King remembers that the image "immediately politicized her" despite her young age at the time. Gov. Michael Dukakis agrees with King's analysis. Prior to the arrival of computers in daily life, the technology that had the largest impact among African Americans in Boston, and in the United States, was television.[36]

While the Birmingham incident made the power of media apparent to many community activists, King also had another personal moment of realization in the fall of the same year. One night in early October, King, already a believer in the power of media content and images, was watching a World Series baseball game between the Los Angeles Dodgers and the New York Yankees on TV with his African American friends. During a commercial break, they saw a Gillette advertisement in which two African Americans—unlike any other commercials in the past with white actors—were talking about using Gillette blades. "The room got silent," King reminisces. When he asked his peers what had just happened, one of them answered, "[The commercial] makes me feel, 'We can, too.' "[37]

After his experiences at the college theater in Orangeburg, with the commercial, and with his daughter, King was convinced that image production could improve the quality of African American lives in Boston. Once he learned that cable television was to be introduced in his community in the late 1960s, he decided to take advantage of it. By maximizing the quantity of local production by, for, and on people of color, he aspired to boost their self-esteem and develop community.

Foreseeable Advantage of Cable Television in Boston

Sixteen years after the first cable system in the state of Massachusetts appeared in Shelburne Falls in 1953, Boston City Councilor Thomas Atkins submitted the first proposed ordinance that considered adoption of a cable television system in his city. After several hearings and examinations based on Atkins's proposal, Foundation 70 submitted its report to the City Council on April 6, 1971. Foundation 70 was a nonprofit group examining the feasibility of a cable system in the city. Operating under the Educational Development Center that was based in Newton, a suburban city to the west of Boston, the group claimed that the city should be divided into fifteen cable television communities comprising three cable television districts. Approximately one week after the report was submitted, on April 14, the Boston City Council held another hearing. Edward J. Roth, consultant to the Corporation for Public Broadcasting and National Education, made five recommendations, one of which particularly spoke to the later contested issue of control, claiming "[c]ity [should] not attempt to operate the system or direct its programming." To address his other recommendations as licensing, ordinance drafting, and others, the committee held further hearings on May 27 and May 28, 1971.[38]

The twelve-page ordinance proposal sat on the desk of the council committee for two years until the process went public in 1971. Although the draft of the second ordinance brought "no definitive action" to the city, the proposals and hearings pertaining to the possible introduction of cable television to Boston encouraged closer studies in the 1970s. Coincidentally, Massachusetts became the first state to have an independent cable commission to determine the cable service provider on the local level. This regulatory shift was significant for Boston. It was no longer the City Council but the mayor who was to officially determine who the provider would be.[39] As later examples will suggest, this is why Boston's African American community leaders tried to influence not only council members but the mayor and other decision makers.

On June 25, 1971, the Finance Commission of the City of Boston, led by Chair Lawrence T. Perera, submitted its statement regarding the introduction of cable television to the city to the Committee on Ordinances and Resolutions of the City Council. The statement recognized that cable television could bring "educational and other public benefits which are not presently offered by the television industry." Despite the promising positives, the commission claimed that the lack of federal and state regulations and coherent policies might compromise the potential benefits of the new technology for the city's residents. The commission's emphasis naturally centered on the formation of regulation. Although the statement did not clearly foresee the type of benefits the new technology could bring to the city, it accurately identified the risk that the young technology possessed. It argued that only after federal and state regulations were clearly set and met that the public could "be assured that this industry [would] indeed operate in accordance with the public interest and it [was] the responsibility of both [federal and state authorities] to see to it that the public interest [was] protected in a manner of such importance and public impact."[40]

A very extensive study about the use of cable television for Boston's development took place during the time between September 1971 and May 1972, as a part of the Boston Development Strategy Research Project. Led by Lloyd Rodwin, the head of the Department of Urban Studies and Planning at MIT, and Lawrence Susskind and Antony A. Phipps, who respectively served as co-director and associate director, the project examined numerous ways to revitalize the city of Boston. Cable television technology was one such means for development.[41]

The report prepared by Konrad K. Kalba, from the Department of City and Regional Planning at Harvard University, recognized the importance of communication systems. He opens his report claiming that "[c]ommunication systems are the infrastructure of human activity. . . . The structure and effectiveness of communications become critical particularly as human settlements become more organizationally complex and more extended over space."[42] Although Kalba did not directly mention race, he recognized the potential danger that communication systems possess. He cautioned that "access to all but the most mass of media is differentiated by income, education and social or occupational status. Geography also continues to affect communications activities."[43] As Kalba recognized, communication systems including cable television had obvious benefits for communities in Boston once it was implemented. As later scholars and community organizers would argue, the key was to ascertain that decisions made prior to the implementation of the new technology was not going to reinforce preexisting biases.[44]

Boston was far from a forerunner in citywide cable television services at the time. The technology had been invented almost twenty-five years prior to the studies that were performed in Boston. However, this late start was advantageous for Boston as it was able to analyze the experiences of other municipalities, particularly in terms of some of the difficulties anticipated by city officials and organizers. In his study, Kalba focused on New York City's cable television system. He identified three strengths and two weaknesses in Manhattan's system. The report first praised the diverse kinds of programs that cable television could bring to its audience. He also explained that for approximately $60, subscribers in small communities were able to watch a variety of television programs, as if they lived in a major metropolis.

The other two positives Kalba enumerated were more consequential for the later development of the cable system in Boston, where issues concerning African Americans and other racially minority populations would arise. Kalba explained that one of the new concepts introduced by cable television was the idea of local origination. He stated, "[it] has emerged because cable television typically serves a community much smaller than the one reached by conventional TV signals. Consequently, much more localized coverage of political, social and cultural events becomes possible on an unused channel." Reflecting on how this might affect the city, he continued, "[u]ltimately, as penetration increases, the local politician will no longer have to buy a metropolitan audience in order to reach his constituency by television. And newscasts will be able to focus on the problems of Wellsley [sic], Roxbury or South Boston rather than of the entire region (e.g., *New England Tonight*)." This idea of local origination and its capacity to have specific audience targets would later attract city and community developers. Due to this narrowcasting capability, cable television could possibly satisfy the needs of a small number of community members, about which the majority of the population in the surrounding communities seemed indifferent.[45]

Kalba also found public access to be strength. In the Manhattan systems, there were two channels reserved for public access, or "for materials prepared by any individual or nonprofit community groups" with no or very minimal fees. He was optimistic about this programming concept. He elaborated:

> This approach places the initiative on dispersed groups, making it difficult to organize the effort, but it does liberate the programming product from the community-relations bias and frequently limited imagination of the system operator. At the same time, in the New York demonstration, the innovation problem is gradually being overcome by the establishment of intermiediary [sic] organizations such as Open Channel and the Alternate Media Center at New York University. These resources centers have succeeded in assisting a variety of groups to produce their own videotapes for playback on cable television.[46]

As explained later, for Mel King, this idea of public access was one of the most attractive characteristics of cable television. Unlike mainstream television programs in which predominantly white producers portrayed African Americans according to their stereotype-ridden image, public access was going to give many African American residents access to the production side of media structures.

The two weaknesses cable television possessed in Kalba's mind were economic feasibility due to the small audience size and the lack of technological development. Kalba problematized the fact that specialized or neighborhood programs might require substantial financial support to maintain their viability. Even in a large city like Manhattan, he estimated that "the audience for these programs is miniscule now and will remain small even should penetration reach most of Manhattan's households." His second concern was the absence of a clear vision of where cable technology was headed. By the early 1970s, interactive systems between the producer and the consumer were nothing new. He was uncertain, however, how households and institutions balanced their control and ownership over media services.[47] Interestingly enough, as later examples show, such concerns would soon disappear. The two major positives that Kalba correctly identified attracted Black community organizers.

In relation to the potential benefit for the city of Boston, Kalba's report used Larry Susskind's framework, which examined the type of urban development cable technology might bring. Out of six objectives, two were particularly significant for African Americans in Boston. First, the report claimed that the technology could create "more equal access for all individuals and social groups to qualify [sic] education, shelter, and jobs; more efficient and equitable distribution of public services throughout the entire region with particular attention to meeting the needs of the disadvantaged."[48] Although Kalba did not make it clear in his study, this concept was only valid as a result of the aforementioned capability of cable systems to narrowcast. Because the information aired on cable channels could have specific targeted audiences, viewers would be able to receive housing, employment, or other information pertaining to their own local communities. The outcome was what Kalba called "equalization of information resources." Regardless of the audience's location or income, access to information would be more equal than what was available from the conventional television system.[49]

Public access also was going to help realize Kalba's prediction. Equalized access to information referred not only to the retrieval of information but also to its distribution. Kalba claimed that "low-income communities could utilize the medium for self-expression purposes and for political and economic development."[50] Although the author did not make a direct reference to the issues of racial disparity in his report, this was an idea that many later developers would share. With Boston's African Americans disproportionately confined to the lower-income group, providing them with a means for self-expression without relying on costly broadcasting media was going to be an effective tool for self-representation and improved self-esteem.

The second framework was related to a decision-making process on the municipal level. Kalba argued that cable television would allow "more direct involvement for all individuals in the administration and decision-making processes affecting the allocation of resources, the delivery of services, and the formulation of social, economic, and environmental policies."[51] The author anticipated that, thanks to local origination and public access features, it was going to be easier for decision makers to conduct opinion polls and find the opinion of citizens, while the same was true for citizens to provide feedback and suggest policy alternatives.

The most significant message of Kalba's paper existed in his preference of the interest group–driven approach of cable system development to the marketplace economy–driven model. By the early 1970s, the cable industry had undergone various policy changes on the national level through regulation and deregulation. In what he called a "development-planning strategy," public ownership allowed more substantial regional development and problem-solving procedures.[52] As Mel King noted, ownership was a key concept in African American involvement in the cable system. Kalba's suggestion challenged the model that had been followed by the existing media industry. It functioned as an authority and controller of information. Market-driven corporations determined what information deserved air time. In order to materialize the potential of cable television, however, Kalba correctly identified that public ownership was indispensable.

The idea of race was absent from Kalba's report. This omission, however, was not a surprise or anomaly because most scholarship and reports published at that time did not include race as a category of investigation variables. This fact, however, should not negate the value of such research projects. As mentioned earlier, many African Americans in Boston belonged to the lower-income group that had historically been deprived of access to media and information. Although "lower-income population" and "African Americans" were not always synonyms in Boston during the 1970s, it was a rather unfortunate but honest fact that development programs and projects like Kalba's that focused on the lower-economic status population were beneficial to more African American residents than those in the middle and upper classes. Additionally, the economy-based discourse eventually established the foundation for race-based programs.

In November 1973, the Boston Consumers' Council submitted its report on the development of cable television to Mayor Kevin White. The council consisted of eight members. Maureen F. Schaffner led the council as the chair. Richard A. Borten served as executive director. Two public members, Harold Fennell and Ruth Strauss, joined as city residents. Other members included Richard G. Huber, dean of the Boston College Law School; Edward F. Lownie from Sealer of Weights and Measures; Leonard Pasciucco, deputy commissioner of health and hospitals; and Michael Tarallo, labor representative. The council also received support from various scholars at MIT, Harvard Graduate School of Design, Harvard Business School, the Cable Television Information Center, and the Urban Institute, just to name a few.[53]

The report was an outcome of the eighteen-month–long study regarding the development of cable television in the city of Boston. Although it predicted that cable television "may someday develop into a consumer service comparable in scope to the three major public utilities (gas, electric, and telephone)," it expressed a sense of hesitancy about the implementation of the technology. It argued, "cable television has yet to provide solutions to the communications problems of the nation's cities, or to fulfill its promise of creating a new communications era."[54] Due to this ambiguous reflection about the technology's "exciting" but "undemonstrated" potentials, the council attempted to identify both promise and risk that not only city officials but also city residents felt that cable systems held to attain beneficial presence in the city of Boston, including special programming services for minority groups and community programming.[55]

The report cited citizens' concerns expressed at hearings held during the first two weeks of October 1972. In total, four hearings, open to the public, were held on October 3, 5, 10, and 11. Although these hearings were not required by state law, they took place "as an opportunity for all interested residents to air their views on cable television, before decisions on franchising [were] made."[56] As Maureen Schaffner, chair of the Consumers' Council responded to the question posed by Irene Burns, a resident from South Cove, it was a forum where residents could voice their concerns, ask questions, and have their opinions integrated into the report the council was preparing for future submission to the mayor.[57]

Schaffner opened the hearing at 3 p.m. on October 3, by explaining that the mayor had "assigned the initial planning for cable television to the Consumers' Council."[58] One of the major highlights of the hearing was the discussion of race. Although it rarely had been a major topic of discussion unlike class and economy, race appeared as an issue toward the end of this first hearing on October 3. An unidentified person spoke up from the audience. He problematized the fact that there was no commercial station that was willing to program jazz music, which he argued would benefit certain racial groups that had been underserved. He lobbied for such a program and argued that if African Americans could manage 25 percent of the available channels, it would be beneficial for their communities.[59] This hearing was one of the earliest examples in which a Boston citizen connected African American community identity and its formation and the use of cable television.

At the second hearing, held on the night of October 5, Charles Grigsby, president of Roxbury Cablevision, enumerated five issues that the city needed to consider as it decided on the service provider. The five points were: "diversification of ownership," "amount of time owners are willing to put into system management," "innovations proposed by the applicant," "degree to which applicant will provide jobs for residents," and "degree of local and minority ownership and participation."[60] David Smith from Roxbury and Thomas Rivera from Cable Antenna Television Information Service emphasized the importance of local ownership and control to bring about "a franchise capable of meeting community needs and satisfying the public interest."[61] Rivera continued to express his wish to see local ownership of a system that takes "the bi-lingual, bi-cultural, and the bi-cognitive needs of the community" into consideration.[62] The fact that he did not clarify which two linguistic, cultural, and cognitive groups he was referring to in his statement is a minor issue here. What was important was that he recognized cable television's capacity as a media outlet that could provide citizens of various backgrounds opportunities to voice their opinions.

On the other side of the serious concerns demonstrated by residents and their groups was a relative indifference among other Bostonians. The report highlighted that although the hearings had been announced on television, in newspapers, and via other media, citizen participation was limited. Those who took part had vested interests in cable television.[63] The aforementioned two figures, Charles Grigsby and Thomas Rivera, had more interest in cable television than other citizen participants. Rivera argued that the public was still unaware of cable's potential.[64] The council suspected that the lack of information among citizens was worsened by a sense of distrust of the city's motives. The solution it offered was educating Bostonians about the benefits cable television could bring, by distributing videotapes, as suggested by Rivera.[65] Gloria Gibson, a graduate student from Boston University, agreed with Rivera and voiced her idea that "someone needs to inform the community before it can come before the Consumers' Council and present its additives."[66] Robert Coard, director of Action for Boston Community Development argued at the fourth hearing, that a cable system could provide jobs and information about jobs with residents, as well as educational opportunities and low-cost advertisement opportunities for small businesses.[67] In the council's view, few Bostonians were aware of these positives.

Although the council acknowledged and was convinced of the positive impact that the cable system could bring to the city, it made a conservative recommendation at the end of the report, by stating that "postponement [of cable development] is a laudable approach," because both the city and its residents appeared unprepared for its use. As for the city, its role in the context of federal and state regulations remained unclear. Such legal and regulatory uncertainty could bring a lack of control or too much control over the medium. Without well-prepared residents, the technology could not achieve its full potential. Additionally, too many citizens were unaware of the difference between cable television and regular television.[68] People knew too little about cable. A female participant, identified as Mrs. Henderson, from Roxbury was quoted in the report asking how cable television could "free" television, how her utility bill was affected, and if she would have to pay more in taxes. On October 11, she stated, "The poor cannot pay any more payments; the potential increase in electric power costs would impose a new burden on the poor. . . ."[69] Reflecting the discussion about the involvement of Northeastern University and the Boston Public Library in the development of the future cable system, she questioned, "Who will pay for all the educational programs? Will all taxpayers pay, even if they don't have cable?"[70] Regarding the technical development, she continued to ask:

> Does this mean on existing telephone poles or will we have to pay the telephone company new higher bills for the use of their poles? Who pays for laying the cable underground? Will this lead to an increase in taxes? . . . Will cable interfere with regular television? What will cable do to the environment? Will it add more to the already polluted environment? Who will own the cable system; will it be owned by utility companies? By outsiders? What is this discussion about districting the City into separate cable areas? To me, this sounds like separation instead of integration. I haven't heard any answers.[71]

Sam Sawtelle of the Dorchester United Neighborhood Association also seemed to have the same concerns when he asked, at the first hearing if consumers would end up having to pay more fees upon the implementation of a cable system.[72] In 1973, the city of Boston in general was not ready for cable implementation.

In 1974, the research firm of Whitewood Stamps issued *Cable in Boston: A Basic Viability Report*. This study was released in response to Mayor White's statement in the previous year. In his announcement on November 29, the mayor noted that "cable television [was] too expensive and technically too immature to function successfully in a highly complex, heavily populated city" like Boston.[73] This announcement meant that "cable television development in Boston would be postponed indefinitely." Mayor White's decision affected not only the lives of residents in Boston, but also those in Chicago, Kansas City, Dallas, New Orleans, and Philadelphia. After his announcement, these cities followed suit. In the eyes of decision makers in these municipalities, cable systems as they existed in the early 1970s were insufficiently developed to be of beneficial use.[74]

Although the focus of Whitewood Stamps' study was on the economic feasibility of the technology and did little to examine the racial and ethnic composition of the city, it also showed the importance of the African American community in the city of Boston. It claimed that "Roxbury, the center of Boston's black community since the turn of the century, is the geographic heart of the city of Boston."[75] Its neighboring communities such as the South End, Jamaica Plain, and Dorchester had a large Black population, as well. Taking note of the African American concentration in the geographical center of the city, this study confirmed the significant role African American leaders played in the following development years.

May 1979 marked the establishment of the Cable Television Review Commission. The *Resident's Attitude Study* submitted in March 1979 showed that about half of the respondents were interested in cable television. An additional 5 percent to 10 percent were interested in various access opportunities. White's appointment of the commission was to realize such needs of the local community. The commission consisted of fifteen members who represented the local religious, academic, consumer, and other interest groups. The members included Chairperson Peggy Charren and Vice Chairperson Bill Owens. Charren was the president of Action for Children's Television. Owens was the state senator from Mattapan. Other members included James P. Collins, a community leader from Charlestown; Joseph C. Dimino, general manager of WSBK-TV; David O. Ives, president of WGBH; David C. Knapp, president of the University of Massachusetts; Michael LoPresti, state senator from East Boston; and Ithiel De Sola Pool, director of the Research Program on Communication Policy at MIT, who had published substantially on cable television. The commission was supplemented by five staff members. Commission members met biweekly to review relevant papers, listen to presentations given by experts in cable television, and meet with interested parties.[76]

Around the time the commission was formed, a coalition of various local community groups also was established. The Cable Television Access Coalition had researchers, community organizers, and producers to check the proceedings of the commission and to raise public interest and awareness in cable television. When the coalition spoke in front of the commission, it emphasized the importance of community involvement during the planning stages, even before the actual franchising process.[77]

In October 1979, the Cable Television Review Commission issued its first report based on its five months of research. The aim of the report was to identify concerns about cable television in Boston and to make recommendations. As was made clear by Mayor White's decision against proceeding with the potential introduction of cable television to the city, there were several major concerns that were repeatedly discussed. First, it was clear that regardless of the type of franchise decision the city made, it would affect the city on a large magnitude. Second, considering the significance of the technology, the city would have to conduct a "thorough planning process" to determine "what [the city] wants from a cable system." Third, the technology would no doubt affect the city's "cultural, economic, and social fabric." Finally, cable policy regarding public interests, technological advancements, and economic incentives had to be clearly outlined.[78] By this time, albeit implicit, race had become a part of the discourse. As this report claimed, public interest was a major part of the cable commission's concern. The report from 1974 showed that African American communities existed in the middle of the city of Boston. Although race might not appear explicitly as a part of their listed concerns, it had attracted significant attention by policymakers and community leaders.

Some of the commission's recommendations were particularly important in terms of knowing how the system was going to be developed in relation to the interests of African American residents and communities. The commission's first recommendation was to reconsider Mayor White's earlier decision. This signaled the beginning of the city's serious and more concrete step toward the introduction of a cable system. It claimed that "the City is now in a far better position to provide Boston's residents" with a sophisticated and advanced cable system. Before this report, no official document had been issued by an entity that had authority over the mayor, who was the only decision maker in the cable system policy as stipulated by the state of Massachusetts in November 1971. The report concluded its first recommendation by stating that "We, therefore, recommend that the City begin the process of awarding a cable television franchise in Boston."[79] Approximately five months after Mayor White officially appointed the Cable Television Review Commission to re-examine the foreseen consequences of the introduction of cable television to the city and to urge policy recommendations in May 1979, this recommendation represented a decisive moment in the history of cable television in Boston.

The fourth recommendation was particularly important for African American residents. The commission recommended that "applicants for the franchise be carefully scrutinized with respect to their plans to hire minorities, women and City residents, and meet the City's needs for foreign language, children's locally originated and public access programming." This franchise candidate review process recommendation was to make sure the two major advantages of cable television, namely public access and local origination, would materialize for the benefit of local residents, including African Americans.[80]

Behind the city's decision to encourage the commission to re-examine the feasibility of cable television, existed several promising developments that exemplified what cable television could achieve. Between 1973 and 1979, cable technology advanced substantially with more reliable satellite interconnection, relaxation of the FCC's technical ruling, and so on. Cable television itself had become a popular service for many households across the nation. Most importantly for Black community leaders, after 1973, cable television began to show its promise for social, municipal, and other public service purposes. The commission summarized:

> [Many cities were] excited by the possibilities. Special programming for the handicapped, non-English speaking citizens, professional groups, and the elderly; home instruction for shut-ins and working people; and community programming through local origination and neighborhood access facilities captured the imagination of potential providers and recipients alike.[81]

Reinforced by successful experiments by the National Science Foundation that affirmed the technology's usefulness for public service purposes, cable television was becoming increasingly more attractive as a tool for African American community leaders.

With regard to topics that might affect Boston's Black communities, the commission showed most concern over the issue of ownership. The cable system could have one of two forms of ownership: municipal or private. Between these two choices, it argued that the former would provide "greater public service benefits." The commission, however, acknowledged the financial burden on the city in municipal ownership. It claimed that although municipal ownership was "an interesting concept, . . . such a policy would constitute an impractical and inefficient approach to cable development in Boston."[82] It concluded to recommend private ownership for this reason. This recommendation did not mean that the commission abandoned local participation in the cable system solely because of the economic reason. It recognized the importance of local economic participation by mandating 30 percent of the system in Boston to be owned locally. The report summarized:

> The Review Commission hopes that a substantial part of the local investment will be by neighborhood economic development groups whose primary activity is to promote and make investment in neighborhood enterprises. We recommend that the directors of the Boston cable television system be elected so that proportional representation of local investor/owners is assured.[83]

Although 30 percent local ownership was far from what municipal ownership could attain in terms of true "ownership" by the local political entity, this represented a major difference between mainstream private media corporations and cable television.

Establishing a new infrastructure means an increase in the number of available jobs. A major contribution that a cable system could bring to Boston was such job opportunities. Identifying this occupational development in the city of Boston, the commission called for "specific and meaningful employment opportunities for minorities, women, and City residents." What was significant about this recommendation was that the commission tried to urge franchise winners to go above and beyond what the FCC required it to do. The report explained that "Cable companies are already required by the Federal Communications Commission to meet minimum standards in employment of minorities. Furthermore, City policy now specifies employment patterns favoring City residents and minorities." For franchise candidates, this meant that it would have to clearly outline their employment policy in the document at the time of bidding. The commission concluded this recommendation dossier by stating that "It is most strongly urged that the applicants' attention to these specifics weigh heavily in determining award of the franchise."[84] The report later included "Employment policy which uses present FCC and City of Boston standards as a base and includes specific provisions for minorities, women, and City residents" as a mandatory undertaking of applicants.[85] As the history of African Americans and their visual representation attest, African Americans had been excluded from the production of television programs. This type of policy was a positive turn from this century-long trend of Black exclusion from American media.

The idea of minority employment, however, was not free from problems. As David Brundnoy summarized in his report, "A Dissenting Opinion, Differing in Some Particulars with the Review Commission Report," there was a lack of clear definition of what "minority" meant. He suggested that the use of such a vague word had a "politically motivated reason." He was concerned that such ambiguity would result in "intellectual bankruptcy" only including "Negroes but not Jews, or people with Spanish last names but not Portuguese last names, or American Indians but not homosexuals, or Chinese-Americans but not Slavic Americans," and so forth.[86] Although he was correct in that the report lacked a clear definition of "minorities," for the purpose of understanding African Americans' relationship to cable television, Blacks who constituted the geographical and demographical core of the city of Boston were a part of this "minority" population. This ambiguity in definition reflected the fact that the RFP was far from perfect. But it also addressed issues that mattered to many underrepresented people, including African Americans.

To make sure that the residents' benefits were respected, the commission recommended the establishment of the Citizens' Advisory Board that would be involved in the franchising process, policymaking, and advising. It could also help establish an access policy to ascertain equal access to the system, guarantee the use of the system for educational purposes, and other public service reasons.[87] Although 1979 did not see much development toward the introduction of cable television, these recommendations made by the commission were precursors to the future negotiations among the franchise candidates, city officials, and community leaders.

Detroit's Social and Historical Background in the 1970s

By the 1970s, Detroit was one of the most segregated American urban areas, along with Gary, Indiana, and Newark, New Jersey, suffering from various symptoms of social decay that ranged from poverty and lack of education to high instances of single parenthood and unemployment. The city known earlier in the century as the "arsenal of democracy" and "dynamite," with its prospering manufacturing industry led by automobile factories, had undergone white flight, an influx of economically disadvantaged African Americans, deindustrialization, and two major unrests in 1943 and 1967, just to name a few factors. In the mid-century, Detroit was "a colorful patchwork of communities" where "music and the culture of other lands come together."[88] Systemic discrimination since the pre-World War II period continued to persist. In the midst of the decline, Coleman Young's victory in the mayoral election of 1973 suggested that African Americans in Detroit, constituting approximately 45 percent of its population in 1970, began to have their voices heard in the city's political arena. Electing city officials who were sympathetic to African American values and life was one solution to gaining political power in a difficult time.[89]

Thomas J. Sugrue and other historians of Detroit argue that demographic change, industrial change, and increasing poverty account for the major shift the city experienced in the mid-century.[90] Since the turn of the century, Detroit witnessed a large influx of African Americans. The trend continued in the post-war period. The Black population in Detroit increased from approximately 150,000 in 1940 to 480,000 in 1960. Although Blacks represented less than 10 percent of the total population in 1940, it accounted for almost 30 percent twenty years later.[91] Many African American migrants hoped that by moving to northern cities, they would be able to get away from the harsh segregation and discrimination in the South.[92] By 1970, the Black population had expanded to more than 660,000, comprising almost 45 percent of the total population of the city.[93] These figures suggest that even though segregation and discrimination existed in the North and African Americans did not have an easier time finding jobs or places to live,

Detroit still remained as a destination for many African Americans moving from the South during the post-World War II era.[94]

While Detroit attracted increasing numbers of African American migrants and residents, the city was concurrently experiencing a major industrial change and a shortage of available work. These changes negatively affected the incoming Black population. Throughout the first half of the century, African Americans largely worked in the manufacturing industry, particularly in the automobile industry. When European immigrants were called on to fight in World War II, they replaced the absent labor force. It was no longer the case by the mid-century. The labor market in Detroit was hostile to African Americans. Detroit saw an increasing number of discriminatory job advertisements.[95] The automation, deindustrialization, and decentralization that started in the late 1940s also disproportionately impacted African Americans and immigrants, whose job opportunities had been limited to such areas. As manufacturing industries left the city to seek lower taxes and larger plant space, the job competition became harsher.[96] Both the total manufacturing and production job positions had been halved by 1977 compared with the conditions in 1947.[97] Even in the steel industry, which had seen an expansion in Black labor participation during the war, the trend stagnated in the 1950s and later.[98]

As a natural consequence, post-World War II African American joblessness increased. The unemployment rate for African Americans in Detroit increased from 11.8 percent to 22.5 percent between 1950 and 1980. The ratio of those not in the labor force increased from 18.7 percent to 43.9 percent during the same period. Black unemployment doubled from 28.3 percent to 56.4 percent. Joblessness brought poverty. Sugrue explained that poverty brought a concentration of poor people in the city, detaching them further from the labor market. He used statistical evidence to show that between 1970 and 1980, the unemployment rate in Detroit's high-poverty area doubled from 14 percent to 28.4 percent.[99] Although these industrial trends were not unique to Detroit, these shifts heavily affected the lives of African Americans in the city.[100]

The unfavorable living and occupational conditions described earlier often lowered the self-esteem of many African Americans. Although many of the achievements of the Civil Rights movement may have positively affected their lives, systemic racism persisted both at work and in the community. The lack of access to jobs, income, and wealth negatively affected African American perception of self-worth. Sociologist Elliot Liebow surmised that these factors not only had an economic influence over African American individuals and their families, but also affected the level of self-esteem and self-respect.[101]

As was the case in Boston, the use of media, particularly narrowcasting media in this social context appeared as a potential remedy for lowered African American self-esteem. The beginning of the city's interest in cable television coincided with a hopeful sentiment regarding future views shared by many African American Detroiters. *The Detroit News*' 1970 survey revealed, for example, that more than 40 percent of African American residents felt jobs and job opportunities were better than five years before. Forty-seven percent responded that there had been an improvement in housing conditions. Forty-nine percent claimed that there were reasons to be optimistic about their future opportunities. Although between 15 percent and 30 percent of respondents also explained that their job situation, family income, neighborhood safety, education, and other domains of life had worsened, it was worth noting that 69 percent still expected positive change to arrive over the next five years.[102] It was out of this social sentiment that the cable project was to become a part of Michigan's African American self-help tradition.[103] Recognizing the potential of cable television, many African American community organizers predicted the arrival of an African American–driven and –oriented solution to negative Black images on television. Similarly to Boston, the city of Detroit began its feasibility study of cable television in the 1970s and started planning for franchises in the late 1970s. Throughout the process of the feasibility study, drafting of the RFPs, evaluation of proposal, and the cooperation and struggles encountered with Barden Cablevision, America's largest African American–owned cable service provider, the ideal of fair African American representation and participation for the betterment of African American self-esteem was always present. Residents and city officials of Detroit, in other words, considered cable television a potential locus of African American social uplift.

Detroit's Twenty-Year Period of Feasibility Discussions and Study

Detroit's interest in cable television began in the early 1970s. Two publications marked this period. The first was *Southeast Michigan Community Needs '70* published by WJBK-TV2 Storer Broadcasting Company. The other was *Cable Television in Detroit: A Study in Urban Communications* prepared by Detroit's Cable TV Study Committee. Issued in January 1970, the first document conducted a formal study to identify the concerns and needs shared among residents and leaders in the city of Detroit, especially in relation to television news and informational programming. Although the report itself did not explicitly suggest how a cable television service could help local communities, the simple fact that a local cable television company conducted the survey to investigate local issues exemplifies the role locally originated media possessed.

The study consisted of a threefold investigation. First, WJBK-TV2 interviewed hundreds of Detroit area residents who had been selected randomly to yield a statistically representative sample. The interviews identified problems residents perceived as imminent and changes they wished to see. Some interviews took place at local homes and lasted several hours. The second type of research was conducted through the mail. Detroit and its suburban residents received a questionnaire and were asked to "rank the most pressing community problems according to their seriousness and urgency." Additionally, business people, government officials, educators, church and civil leaders, and other community leaders were similarly asked in a questionnaire to list problems that the greater Detroit community was facing. The survey took place in November 1969 as the base for the publication that came the following year.[104]

The report revealed that Detroiters, both regular citizens and opinion leaders, considered race relations, crime, and education to be the three areas of highest concern. Of residents who responded, 47.9 percent mentioned race relations as a community problem. Two years after the 1967 "riot," crime came second with 42.9 percent of residents perceiving it as an imminent problem. Race was a concern for both African Americans and whites. Several upper middle-class white residents are quoted as saying, "The race problem is very bad. I live in a community that is all white and a lot of people in my area feel that once a colored person moves in, it's ruined, and myself, I feel that they are as good as we are. Some are good and some are bad." A male resident responded, "The racial situation is impossible. The state of racial turmoil in this area is a constant threat to the safety of the community. I don't know what the Blacks want, but something better be done about it soon." Other residents agreed by saying, "The racial problem is number one. I feel that there's going to be another riot," and "There's unrest with the Negroes. . . . It's the Black power ones that's the problem."[105] African American residents also expressed their concerns about racial division. An African American woman identified as a member of the lower-lower class argued, "The big problem is not having togetherness." An upper lower-class man stated, "There's the race problem. A whole lot of people don't want to get involved. They like to sit and talk about it, but do nothing about it." He continued to express his concern about unemployment and economic difficulties among African Americans and suggested a more active municipal involvement to mend the problem.[106]

Published two years later, the report prepared by the Cable TV Study Committee on the City of Detroit advocated an introduction of cable television to the city. The committee, established by the city's Common Council, aimed to examine and advise the council on the feasibility of bringing cable television to the market. Lois P. Pincus served as project director. Rev. James W. Bristah, superintendent at the United Methodist Church in Detroit served as chairperson assisted by Ola Nonen as vice president. David Harper, president of First Independence National Bank was treasurer. Dwight Havens, president of the Greater Detroit Chamber of Commerce sat as secretary. Twenty-three members included Esther G. Edwards, senior vice president of Motown Record Corporation; Erma Henderson, executive director of Equal Justice Council; Otto J. Hetzel, associate director of the Center for Urban Studies at Wayne State University; and other locals who had educational, political, business, and other stakes in the city.[107] Despite the participation of media studies scholars such as Morton Miller from Wayne State University and Hazen Schumacher from the University of Michigan, few members "either understood or appreciated" their responsibility to study and make recommendations regarding cable television in Detroit.[108] The outcome of a year-long investigation with eighteen committee meetings that extended over hours and many more subcommittee meetings was report in excess of 160 pages refined by external consultants.[109] By the end of the study, the committee recognized cable television as "too important to Detroit, and its impact on the daily lives of Detroit citizens will be too dramatic and pervasive, for cable system operators to be regarded as anything less than gatekeepers to the mind."[110]

The committee began its study by holding various meetings with experts from different fields to learn about the problems. This early phase was a learning opportunity for both committee members and local residents. One of the consultation sessions was videotaped and put on air for Detroit's audience over a two-week period. A series of public meetings followed the television session. At the meetings, residents and representatives from any of Detroit's local organizations we allowed to share their perspectives about the city's cable system development. These meetings were held at different locations and at different times of the day to facilitate and encourage greater participation.[111]

The study predicted that within a decade, almost all of the cities in the country would have cable systems. The question was no longer if the city should consider establishing a system, but rather, when and how. The study claimed that to realize "an innovative and comprehensive . . . use of cable television as a communication medium . . . calls for a sophisticated system and a diversified structure with a high level of commitment to local origination, maximum access, and a wide range of municipal and educational services."[112] The study saw the potential of cable television as something that would allow "access to telecommunications to many individuals, organizations and institutions previously denied all but token access."[113] After identifying several technical advantages of the system, such as the improved signal reception and more diverse radio signals, the study suggested ways to make the cable television system beneficial to Detroit's minority community particularly through public ownership, financial distribution, channel allocation, employment, and public access.

The committee strongly advocated public ownership of the cable system over private ownership. The former style of ownership was preferable for Detroit because it guaranteed high-quality public service and encouraged involvement by local minority entrepreneurs in the system's development. Because the size of the city's financial investment would be immense—anywhere from $30 million to $120 million—the system had to be beneficial to the public as a whole. To ascertain the system's role as a "public utility," it also was important to avoid the competition between profit generating and public service that would be more likely to happen if privately owned. The study, therefore, recommended the cable system should not be established "until and unless a system is devised that will protect the public interest in this new telecommunications medium and guarantee that public needs will be met. The matter is too vital to be left to promises and good intensions." The study claimed that without responding correctly to what the local residents need and want, "the flow of information, entertainment, news, social and commercial services to public" would be disrupted.[114] This was one of the reasons why the report recommended a single entity be charged with distribution of service. Recognizing that racial and ethnic groups were very unequally distributed in the city, separation of distribution by area would result in a *de facto* race-based system, because "increased communication within the City is seen as a major potential of cable."[115]

The committee's concern in terms of ascertaining public service to guarantee residents' access to quality of information as their right, invokes Marshall McLuhan's idea of media as "the extensions" of a person and Carter G. Woodson's notion of mis-education. The report made reference to

McLuhan who claimed, "Once we have surrendered our senses and nervous systems to the private manipulation of those who would try to benefit from taking a lease on our eyes and ears and nerves, we don't really have any rights left."[116] Approximately thirty years prior, Woodson argued that African Americans' mis-education resulted in reinforcement of their disadvantaged living condition. His famous axiom reads: "when you control a man's thinking you do not have to worry about his actions."[117] He further continued:

> Lead the Negro to believe [they have no worthwhile past and they will achieve nothing] and control their thinking. If you can thereby determine what he will think, you will not need to worry about what he will do. You will not have to tell him to go to the back door. He will go without being told; and if there is no back door he will have one cut for his special benefit.[118]

The Study Commission's objective was to make sure that such a history would not be repeated with cable television in Detroit.

The committee suggested that financial allocations for community programming and lower access barriers would benefit Detroit residents. It argued that "the Community Cable Board [should] allocate programming funds to grantees for local programming" so that community channels would be able to secure programming funding through advertisement or system revenues. Similarly, according to the report, such funding should enable free-of-charge community and public access channels for local communities, including African American communities. Acknowledging that the FCC only allowed the first five minutes of live presentations to be free of charge, the committee recommended that the city petition for a waiver of such FCC regulation for the benefit of citizens of different financial backgrounds.[119]

The study also emphasized the importance of developing a sufficient number of channels for the needs of local communities. Of all thirty-six channels, or the minimum number of channels that the committee considered necessary, the committee felt ten should be reserved to serve communities. That number was larger than what the FCC required. Explaining that the purpose of such channels was "to encourage programming that accurately reflects the needs and interests of the community," the report argued that the realization of community channels would "provide the basis for heightened community consciousness, increased communication, strengthened ethnic identity, and improved community problem solving." It also stated that such systems would "facilitate the flow of information of particular interest to the community . . . and stimulate much needed intra- and inter-community communications."[120] The report suggested that establishing a cable television system with an appropriate channel allocation plan would alleviate and help improve some of the concerns that *Southeast Michigan Community Needs '70* had revealed—a lack of intercommunal understanding, racial tension, and others—two years earlier.

Decentralization was a key to realizing the capacity of community access channels to their fullest extent. By creating five separate cable districts in the city, with at least two channels for each district, the cable system would have programs directly serving the interests of the local residents. This structure would provide "a mechanism of community control of the medium and community access to the medium" without sacrificing local and communal needs and concerns. Responding to the interests of community was particularly important in urban areas where many African Americans resided, including Detroit. These residents used telecommunications differently than, for example, white working-class residents. In order to better serve African Americans living in inner cities, community channels needed to cover issues such as "community news, health delivery programs, use of community resources, courses in home management and child care, job training and job availability."[121]

In addition to the community channels, the breakdown of the thirty-six channels included two public access channels, three municipal channels, seven educational channels, eight local television broadcast signals, two distant television broadcast signals, three commercial channels, and one program guide. For more than sixty ethnic groups in Detroit, cable television would be a medium to share their culture. The report claimed, "Their music, their dances, their festivals all are a part of our pluralistic culture. But too seldom do these ethnic groups have the opportunity to share their cultural heritage. A cable system with open access would give such groups wide opportunity to produce their own programs."[122] This proposed list showed the importance of community involvement that the committee saw as being so important for the development of the medium.

The introduction of cable television to the city was expected to bring professional opportunities. Residents hoped that they would be able to contribute to the improvement of the lives of African Americans in Detroit. Carl Edwards, a minority recruiter at the Automobile Club of Michigan explained that although white industrial decision makers had begun to acknowledge the important roles of minority recruiters, there remained discrimination when it came to hiring. Edward A. Benford, president of Ludet Personnel Services, explained that discrimination was underestimated among African

Americans because they sometimes did not apply to what were considered "white jobs." Because corporate executives felt African Americans were less ideal job candidates than whites, training programs for African Americans had been reduced and eliminated. Cognitive maps of racial divide affected African American lives in Detroit, as well. To amend this situation, the "minority group clearing house" was born. If Edwards, for example, could not find a job for an African American who was looking for a job, he would refer the candidate to the clearing house, which would match the person to a job at another company. This was one of the initiatives by African Americans to collectively improve their employment situation. Additionally, job training and education were important to make sure African Americans could obtain and maintain a job. Twenty-six percent of Black Detroiters agreed that the lack of education and training was the largest barrier to finding a job. James Banks, an African American urban affairs officer for City National Bank claimed that the lack of training accounted for 60 percent of cases in which African Americans could not obtain a job. Banks, therefore, had offered bank teller and other kinds of job training to local African American residents.[123]

The introduction of cable television was going to supplement these existing efforts. As an immediate benefit, Detroit's large African American population had the potential to become an integral part of the labor force in the cable television business. From construction to operation and maintenance, local residents would play "the leadership role in regional development of cable television." In the long-run, the cable television industry could not move out of the city as long as the industry served its communities. The report explained "[u]nlike many other industries, [the cable television industry] cannot move to the suburbs; it will be primarily located within, provide service to and derive its income from within the city limits."[124] This was significantly different from manufacturing industries such as automobile and steel, which had moved out of Detroit seeking better business locations, resulting in the economic downturn the city experienced in the mid-twentieth century. The committee emphasized establishing the cable system in such a manner that the residents of the city would truly be able to benefit from it, not only by receiving television content, but also by supporting the industry and obtaining occupational opportunities.

Reginald D. McGhee, a Detroit resident and member of the Cable TV Study Committee, expressed his agreement with the idea of public ownership to promote minority involvement. He argued that if cable television was owned and controlled by the hands of a few, rather than many, such a concentration of media might result in a lack of attention to local interests.

He explained that by adopting a public ownership model, the cable system would encourage voluntary engagement by minorities in the city. Such a system "would create literally hundreds of jobs for the trained as well as the untrained." If successful, McGhee claimed that minority residents of Detroit could be employed as "actors, actresses, technicians, equipment operators, repairmen, and . . . service people."[125]

Reflecting McGhee's belief in local minority employment, the recommendation in the study report explained, in terms of the construction of the system, "the work force of the system [must reflect] the racial, sexual and ethnic group composition of the population of Detroit." For maintenance and operational personnel, it argued that "all qualified applicants will receive consideration for employment without regard to race, creed, color, sex, or national origin." These recommendations were the committee's response to what it called, "the historic pattern of job discrimination" through which many African Americans experienced limited occupational opportunities. In order to stop these racial filtering processes in the job market, the committee made sure that the racial and ethnic, in addition to gender, composition of the city would be reflected in the employment of the cable television project.[126]

The employment of city residents also would benefit the city through active financial transactions. The economy of the city had long been stagnant. The long-term unemployment rate within the city of Detroit had been much higher than that of the Detroit metropolitan area. By giving residents jobs, the cable system would alleviate the financial difficulty suffered by African American Detroiters. Consequently, direct income taxation, increased purchasing power of residents, and capital investments within the city were expected to revitalize the economy of Detroit. McGhee argued that an increased employment rate in Detroit would "create a new work force, replete with a purchasing power to bolster the sagging economy of the City of Detroit."[127]

Additionally, local minority entrepreneurs would face a lower entrance barrier to the new system when owned by the city rather than by a corporation. The report explained, "[public ownership] would provide significant opportunities for small business and minority entrepreneurs through contracts for regional or system-wide construction, maintenance and sales."[128] The committee examined what was good for the residents of Detroit and local minority businesses in the process of making its recommendation for public ownership.

According to the study, African Americans in Detroit could benefit from the concept of access in two ways: programming access and viewer access. Programming access, which the report considered one of "the hallmarks of cable television," was the opportunity for people "who want to transmit programs or offer services over cable systems to do so." It was a locus of self-expression and cultural diversity, rather than of "mass tastes." Because cable television could offer a large number of channels at lower cost, unlike broadcasting television that had to appeal to as wide an audience as possible, programming access, which was difficult to attain with a broadcasting service, would be improved. Additionally, unlike a thirty-minute network television show that cost in excess of $50,000 to produce at the time, local programming would not cost more than several hundred dollars at maximum. In many cases, the cost was only a few dollars.[129]

Cable television could develop new types of programs intended to cater to the interests of minority audiences. In the case of Detroit, the report suggested, such programs could deal with "school plays, athletic events, health care programs . . . [and] neighborhood news." These locally produced programs could function as "a problem solving mechanism" to tackle social issues such as welfare, unemployment, health, and so on.[130] Additionally, as the report prepared by the Rand Corporation in 1970 argued, "local origination in the ghetto can perhaps be used to interest local residents in solving their own problems rather than being totally independent on outside help."[131] These recommendations suggested that bringing a cable television system to Detroit not only would be an introduction of a new media format, but also a new opportunity for minority community development.

Programming access was particularly important for African Americans. By 1970, African Americans disproportionately relied on information retrieved through television. In 1967, 61 percent of African Americans trusted television, whereas only 15 percent trusted newspaper, 6 percent radio, and 3 percent magazines. The number for television had doubled since 1960 but halved for newspaper. Whereas the figure remained almost stable for magazines over the seven-year period, in 1960, 22 percent of African Americans trusted radio as a source of information. Despite their reliance on television, images on television were "not reflective of the needs or interests of the minority viewer. Although some efforts have been made to correct the portrayal of minority groups as stereotypes, significant minority programming is still non-existent." This explained why it was important that the programming accessibility prevented the cable television from duplicating the history of television.[132]

The study also made sure that the minority communities of Detroit would benefit from viewer access. It argued that because programming access would bring problem-solving mechanisms and empowerment opportunities to the city, programs with such potential needed to be available to all viewers. The committee claimed that "access to the cable system [must] be available to every person in the city wishing to subscribe, on a non-discriminatory basis, and within a reasonable and specified period of time," and "the cable system authority [should] develop an appropriate mechanism to permit access to viewing for low-income families." The report called attention to the fact that for many African American residents living in poverty-stricken areas of Detroit, even a $5 monthly subscription fee could be difficult to afford. Moreover, low-income households were highly dependent on television to obtain information and entertainment. Although the study did not make a clear suggestion of how to realize viewer access among low-income families, it did state that such a consideration was "a matter high on agenda of the cable system authority."[133]

These two studies taking place in the early 1970s reflect the city's interest in the cable television system. Race was a major point of emphasis for both studies. Lead by the local television station, the first report identified race relations as one of the most controversial issues in Detroit in the minds of its residents, business people, community leaders, and others. The Cable TV Study Committee especially attempted to make sure that when a system was introduced, African Americans and other minorities would not be left behind in production, channel allocation, employment, and access. It sought a way for African Americans to be not only consumers of cable television service, but also active participants in the industry.

Despite these two studies embodying the relative enthusiasm concerning cable television during the first five years of the 1970s, the rest of the decade did not see much advancement in the development of the technology in Detroit. This was not unique to Detroit. The general understanding of the cable television market in the late 1970s was that the size of the urban market for cable systems was too small. Summarizing the changes that took place during the second half of the 1970s, the Cable Television Advisory Committee of Detroit explained in March 1981 that "it has only been in the past two years that the market for cable television has become strong enough to sustain systems in major cities such as Detroit."[134]

Even within the Executive Office of the City of Detroit, cable television was not a part of the top priority agenda at the end of the decade. Despite City Clerk James H. Bradley's concern about the lack of commitment by the city office to the study of cable television, Mayor Coleman Young explained in his letter to the City Council on February 15, 1979, that the Media Council that was going to be established would not deal with cable television–related issues. Its aim was to study how to attract industrial and commercial television and film programs to the city and how to nurture talents for mainstream media. He wrote, "it would not seem that the objectives of the Media Council would be compatible with the issue of Cable T.V. for the City of Detroit nor with dealing all media as it related to the Federal Communications Commission."[135] Later in the year, the city started in-depth discussions on the establishment of a cable television system in the city by drafting an ordinance and an RFP. But the process was not as swift as many residents wished.

Recognizing that the City Council was once again interested in the potential introduction of cable television to Detroit in summer 1979, the City Planning Commission, led by Jack L. Korb, submitted a copy of the August 1979 issue of *Planning*, a publication of the American Planning Association, to council members.[136] Written by William F. Rushton, the director of research at the Center for Non-Broadcast Television, Inc., the article elaborated on the benefit that cable television could bring to city planners. It claimed that, "with cable television and other communication systems that are available right now, planners can get more citizens involved and expand their efforts in directions they might not have thought possible a few years ago." Rushton argued that such technologies "represent the fastest way to send masses of information to masses of people and get a quick response."[137] The article emphasized the potential of cable television to connect residents in a city and bring them together to discuss many of the issues that matter to the community.

The second half of the year saw a sudden rise in the interest in cable television within the city office. In November, the Detroit Cable Television Advisory Committee submitted its status report to the City Council. The report turned out to be controversial. James Bradley, who was concerned about the absence of serious commitment to cable television by the City Council about six months prior, found himself anxious that the Detroit Cable Television Advisory Committee had not turned in its report by the deadline. On November 2, he wrote to Lois Pincus, the chairperson of the committee. The letter read:

> The City Council requests that you kindly furnish immediate information as to when the status report on Cable TV will be submitted. The Council was informed some time ago that the status report would be submitted prior to November 1, 1979.
>
> Please respond.[138]

The letter was short but sent a very strong message to the committee. The City Council was eagerly waiting for the status report. The committee did not submit its report until November 13, approximately two weeks past the deadline. It summarized the status of its work by explaining:

> The Detroit Cable Television Advisory Committee is continuing the work of developing a draft ordinance and a Request for Proposal which will define the basic structure of the cable system for the City of Detroit. In this regard, the Cable Television Information Center was hired to assist the Committee in the development of an ordinance which would be designed to secure the best possible system and the highest level of service for the City.[139]

Although the committee's ten study points included minority entrepreneurship opportunities, local regulation, franchise fees, and other issues that were relevant to Detroit's residents, it did not contain concrete plans for the upcoming ordinance or the RFP.

As the frustration within the City Council rose, the Detroit Cable Television Advisory Committee continued to not produce detailed plans for the RFP. Six months after the status report, the committee was only "nearing the conclusion of the first phase of the franchise process." The new status report explained that drafting the RFP had been a "complex job." The ordinance had not been finalized either. The only good news was that the report promised that it would "inform Detroiters about what a cable television system might mean to them and how they can use it for education and information as well as entertainment." It would also "develop the education, municipal, and health consortia and involve those interested in community access programming." Although the committee had four drafts by the time the report was issued, it still had a long way to go before it would issue the ordinance and the RFP.[140]

The year 1980 was not one of progress for Detroit and its future cable system. The committee failed to deliver the ordinance by the end of June and the RFP by the end of the year, as the status report issued in May promised. Furthermore, Lois Pincus, chairperson of the Advisory Committee offered her resignation. Although she did not explain the reason behind her decision, she suggested that she would return to the cable television–related business in the future. She emphasized in her letter to the City Council members that it was a "difficult decision." She also said that she "may assume new responsibilities offered to [her] in the cable television industry." In a response to her letter, Maryann Mahaffey, a council member, wrote, "I know with your capabilities that someone would grab the opportunity to get your expertise, sooner or later."[141] Although the following year was to be one of progress, 1980 was, if anything, a year of stagnation for the city and its future with cable television.

The concerns that rose during the meeting between city officials and nineteen residents of Detroit on December 16, 1980 showed the effects of the lack of a concrete proposal by the committee. Taking place at a restaurant in the city, the meeting produced almost nothing more than simple reiteration of and re-emphasis on the importance of local citizen participation, public access, and other issues that had been mentioned in the reports published ten years earlier. Although the discussion raised several important questions, such as what would happen to left-out communities where African Americans lived, what kind of "dedicated" channels communities would have, and how public access would be guaranteed, the final conclusion of the meeting with nineteen local attendants was that "the group [meetings] should continue" in the future.[142]

Final Draft of the Request for Proposals

On March 6, 1981, the Detroit Cable Television Advisory Committee finally submitted the final draft of its RFP, a franchise agreement, and the ordinance to Coleman Young, who three days later turned them in to the City Council. The document package was the fruit of the committee's seven-year effort. After reviewing various cable systems from around the country, it concluded that "the City should act expeditiously in the granting of a franchise in an effort to insure the integrity of the process and so that the City might enjoy financial benefits associated with CATV as soon as is practical." In its cover letter, the committee made a bold timeline suggestion. Its plan stated that the RFP would be released three days later, on March 9. Bidders would respond to the RFP by June 15. The soon-to-be-established Detroit Cable Communications Commission would finish its review process by the end of July so that the applicants could submit their clarifications by August 6. By August 16, the mayor would make a recommendation as to which provider would offer the service, to the City Council. The council would spend the rest of the month reviewing the recommendation. According to this schedule, the franchisee would be determined by the end of the year.[143]

The creation of the Detroit Cable Communications Commission was deemed necessary for both practical and legal reasons. On the one hand, if the commission was established as would be requested in the ordinance, it would serve as "an arm of the executive branch of the government." It also would allow the separation of power, as requested by the Michigan Constitution of 1963, between the Advisory Committee that created the RFP and relevant documents and the entity that reviewed proposals. As a result, it would be the Cable Communications Commission and its associated consultants that made the recommendation to the mayor. This would secure the separation of power between two pertinent entities: the one in charge of drafting the RFP, and the one that evaluated the response to the RFP.[144]

In 1982, the Michigan Library Association's Telecommunications Committee issued a resource guide, clearly marking the commission's expected roles. The document also educated those who were not particularly knowledgeable about cable television. Led by Bruce Schmidt, chairperson of Southfield Public Library, the committee provided Michigan residents with a basic understanding of cable television and its franchising. The group also had six additional members: Barbara Buckingham from the Macomb Public Library, Clara DiFelice from Oakland University, Shelly Gach from the Huntington Woods Public Library, Jim Hibler from the Independence Township Library, Shelagh Klein from the Sterling Heights Public Library, and Linda O'Donnell from the Oakland Schools. The document listed what the committee members thought the residents of Michigan expected from a cable television.[145]

Discussing the responsibilities of the commission, the Library Committee clearly necessitated its independence from the other branches of the city office. The committee, as a separate entity, was expected to advise the city about the basic requirement for the franchise. Its responsibility also included giving advice on rates, ordinances, general policies, and other issues. Representing the residents of the city, the committee would make sure that access channels were truly accessible to the institutions, organizations, and individuals, including ethnic minorities of Detroit.

The proposed RFP drafted by the Advisory Committee emphasized the importance of minority participation. It argued:

> the city find that there has been an historic exclusion of minorities from participation and ownership in the communications field and therefore it has included provisions in granting the franchise for the city that are intended to provide opportunities for minorities to remedy the discrimination that has prevented them from access to the communications media, its operations, management and policy boards.[146]

Although it was far from a surprise for the RFP to contain such an explicit statement regarding equal participation considering the continuous discussions on the same topic dating to the early 1970s, it was noteworthy that the city of Detroit, containing a large African American population, expressed its interest in minority inclusion. Particularly in employment and job training, the RFP draft ascertained that local minorities would be hired for the project. It argued that "the city expects that the selected applicant will employ Detroit-based labor, specifically including minorities and women, to the maximum extent possible in the construction and management of the system." Similarly, regarding professional development, the franchise candidates would be asked to:

> describe in detail for each year of the term of the franchise plans and policies for training programs to provide personnel for the various aspects of installing, managing, operating, marketing, programming and maintaining a cable television system, including specific plans and policies that will assure representative participation of minority persons and women business enterprises in such programs.[147]

Along with the proposal's request to contain two lease channels "reserved for minority individuals, groups or institutions," the draft constantly reiterated the importance of minority involvement.[148]

Outside of the city office, the frustration among Detroit residents who wished to have cable television sooner began to surface. On March 20, 1981, Detroit resident Sharlan M. Douglas urged the City Council for quicker development. She wrote, "For heaven's sake, stop bickering and get on with the business of approving cable television for the residents of Detroit. . . . Consider the needs of your constituents and not your delicate egos and get on with the business of wiring Detroit." In response, Maryanne Mahaffey expressed her gratitude for Douglas' honest comment while emphasizing the importance of being careful and thorough while examining the ordinance draft. Douglas, however, was not alone. Detroit's local newspaper also surmised and reported frustration and concerns about the slow process.[149]

The first sign of the delay in the cable system introduction to Detroit appeared in the memorandum dated April 15, 1981 by Michael E. Turner, director of the City Council Division of Research and Analysis. In his three-page note addressed to the City Council members, he shared his doubt that the time schedule as proposed in the Advisory Committee's report would be met. He explained, "it must be noted that if the RFP is not released in mid-April 1981, it will be essentially impossible to award the franchise in August, 1981." Already more than one month behind the original schedule,

Turner proposed a very tight timeline. He suggested that instead of allowing bidders to spend ninety days to study community needs and prepare the proposals, the council could only allow seventy-five days to do so. Similarly, the Cable Communications Commissions would subsequently have forty-six, rather than sixty to ninety days in the case of an average franchise plan, to review the proposals. If the bidders could cut the time to prepare their response and clarification to a week, rather than two, and the commission could review the response in eight days rather than in two to four weeks, the mayor would be able to make a recommendation to the City Council on August 16, 1981 as planned. In this case, the City Council would have sixteen days to review the recommended proposal, conduct hearings, and approve the agreement by the end of the month.[150] Despite this bold plan, due simply to the fact that the timeline was already a month late and that what issued on April 15 was this memorandum rather than the RFP, it seemed to signal the major delay that the city would experience during the development of its cable television system.

By the end of April 1981, the structure of the proposed Cable Communications Commission had become a topic of heated debate. On April 22, E. Barbara Wilson, chairperson of the City Council's Cable Television Citizens Review Panel, wrote that the mayor possessed too much power over the appointment of the commission members. She wrote to the council, "As presently proposed, the members of the Commission would be appointed by the Mayor, subject to City Council approval. . . . Due to the importance of cable television to the city of Detroit, the Review Panel's consensus was that the power of appointment of Commission members should be shared by the Mayor and the City Council." Additionally, the panel recommended that the commission be composed of five, seven, or nine members, a majority of whom would be appointed by the City Council.[151] This represented a dramatic change from the existing plan in which the mayor was the primary appointer of commission members. The panel, however, claimed that its recommended plan would "represent the broad diversity of [the] community, including considerations of geography, race, gender, age, organizational affiliations, and physical handicap."[152]

Two days after the panel's recommendation, Michael E. Turner issued a list of changes that the City Council deemed necessary for the pending ordinances. Reflecting Wilson's logic, the memorandum claimed that:

> Council directed that the ordinance creating a Cable Communication Commission be amended to provide for a seven-member

commission, with four commissioners appointed by the City Council and three commissioners appointed by the Mayor. The ordinance as submitted to Council provided for a five-member commission with all commissioners appointed by the Mayor, subject to City Council approval.

Under this new provision, the mayor would have the authority to appoint a new commissioner in order to replace one that the mayor had previously appointed. Similarly, the council would appoint a new member when a commissioner appointed by the council decided to leave. The council discussed these changes over two days and issued the memorandum for the approval by the Law Department.[153]

Local newspapers were aware of the negotiation of power relationships between the mayor and the City Council. On April 30, 1981, Nolan Finley wrote in the *Detroit Free Press* that the power play between Mayor Young and the City Council would possibly hinder a smooth decision-making process in the franchisee selection. Although the City Council had passed the ordinance a day earlier, the article reported that the mayor opposed the post-amendment ordinance that allowed him to appoint only three commissioners out of seven, instead of all of the five as originally planned. Young's aide, Carol Campbell, was quoted as saying, "a veto [by the Mayor] on this matter is not a bad guess."[154] Similarly, *Detroit News* cautioned that the power struggle between the council and the mayor might "stall the efforts to bring cable television to the city." It also cited Charlie Williams, the head of the Advisory Committee, saying, "If I had to make an educated guess, I'd say the mayor will veto the ordinance." Although the council had the power to override the veto, the article continued to express its concern that the proposed deadline set for August 15 set by Young to select the franchisee might not be possible.[155]

Campbell and Williams' prediction came true and fueled the intensity of the debate. After Mayor Young vetoed the ordinance, the City Council decided to make an amendment on May 13. While keeping the number of commissioners at seven, the mayor would appoint four and the City Council would appoint three, according to the suggested new draft. This amendment would give the mayor more power over the Cable Television Commission.[156] At the formal session on May 14, the City Council did not support the new draft. Instead, it decided to override the mayor's veto from the previous week. Not willing to adopt the five-commissioner plan that Mayor Young had proposed, Erma Henderson, council president, commented that she needed to examine which entity—the mayor or the council—would have control over the issues concerning cable television. Williams, on the other hand, suggested that the five-commissioner plan in which the mayor would appoint the commissioner and the council would approve the appointments was the most reasonable solution, because "[the council members] don't have to approve them."[157] What this power struggle meant for the residents of Detroit was an even greater delay of cable development in the midst of this power play.

Mayor Young's political interests also interfered with the development of cable television. Although he continued to push for an early cable franchise selection by the hard-to-achieve deadline of August 15, 1981, it became clear that the reason for such a push was "to avoid the possibility of cable companies doing election lobbying."[158] Because of such political pressure, Erma Henderson and his colleagues at the council began to consider postponing the introduction of the cable system. Nicholas Johnson, former member of the FCC and chairperson of the National Citizens Communications Lobby in Washington, DC agreed with her. He explained that "a lot of communities in America are rushing into [cable] like lemmings into a barbed wire fence."[159] Although the discussion on cable television continued despite Henderson's reservation, it added more reasons for Detroiters to expect a delay in the delivery of the system.

Despite the slowdown caused by the power play, two seminars took place in 1981 to provide opportunities for residents, policymakers, and those involved in cable development to learn from the experience of other municipalities that were more advanced in the process. The first of such events was called "Cable Television: In the Public Interest," sponsored by WTVS/Channel 56, held on April 24. In one of the panels, Jeffrey Forbes, Massachusetts State cable commissioner, discussed the licensing process. Forbes shared the draft of his more than thirty-page document entitled *The Cable Licensing Process: A Practical Guide for Municipal Officials and Cable Advisory Committees*.[160] Similarly, seven months later, another seminar on cable television was held. Sponsored by the Benton Foundation and held on November 28 at Cobo Hall in downtown Detroit, it discussed the franchising process, ownership issues, access issues, and regulatory and administrative issues. Speakers came from Seattle, Washington; Washington, DC; Tucson, Arizona; Bloomington, Minnesota; New York City; Brea, California; and other municipalities.[161] These events were rare but were certainly existing signs that kept hope for the residents of Detroit.

Compromise concerning the Cable Communications Commission appointment came on February 10, 1982. The revised ordinance read:

> There is hereby established a Detroit Cable Communications Commission. The commission shall be composed of seven commissioners, four of who shall be appointed by the Mayor and three of whom shall be appointed by the Mayor from a list of four names submitted by the City Council.[162]

This was a major concession of power for the City Council. During and right after the battle between the City Council and the mayor to obtain control over commissioner appointment, the council received applications and recommendations from Detroit residents for the position of commissioner in late summer. In a recommendation letter dated September 18, Morris H. Goodman, a Detroit-based attorney, recommended Amy Davis, executive director of the Sundial Center, as a commissioner. He claimed that Davis "is a pioneer in the development of production of community programming for cable televisions and is actively promoting the protection of the public interest in the communication's industry."[163] Donald Snider also recommended himself for the position. He believed that his experience at various cable-related seminars in the past would be beneficial for the city.[164] Similar letters continued to flow in to Mahaffey and other council members' offices for the following six months.

As an increasing number of résumés, applications, and recommendation letters began to arrive, council members began to review the documents and prepare their nominations for the interviews slated to take place on March 15, 1982.[165] Some candidates received positive evaluations for their previous engagements in local communities or in education, key fields for African American empowerment through cable television. For example, council member Maryanne Mahaffey was intrigued by the teaching experience of Alta Harrison, a master's candidate in communications at Wayne State University at the time, in a faith-based context to enhance "awareness, sensitivity, and consciousness" and encourage "group interaction."[166] Similarly, she showed her interest in another candidate, Joseph R. Jordan, for his technical background and "ideas about involving youth in training and employment."[167] She wrote in her personal note that Jordan had "strong support for [a] minority operator" and was a reasonable negotiator. He had the potential to become a "minority representative at technical level."[168] These were some of the key methods through which cable television was believed capable of enhancing the self-esteem of African Americans.

Due to the large number of applications received from thirty-seven residents, the interviews took place from March 15 until March 22. The council's interests in minority empowerment were apparent in interview questions. "Cable has been characterized as one of the last opportunities for significant local and minority control, and ownership. Would you share with us your views on this subject?" "What are some of your suggestions to implement . . . local economic development?" "Do you have any suggestions for implementing plans to train Detroit citizens for jobs in the CATV industry?" "Have you given any thought of ways to provide cable television service for low-income citizens?"[169] These questions were designed to make sure that selected cable television commissioners and commission members would be capable of executing the plan for minority community involvement in production and consumption.

By the time all the interviews took place, the candidates for the Cable Communications Commission had been narrowed down from thirty-seven to nine. Of those nine, four would be selected and their names would be sent to Mayor Young, who would choose three of the four. The nine nominees were Kenneth Cockrel, former councilman and partner at a law firm; Amy Davis, director of Sundial Center, a nonprofit educational center; Alta Harrison, a preschool teacher planning to obtain a master's degree in communications from Wayne State University; Jim Holley, pastor of Little Rock Missionary Baptist Church; Derrick Humphries, a lawyer and media director for the Congressional Black Caucus; Joseph Jordan, pastor of the Corinthian Baptist Church; Helen Love, producer of WJBT-TV's PM magazine; Myrna Webb, administrator for the Medical Center Citizens District Council; and Barbara Wilson, an art teacher for the Detroit public schools and Wayne State University.

While the Cable Communications Commission selection process was under way, the city's decision to make its cable television system reflective of its minority composition remained strong. In April 1982, council person Barbara-Rose Collins shared an article on African American involvement in cable television published in *Sepia* magazine. Explaining that "Detroit [was] at least 63% Black," Collins urged her colleagues to remind themselves of the significance of their work for the city's African Americans.[170] Ken Coleman opened the article with a positive tone claiming that "the picture is getting bleaker for Blacks in cable. But while the established networks are attempting to muscle out those who try to enter the market, there are African American companies who refuse to give up the challenge."[171] Coleman, however, was not overly optimistic. He recognized that of 4,300 cable systems in the United States only 6 were African American-owned. Those systems were KBLE in Columbus, Ohio; Telecable Broadcasting in East Cleveland, Ohio; Connection Communication in Newark, New Jersey;

Video Vision in Tuskegee, Alabama; Bardon Communications in Lorian, Ohio; and Small Cities Communication in Carrboro, North Carolina.[172] The rest were owned by non-African Americans. The article closed by identifying ten ways to make sure that a new cable system would be beneficial for the Black community, such as insisting on public disclosure of hearing proceedings, encouraging minority participation.[173]

On May 20, 1982, Mayor Young announced the appointment of seven members to serve as the Cable Television Commission members. The four commissioners he appointed were James P. Robinson, pastor of St. Catherine and St. Edward churches; Otto J. Hetzel, professor of Law at Wayne State University; Jean Leatherman-Massey, administrative coordinator of the Detroit Board of Assessors; and Lester Morgan, instructor at Washtenaw Community College. He also appointed Joseph Jordan, Myrna Webb, and Alta Harrison, whom the City Council recommended. The mayor emphasized that the members represented the "cross-section of [the city's] citizens."[174] The first major responsibility of the commission was to review the draft RFP prepared by the Advisory Committee.

As the commission revised the draft of the RFP, Mahaffey received a personal letter from Geraldine L. Daniels, assembly member from the seventy-first District of New York County.[175] Daniels shared statistical information released by the FCC approximately one year previous. The FCC's report, published in 1981, dealt with minority participation in the national cable television industry. The report showed that in 1980, more than 87 percent of cable employees were white. Only 12.8 percent were non-white. Additionally, the FCC found that minority employees who reached the cable official and manager level only made up for 0.6 percent of all cable workers. Approximately 30 percent of non-white workers were technicians. Another 30 percent were engaged in operative works. Minority concentration in lower ranking jobs was obvious. Approximately 90 percent of the upper four job categories were dominated by whites.[176] If the city of Detroit successfully introduced cable television to the local community and was able to reflect the racial composition of the city, the feat was going to create one of the most racially integrated cable television workplaces.

Independent from the city's policymaking process, ELRA Group, Inc. conducted a study on public interests in cable television for the Center for Urban Studies at Wayne State University, and published its outcome as *Cablemark Survey: Public Interest in Cable Television Detroit, Michigan* in June 1982. The study aimed to "identify the types and numbers of entertainment programs, services, and channels of information that Detroit residents would like to have cable TV supply." The project successfully interviewed

405 people and gathered information regarding their expectations about cable television. 63.5 percent of the respondents were African Americans and 31.1 percent white. The ratio of African Americans in comparison to the subject pool equaled the proportion of Blacks in the city, 63.1 percent according to the 1980 census.[177]

The study revealed that 50 percent of Detroiters could only watch five to seven channels. Only 4 percent of the population had eleven or more channels. Almost 20 percent had four or less. These numbers showed that the diverse programs that cable television promised to bring were something new to most residents of the city. Although limited to a small number of available channels, 97 percent of the respondents were satisfied with the quality of pictures. Furthermore, 87.7 percent did not have a paid television subscription.[178] These statistics suggested that although the potential of the television technology had not been fully taken advantage of, people positively received the technology in their life. Additionally, due to the high unemployment rate, cost would be a significant factor to consider during the planning of a new television service.

The survey revealed that approximately 70 percent of Detroiters wanted to have a channel "with special programs featuring Black entertainers." There was more interest in a Black channel than in a cultural, religious, or sports channel. With in excess of 60 percent of the city's population being African Americans, this was not a surprising figure.[179]

Communication among the Detroit Cable Communications Commission, the City Council, and Mayor Young's office was very frequent during summer 1982, a few weeks ahead of the submission of the RFP. On July 30, Michael E. Turner, then-interim executive director of the commission submitted the City Council its final drafts of the Cable Communications Code of Conduct and RFP.[180] On August 2, six minor definitional adjustments were requested for the Code of Conduct.[181] Two days later, the revised final draft of the RFP was resubmitted to the City Council to be voted on at the Formal Session.[182]

Chapter 4

Drafting of Democratic Communication Media

The process of drafting a scheme for a democratic communication medium had two major phases: issuing of the RFP and the submission of cable application. Although this section covers a very short period of time—from August 1980 to February 1981 for Boston, and from August 1982 to December 1982 for Detroit—it encompassed both of these significant events. First, the RFP was an outcome of city officials' commitment, local residents' participation in decision making, and various study projects' months and years of long work. It summarized exactly what the city wanted to achieve through cable television. It was a summation of discussions that had taken place during the incubation phase.

Second, cable service providers submitted their applications several months later as their response to the RFP. None of the applicants were new for either Boston or Detroit. They were all corporations with some, and at times, extensive experience in the cable industry. Tailoring their service to match what the cities of Boston and Detroit expected, candidates submitted an application explaining how specifically they would meet the needs of each city. Until the two cities made their final decisions about the service provider, they kept in close contact with each corporation asking questions and receiving answers.

Analyzing what the city expected and how candidates strove to satisfy the needs reveals one of the most dynamic and fundamental moments in the history of cable television development in Boston and Detroit. Issues that concerned African American residents such as local origination, access channels, employment, show that African American interests were not marginalized during this period. Additionally, the study of applications also demonstrates how much candidates tried to live up to the expectations of Black Bostonians and Detroiters.

The focus of this chapter is on how discussions before the issuing of the RFP, the RFP itself, and applicants' responses to the RFP dealt with issues concerning the residents of Boston and Detroit, particularly African Americans. In one way, such a focus risks paying too little attention to issues such as cable suppliers' financial condition, technical capabilities, and others that were highly consequential in the cities' decision making regarding which candidate would win the franchise license. All of these issues, without a doubt, were major concerns for city planners and minority future customers. Examining the history of cable television as a community medium and tool, however, reveals that ideas such as community access, local origination, minority involvement, and minority employment, were just as significant as more business-oriented topics. Therefore, the need to investigate technology with a special emphasis on race is imminent. To do so also enables a more comprehensive study of cable television because the new medium is put in a larger social context.

It also is important to identify some of the arguments and claims that were made during this period that did not seem to have any impact on African Americans in Boston and Detroit, but that were quite consequential in reality. For example, many city offices stated that the cable technology would be useful to ensure and improve the security and safety of the neighborhood. Although such a statement itself does not contain any race-related language, it must be read within the context of municipal control of a certain neighborhood or that of positive or negative relationship between residents and the police department. A close reading of the RFPs and applications can reveal these examples of tacit race politics surfaced, highlighted, and mediated by cable television.

Finally, as had been the case in previous chapters, residents, city officials, and companies' discussions on African American viewers often had been grouped simply as a minority issue. RFPs, for example, refer to minority issues more frequently than African American issues. The category, of course, includes ethnic minorities such as Blacks, Latinos, and Asian Americans. It also includes women and people with disabilities. Despite the diversity within this category, African Americans were most affected by such minority issues, because 75 percent of the minority Bostonians and 96 percent of minority Detroiters were African Americans, in 1980.[1] Therefore, a special focus on African Americans does not ignore other minority groups. In one way, it allows us to understand how the largest minority group in Boston and Detroit was affected by cable television. In other words, if the African American voice was not going to be heard, the likelihood of other minority groups' voices heard was smaller. Second, it recognizes how the policies of cable companies toward the minority population heavily affected African American communities in these cities.

Drafting and Issuing the Request for Proposals in Boston

The shift from the incubation period to the drafting period was a dynamic moment in the history of Boston's cable television. By the early 1980s, many Bostonians spent their free time watching television: 56 percent watched more than two hours of television on a typical day and 27 percent said that they watched more than twenty-five hours of television per week. Only 26 percent watched television less than ten hours a week.[2] In such an environment where television had become a significant part of Bostonians' free time, the cable television decision-making process faced a new phase. Although residents and community organizers continued their discussions and studies, Boston's political leaders made little progress with their discussions for the introduction of a cable television system. The Boston Cable Television Review Commission presented its recommendations to the city on October 17, 1979, explaining that the franchise must engage as many residents as possible.[3] Reflecting the recommendation, Mayor Kevin White promised during his inaugural address in 1980 to have more community involvement in policies concerning the city government and media. In reality, however, nothing was taking place. Mayor White failed to show a serious commitment to keep his word.[4]

Despite the commission and the Access Coalition's recommendation to engage more residents, Mayor White did not make any plans. This was contrary to his self-appraising statement that "there are few decisions, if any, I have made as Mayor which will have as lasting effect on the people of Boston as the installation of a city-wide cable TV system." In reality, he even postponed the development indefinitely. It was only after late 1979 that he announced that concrete project planning would be resumed in early 1980. Even so, he did not comment on citizen involvement. But the promise was broken once again. Boston's plan remained frozen during the first part of 1980. The city had to wait until the summer to witness progress in cable television.[5]

Dan Jones and Martin Kessel, two members of the Boston Cable Television Access Coalition, outlined the importance of citizen participation in their summer 1980 *Boston Globe* article. The article was published approximately one week after Mayor White announced on August 11 the beginning of a formal process of choosing an operator to build and maintain

Boston's system. The article stated, "the shape of the Boston cable system and the services it provides will depend largely on the extent of citizen participation during the planning stages." The authors also cautioned that due to the large sum of money involved in the deal, the decision could be made with political motivations. This was particularly the case because only the mayor had the final say as to which company would win the franchise. Jones and Kessel claimed that the city's decision to establish the Cable Television Review Commission in 1979 was a valid one. The authors, however, were cautious. The Citizens' Advisory Board had not been established on a permanent basis yet, despite the recommendation submitted by the Review Commission sin October 1980.[6]

In September 1980, the Mayor's Office issued *A Cable Television "Primer."* It sought to educate its readers who were mostly Boston residents, about the history of cable television and inform them of its significance in their lives. The *Primer* explained:

> Boston looks forward to the contributions cable television will make to a total communications network which opens to its citizens and institutions new opportunities for access to information, increased communication and other economic and social benefits made possible by advancing technology. It is hoped that this primer will stimulate your thinking about ways in which this vast and exciting change in communications can affect the city.[7]

The document offered a brief but relatively comprehensive summary of the differences between conventional television and cable television, technology of cable television, and the history of the cable industry. For those who were aware of the delayed commitment by Mayor White, this publication did little to change their distrust in his words.

Another shortcoming of the *Primer* was that it contained little emphasis about the residents' voluntary participation in the production of cable contents or the industry's strategy to encourage such participation. These were some of the key benefits of the technology. Although the *Primer* discussed how the new technology could be beneficial to the elderly, educational institutions, health service providers, and so on, the focus was on what kind of service was to be offered and how it would be useful for the residents. Discussions on how their participation would ameliorate the service even more were limited. The *Primer*'s conclusion reflects the policymaker-driven style of cable development to which the city office was leaning. It explained that the cable system was expected to:

> interconnect the hospitals, the schools, the police and various centers of government. The uses discussed in this report demonstrate that a city such as Boston could easily utilize a significant number of cable television channels for public services alone. But, if Boston is to take advantage of cable opportunities and receive the maximum benefits of a CATV system, it must begin to plan for the future now.[8]

This type of analysis reveals the difficulty among citizens to remain engaged in the discussions on the development of cable television. On many occasions, policymakers and residents were engaged in separate discussions on the same topic, or led by separate leadership.

The city's interest in cable television as a means of public service with a relatively limited amount of voluntary participation by residents served as a basis for some of the position papers made public in September 1980. The Coordinating Council on Drug Abuse (CCDA), for example, explained that:

> If CATV were available for public service immediately, it would be likely that the CCDA would prepare a schedule to target specific audiences (both high risk and general), at times during the day when those populations are likely to have access to a receiver, with programs designed to deal with the unique drug-related problems experienced by this audience.[9]

Cable was especially useful because its two-way capacity would allow a connection among neighborhood health centers, educational institutions, hospitals, and residents. The CCDA's meeting also could be put on the air. Cable, from the perspective of the council, was a way to control, educate, and network for the prevention of drug abuse.

In order to secure the availability of such public service, "no prospective operator should be considered who did not guarantee channel availability for community-oriented public health and safety issues." The council made a strong recommendation that if the franchisee failed to provide a promised public service, it would be forfeited. Under such cooperation with the cable provider, CCDA staff would coordinate and develop cable information to put public. Ultimately, such an emphasis on drug abuse on cable television's public service channels would make city residents more aware of alcoholism and drug abuse and raise their consciousness about such negative substances.[10]

The council's interest in cable television told more stories about the capacity of the technology than simply as a means for tackling the drug abuse problem. The position paper explained that cable television's role is especially important because "both audience and message content can be more efficiently selected and matched."[11] Other departments and commissions also submitted their position papers on cable television. For example, the Conservation Commission and the Air Pollution Control explained, as the CCDA stated, that a greater public awareness would be the most beneficial outcome of cable television. More residents could participate in public hearings if they were broadcast on cable television.[12] The Economic Development and Industrial Corporation of Boston emphasized on cable's capacity to encourage employment. Cable could provide its audience with the latest job listings and training.[13]

Being able to connect the media content and its target audience efficiently was one of the reasons why the minority spectatorship was a major point of emphasis among cable developers. Because there were more channels available with cable television, programs of various emphasis, topics, and interest would be able to exist. Without the technology to connect people to contents they enjoyed viewing, such media development in production would not be successful, because a large number of channels would not guarantee audience satisfaction. Cable's advantage over conventional television, therefore, was not only its large channel space but also its ability to match contents and audience.

The Boston Police Department's position paper was most eloquent about cable television's positive role in community development. Explaining that police officers spent about 70 percent of their time on non-crime issues in local communities, the police department believed that cable television could be helpful in further improving its relationship with the community and bring a better living environment to the residents. Bostonians could acquire information on how to avoid becoming victims of a crime, to get together to discuss community concerns, give feedback to the police department, and so on.[14] In the department's analysis, cable's advantage was not only to connect the city office to the public, but also to help residents connect with each other.

Applying to Wire Boston: Submitting Preliminary Application

Boston's cable television project started a new development phase on August 13, 1980 when the city issued a request for initial application. Cable companies that hoped to wire Boston were requested to submit their preliminary application by November 3, 1980. By the deadline date, nine cable television companies had submitted their preliminary applications to Mayor White. Two were local companies: Abetta Corporation and Boston Cablevision Systems, Inc. Stephen Mindich, publisher of the *Phoenix*, a weekly newspaper in Boston, owned the former. Clarence E. Dilday, James S. Dilday, and others owned the latter. Non-local applicants were the New York Times Corporation, American Cablevision of Boston, Cablevision Systems Boston Corporation, Rollins Cablevision of Boston, Inc., Times Mirror Cable Television of Boston, Inc., Tribune Cable of Boston, and Warner Amex Cable Communications Company.[15] Since the official RFP would not be issued as the *Authority Report* until February 1981, these applications did not always respond to the particular needs and wants of the city and its residents. These companies based their preliminary plans on the information provided in the *Primer* and other sources that described the type of service in which the city was interested.

Boston Cablevision Systems was one of the local applicants that submitted its application in fall 1980. Its plan was to build a fifty-four channel system with about six hundred miles of cable serving more than 225,000 households. It was also interested in serving the minority population of the city. Agreeing with Boston Cablevision Service's desire to build a system for the minorities, Syndicated Communications, Incorporated, a Washington, DC–based venture capital company, expressed its interest in considering an equity investment of $500,000.[16] As for local origination, Boston Cablevision planned to reserve seven to fourteen channels for public access. Although exact locations were not yet determined, the company promised to establish fully equipped studios for the production of access programs. The company considered access to be "the fruition of cable's promise." As a result, "minorities, women, ethnic groups, the elderly, youth, handicapped and disabled citizens could develop programs to serve their information needs." Similarly, foreign language programs would serve as "a cross-cultural 'window' for all Bostonians." Once access centers were wired properly, residents would not only be able to produce media contents of their choice, but also would be able to interact with other residents from different areas of the city.[17]

Boston Cablevision attached its affirmative action policy to the application. It explained that the company:

> stresses the importance of strong and effective affirmation action programs in employment practices and policies. BC shall afford

equal employment opportunity to all qualified individuals and no person shall be discriminated against because of race, color, sex, religion, physical handicaps, sexual preference, national origin or any other factor not related to job performance. This policy shall apply to all personnel actions including working conditions, recruitment, evaluation, selection, placement, promotion, compensation, training, demotion, lay-offs and termination.[18]

The applicant not only promised that hiring, recruitment and other human resource–related processes reflected this basic policy, but also emphasized that it would draft an annual report to discuss the effectiveness of the policy. Although the application did not make a detailed comment about how exactly to solicit minority participation in work force, Boston Cablevision Systems appeared to have a strong affirmative action policy.

In the New York Times Corporation's plan, there would be at least fourteen hours per week of local origination programs to "answer the needs of the Boston Community for quality programming in all fields—current events, culture, entertainment, practical endeavors and finance." Additionally, 30 percent or more of such programming were planned to take place in Boston.[19] To execute such programs, there needed to be adequate facilities and equipment that met the FCC's standard cable-casting rules. Located in Boston, such facilities should be open from 8 a.m. to 10 p.m. on weekdays.[20]

The applicant attached to the application its affirmative action plan for minorities. Its aim was "to assure equality of employment opportunities for minorities consistent with all applicable laws, regulations, orders and constitutional provisions." It clearly stated that:

> the Times does not intend or commit itself to recruit, hire, promote, select, assign or transfer any person, regardless of sex or race, other than a person whom The Times believes at the time of any such personnel action, is well qualified. The Times does not intend or commit itself to create any vacancies or openings, by establishment of new jobs, by retention of unneeded jobs or by encouragement of retirement or severance of present personnel to meet or satisfy any goals contained herein nor to demote, reclassify, terminate, replace or otherwise displace any non-minority employees to meet or satisfy any such goals.[21]

Submission of such a document was a way to underscore the company's commitment to hiring minorities. A clear statement of mission, if accompanied by true willingness to materialize it, offered a sense of assurance to Boston's African American citizens. However, because this part of the application did not directly refer to the ethnic composition of the city and the company until earlier in 1981 would not offer the details, it was unclear how aware the Times was of the ethnic dynamics of the city on the submission of the application.

Rollins Cablevision also showed its interest in cable television in Boston. It had already wired the largest system in Connecticut, in Branford.[22] Despite this experience, its application was not strong. Although its financial plan and a few other plans were made specific for Boston, most policies remained much more vague than the Warner plan or other weaker applicants. Regarding the channel structure, for example, Rollins correctly took Boston's geographical characteristics to determine what channels to air. But there was little reference to the needs of the local communities such as women, ethnic minorities, educational, and medical institutions. Such an application did not impress racial, ethnic, and other minorities in Boston.[23]

Times Mirror Cable Television of Boston also showed its interest in constructing and operating a cable system in Boston. Its parent company, the Times Mirror Company was the sixth largest cable company in the nation. In his correspondence to Mayor White, Ralph J. Swett explained that his corporation owned and operated five metropolitan and three community newspapers including *The Los Angeles Times*, had a publication division, and owned seven broadcast television stations, as well as many cable systems.[24] Times Mirror predicted that "the development of a broadband communications system touching every corner of the city, could be to the 1980s what the development of the streetcar was to the 1890s."[25] It also recognized the unique nature of cable television that position papers of other companies and city offices had identified. Cable television was:

> not just better reception and more channels. It is also a tool by which a community can better understand its government, its public services, its cultural life, and its people themselves. Cable means access and outreach—it can link schools, libraries and senior citizen centers to one another and to the general public. . . . And cable television can open unprecedented lines of communication between neighborhoods and downtown, and among neighborhoods themselves. If pluralism is the lifeblood of this city, this sharing of knowledge and experience can only strengthen it.[26]

The company also argued in its application that cable is especially useful for Boston because "the city's social diversity and cultural richness require more community channels and programming support than most other cities could ever use."[27] Times Mirror proved that it correctly saw the advantage that the new technology would bring to the city through its preliminary application.

Times Mirror considered community programming as representing a small part of its community services. It claimed that "system design, location of offices and studios, and mechanisms of accountability matter very much as well." It also speculated that successful local programming could only happen after bridging the programs produced by the company and those through more conventional public access. In other words, without the company's involvement in community programming and training, there would be no successful community services.[28] In order to realize ideal community programming, the application promised to dedicate channels for neighborhood access; local origination; various academic institutions; local sports, job training and recruitment; and medical, business, and artistic communities. Unlike Warner Amex's application, it also referred to ethnic minority communities such as "the Afro-American, Hispanic, and Chinese" and others as a significant audience group of its cable system.[29]

Times Mirror also planned to offer two fully equipped broadcasting stations. One of them would be located in the North Station area and other at Roxbury Community College. The former would be a part of the North Station area revitalization program and would be the base for the entire Times Mirror of Boston system. It was expected to serve as the neighborhood access center for half of the city. The latter, on the other hand, was not only for the rest of the city, but also should function as a major studio nearest to cultural and medical institutions, occupational centers, and educational facilities. The city would also have eight neighborhood access studios dispersed throughout the city: East Boston, South Boston, Brighton-Allston, Dorchester, Roslindale-Jamaica Plain, West Roxbury, Hyde Park-Mattapan-Readville, and South End-Back Bay-Chinatown-Bay Village areas.[30]

Personnel, training, and financial support would be offered at studios. Programming staff would be located in the two main studios and the eight neighborhood studios. They would help citizens improve and realize their ideas or come up with new ones. Their responsibility included offering training seminars to those who hoped to become familiar with the cable technology. A part of the Community Access Board's revenue would become a resource for the use of equipment, air time, staff support, and others. To sustain these three types of services and supports, Times Mirror claimed that it is significant to establish the board in charge of local community issues.[31]

Times Mirror recognized the importance of engaging African Americans.

> The Black community will have access to a full channel on the cable system. Programs and news stories of interest to Black citizens will be highlighted on this channel, and local groups can use the production facilities to develop additional specialized programs. In addition, the satellite service, Black Entertainment Television (BET), which features films, entertainment specials, and sports of interest to the Black community will be carried on this channel.[32]

This was a clear contrast to previous applicants that did not mention African American involvement as in much depth as Times Mirror did. The application continued to state that "it is hoped that the diverse . . . ethnic groups within the city of Boston will utilize this channel to state their views on many subjects." Offering BET, of course, did not guarantee fair representation for African Americans. Nonetheless, it was noteworthy that the company studied Boston's African American demography. An absence of reference to African American and other minority communities in other preliminary application did not mean that they were uninterested in African American involvement, especially because those documents had been submitted months before the RFP came out, making it difficult for applicants to make specific proposals. Time Mirror's gesture was a positive one to the city.[33]

The Tribune Company's plan was to establish Tribune Cable of Boston even before it was actually granted a license. Unlike Warner Amex, Tribune's proposal had a strong emphasis on local involvement. Thirty percent of Tribune Cable of Boston would be held by a group of local investors and interested parties.[34] Especially in local origination, the company promised "the commitment and experience necessary to provide high quality and innovative local origination programming." It recognized local origination as "a unique opportunity to provide service to the community which in addition to entertainment, can provide many social and economic benefits. . . . These programs should meet community needs and awaken interests while providing information."[35]

The Tribune Company pledged to offer sufficient personnel and facility resources available for local origination. Professional staff would "collect, organize and disseminate information and evaluate and recommend programming concepts." They were willing to help produce fifty to seventy hours of quality local origination programs every week. Such programs were local sports events such as school and league games from football to soccer, hockey and track and field. Linguistic minorities were not ignored, either. There would also be programs for speakers of English as a foreign language. These non-English programs were to serve various ethnic groups in the city. Information about job openings in both local and national businesses also would be a key factor in local origination. The Tribune Company explained that "community involvement is the key to ongoing high quality local origination." In this respect, Warner Amex and Tribune were very different in their preliminary applications.[36]

Warner Amex Cable Communications Company also submitted its preliminary report in fall 1980. The proposal emphasized technological features and marginalized minority or resident participation. Warner's application had a lot of technical content. This was not surprising considering the fact that it had been known as the pioneer in QUBE. QUBE was a new type of cable technology that facilitated viewers to "talk back" to their television. The audience was able to easily respond to questions or comments that appeared on the screen. This concept was rooted in the system launched in Columbus, Ohio in 1977 that served as the forerunner for systems in Houston, Pittsburgh, suburban Cincinnati, and elsewhere. The proposal promised that the company would bring more developed version of QUBE to Boston.[37]

Boston tended to focus on public service rather than public participation. It also discussed minority and resident involvement in the context of community-based activities. Warner's lack of attention to public participation in comparison to service was not surprising. In the three-page cover letter addressed to Mayor White, the application did not mention even once how to engage local community members or how to make the system beneficial to the lives of local minority Bostonians.[38]

Warner explained that in excess of 80 percent of the Bostonians who had been interviewed were aware of the future arrival of cable television in the city. About 40 percent of the respondents were likely to subscribe to basic services, although more than 50 percent of the population was satisfied with the current television service. Warner also noted that more than 30 percent of the city's population was elderly. On the other hand, interestingly enough, it did not remark on any of the ethnic trends or characteristics of the city.[39] More than 20 percent of the population at the time of the survey was African American.

Warner planned to offer eighty-six channels. Installation would cost $15. Tier 1 had thirty-six channels for $5.95 per month and Tier 2 cost

$7.95 for fifty-two channels. Fourteen additional channels were available with the subscription to QUBE. Ten regular channels were a part of "Boston Elders Service," which would be offered at no cost to senior residents. Warner tried to appeal to the city by focusing also on the needs of elderly Bostonians. There also would be seven optional pay channels featuring movies and entertainment. Thirteen channels were reserved for community-originated programs. In addition to channels, Warner promised to provide the city with a fully equipped local origination studio, seven community communications studios, a municipal studio in the City Hall, thirty-five community communication personnel, three field vehicles, and other services and facilities to maximize the benefit of cable television.[40]

In its brief reference to community programming, the application promised to offer a fully equipped studio near the Boston Theater District. Seven fully equipped studios for access and community uses would be dispersed around the city. Three mobile production vehicles were also a part of the offer.[41] It also listed nine categories of interest groups whose needs Warner would try to meet with special priority. Those groups were government, the elderly, Boston neighborhoods, schools and other educational institutions, health care providers and consumers, cultural centers, women, business, and consumers. Ethnic and racial minorities were not included in this list. Although African Americans and other minority populations were included as a part of each of the category, the lack of ethnic framework in Warner's proposal revealed its specific interests and concerns in the media business, especially contrasted to other applicants in Boston.[42] Potential types of programs in the field of arts and culture did not mention African American communities despite their active engagement in the city's arts and culture activities. Five examples elaborated in the application dealt with Chinese opera, the history of Irish immigration as taught at Boston University's American Studies course, the neighborhood theater at the Strand Theater, reading of Ernest Hemingway at the Kennedy Library, and presentations of the city's oral history by its elderly residents.[43]

Submitted by the future winner of the franchise agreement, the application submitted by Cablevision deserves a closer study. Cablevision was a New York based cable provider that had wired neighborhoods near New York and Chicago, and other cities since the mid-1970s. Communication between the city and this major cable provider, however, had taken place prior to the submission of the application in early November 1980. On October 14, John Tatta and Charles F. Dolan from Cablevision met with Mayor White and his staff. Dolan explained that after the meeting, they were "more convinced than ever that [their] company should make a strong bid to provide cable television service in the City of Boston." The outcome of such a belief was the preliminary application submitted on October 31 and received by the city on November 3, 1980.[44]

As other candidates also explained, the official RFP had not been released at the time of preliminary document submission. Therefore, the document was very general in content. Dolan stated that "we are fully aware that subsequent submissions will be dependent upon your further requirements." Dolan, however, identified that Boston is a unique place because "the depth and variety of cultural, intellectual and human resources located in and around Boston are extraordinary, and will help [them] create a television service that is second to none."[45]

Cablevision planned to offer two types of services: low-cost basic cable service and a range of various entertainment options. Because the company "[believed] that no family should be denied this basic service because of cost," the service had to be very low cost. In other words, Cablevision's plan was to sell the cable with a small charge and attract more subscription to earn gains. This scheme was backed up by its study that indicated subscribers were likely to choose a few additional paying channels. In New York and Chicago, for example, about 90 percent of customers subscribed to an additional service. Keeping the basic subscription fee as low as possible was a corporate strategy to sell more with small gain and eventually attain large revenue.[46]

Cablevision foresaw three-level service structure. The basic service offered thirty-six channels for a cost of $2 a month, after $25 installation fee. Because Cablevision's basic service in other localities cost approximately $4.50, the cost for Boston is approximately 50 percent of that in New York and other areas. This basic plan included CNN, ESPN, BET, C-SPAN, and other satellite channels. The second level of service was the premium service, carrying up to fifty-two additional channels. To subscribe to premium services, it cost an extra $4 a month and $30 for installation. This service type included entertainment channels such as Home Box Office. The third type of service was for institutions. It allowed users to obtain information on traffic control, police and fire department assistance, data transmissions, and other services. The application even predicted that once the system attracted sufficient amount of customers, there would be no monthly charge eventually. Dolan expresses his idea that "we would like to see totally free universal access to the cable."[47]

Cablevision planned to reflect Boston's local characteristics on its original channels. It pledged to offer not only sufficient channel capacity but also facilities and funds to "the creation of original channels tailored to

Boston's neighborhoods, as well as for citywide use." For example, Cablevision proposed to launch a Campus Boston channel that would connect educational institutions in Boston. The channel included a special service called "Extra Help." Through this service, teachers would be able to assist students working on their homework at night. It also would be useful for people with disabilities. Dolan pledged:

> Cablevision has never regarded public access programming as a burden. We have gone far beyond the "electronic soapbox" approach to access. We assist members of the community in production with financing, personnel and equipment. Once a citizen has worked hard to make a program, we work hard to make sure he has an audience by promoting it in newspapers, our printed program guide, and the channel guide. This is our way of assuring that citizens continue to use their public access channels.[48]

These unique features that Cablevision proposed about public access would all be included in the basic service program.

Cablevision considered channels devoted to local news and features as "the anchor of Cablevision's local origination programming." For this purpose, the applicant anticipated to recruit and train local staff to work at studios, mobile units, and news vans. *Cable Report* was thought to be one of the programs that trained people would produce. It would feature both live and taped stories from different parts of Boston. It also was to provide a place for discussions on topics that concern the lives of Bostonians. Also proposed were other programs such as *Perspective*, a documentary program on issues facing the city; *Sounds Good*, a performance channel for local artists; and *Gallery*, featuring Boston residents who had distinguished themselves.[49]

More than one month after nine companies submitted their initial applications, the Cable Access Advisory Committee submitted its report to Mayor White. The committee, established in October 1980 by the mayor, aimed to "focus on the issues of public access and institutional uses as an adjunct to the franchising process." The report had four key issues. First, it offered a basic framework for the public and regulatory policies for the management and control of public access. Second, it examined the way of financing community access. Third, it studied institutional and business use of cable. Fourth, it made ensured that the voice of residents would be reflected in the decision-making process. The message of the commission was clear. The city needed to have:

comprehensive telecommunications services which [offered] opportunities for public involvement and control, a well-integrated system of institutional use, and a responsiveness to the needs of both discrete neighborhoods and the whole community, especially previously unserved segments of the population.

One of the major points of emphasis was the establishment of nonprofit corporation to regulate public access channels. The plan was that the mayor would appoint fifty members. In order to make sure that local citizens would have access to such channels for various interest programming and prevent the city office from monopolizing or taking the direct responsibility of the medium, such a corporation would work as a watch dog. Under this plan, 5 percent of the revenue that the franchisee earned would be contributed to finance the costs of public access. The *Boston Globe* explained that in other communities where diversity was a characteristic, nonprofit corporations had served well to ensure accessibility of access channels. It also was important to hold public hearings, at least on the RFP. Only immediate planning could allow various needs to be met for the communities in Boston.[50]

Mayor White promised to review the proposal very carefully. He believed that the recommendations by the committee allowed cable television to be "more than television" in Boston. Two-way technology could connect Boston Public Library and its users, hospitals and patients, community and fire departments, and various city departments and Boston residents. White explained, "we want every segment of the city to share access to the cable system." What the study presented, as well as what residents had discussed, seemed to convince White that "to achieve this promise [of community involvement], cable television must serve the people." Approximately two months before the *Authority Report* was issued, White promised that through the RFP, the city would be "seeking the best from cable television both in technology and ideas."[51]

As the city offices moved toward the finalization of its RFP, several departments and commissions were asked to take another look at their needs for cable television. As an updated version of what they had submitted in 1980, the Boston Police Department, the Boston Water and Sewer Commission, the Public Facilities Department, and the Boston Housing Authority reiterated their interests in the cable system. The interest of the police department was primarily in sharing information with the public, educating Bostonians on how to avoid becoming a victim of a crime, and to enhance the security and safety of the city.[52] The Water and Sewer Commission hoped to use the system for customer information, through which the commission could read each customer's meter and obtain individual information.[53] The Public Facilities Department found cable television appealing as a security and remote control program. It would also increase accessibility for the residents to the department and enhance its public relations.[54] The Boston Housing Authority, serving 50,000 residents and having about 850 staff members, believed that cable technology would be beneficial in six ways: communications; data processing; staff training; resident education and participation; security; and citywide communications.[55] These examples showed that city offices recognized the benefits of cable television. The residents, however, argued that there was little reference about how they would benefit from improved services enabled by cable television. For interoffice communication among departments and commissions, and the cable office, there was very limited discussion as to how residents could benefit from the technology.

Very early in 1981, the Mayor's Office of Cable Communications submitted its grant application to the National Telecommunications and Information Administration (NTIA). The main objective of the application was to realize Boston's potential to the full extent. The application explained that Boston had the capacity to be "the vanguard of the communication revolution, but in an era of budgetary cutbacks and financial restraint it may not have the resources to do so, without the kind of financial assistance provided by this grant [by the NTIA]." By doing so, the project could achieve eight major goals: "stimulation of community involvement in the access planning process, including potential institutional users and neighborhood based public access users;" "education of public institutions and community groups on cable capability through workshops and seminars;" "identification of existing institutional and community communications facilities and analysis of how these might be included in the citywide system;" "assessment of institutional and community communications needs in order to plan for construction and equipment of appropriate facilities;" "translation of identified facilities and assessed needs into a master plan for the location of new facilities, definition types of facilities to be constructed, and provision of equipment;" "catalyzing formation of coalitions to provide the basis for the nonprofit corporation user councils recommended by the BCAAC [Boston Cable Access Advisory Committee];" "creation of testing of permanent materials to be used in ongoing public education and needs assessment efforts;" and "formulation and testing of a replicable procedures and data collection instrument for use nationally by other cities in their cable planning and construction efforts." The project was to serve more than 640,000 people living in Boston.[56] The application form identified the needs of the local community, how the city foresaw to realize them, and how the discussion on cable television had developed in the city.

As the city prepared its RFP, there had already been approximately 4,200 cable systems in 10,200 communities serving more than 44 million people nationally. This was about one fifth of all the households with television. Many of Boston's suburbs also had been wired. The city foresaw that the rapid growth of the technology would eventually turn itself from "a luxury" to an essential part of communication. To make most of the cable system, the city perceived that there were three major objectives in the planning of the system with the collaboration with the NTIA: "translating needs and interests into the location and provision of facilities and equipment," "organizing Boston's wealth of resources to profit from a comprehensive communications system providing expanded services," and "reaching previously unserved segments of the population, and recommending whether to apply for NTIA facilities construction funding."[57]

The application characterized the city as "a relatively small city with many of the problems endemic to older cities." It also described that the city had "a wealth of cultural, educational and health care institutions, a rich historical and ethnic heritage, and a position as a major distribution and transportation center for New England." It analyzed that the city's weakness had existed in its recent history that had witnessed the decline of local industries, rapid demographic change, and others. Its strengths had the power to serve as a remedy for such problems. For example, despite the economic decline, Boston was still the second largest financial center of the nation, after New York. The twenty-five–year trend of economic decline ended in 1975. Since then, the city had witnessed an increase of four-thousand manufacturing jobs. Boston also was an intellectual center with many universities and colleges, as well as libraries and conservatories, which generated about $1.3 billion in economic activity and whose employees and students spent almost $5 billion in the local economy. It also served as the center of various cultural forms from education to entertainment, and to religion. The city also was the home to many advanced medical centers and institutions.[58]

Reviewing the characteristics of Boston, the application saw the city as a potential pioneer of the use of cable system for positive social changes. It summarized:

> the ideal place for such as model [of using the technology for public service and access] because of its many and varied resources. As a cultural, educational, medical, intellectual, and technologi-

cal center, it has been and will continue to be a source of high quality, innovative ideas and programs for the entire nation."[59]

Because of these traits, Boston's plan could be "the beginning of a statewide and nationwide effort to facilitate the sharing of ideas and programming while retaining local control." The application emphasized the portrayal of Boston's cable system as one of the pioneers of a new use of the technology, or "the vanguard of the communication revolution."[60]

"What enriched the value of Boston?" cable companies asked themselves. It was the people who lived there. Because of its history, there existed at least fifteen distinct neighborhoods that had their own uniqueness geographically, politically, and socially. Between 1950 and 1970, "white flight" hit the city. Residents often identified themselves as being from a part of Boston such as Charlestown, Southie (southern part of the city), or Roxbury, rather than just from Boston. Many Bostonians for the past several decades had their own cognitive map based on their racial stereotypes about various groups of people that were often racist by nature. Although such heterogeneity could bring divides in the city, organizations such as the Boston Covenant for Justice and Racial Harmony allowed "positive exchanges among individuals and neighborhoods."[61]

There were several unique characteristics of Boston's demography. First of all, it contained a large minority population. African Americans, for example, could find their roots in the city in the seventeenth century. Black communities established many of the major institutions such as Freedom House, the Elma Lewis Center for the Performing Arts, and the Museum of Afro-American History. Comprising 20 percent of the city's population, African Americans made up for the largest minority. The Latino population was another group that was rapidly increasing. At the end of the 1970s, its population expanded by about 50 percent.[62]

Another characteristic was the city's large elderly population. According to the national demographic trend, that number would continue to increase, making their living conditions worse as a result of rising housing costs. Additionally, women were heading 23 percent of households.[63] The success of cable television in Boston was contingent on the inclusion of these people living in Boston. To serve them better, "it is not enough merely to offer the telecommunications services. Careful, thorough planning is essential to ensure both that the services are used and that the services meet the stated and to-be-defined needs of the community."[64]

In addition to reiterating interest in the cable system, as explained by city offices and departments in their respective position papers, the application supplements how exactly the residents of the city would gain access to the new technology.[65] One of the major ways to enhance community engagement was through various types of workshops. Some would be offered for community groups and some others for institutions. The city also planned to videotape workshops and to generate immediate outcomes from the workshop.[66]

An institutional workshop, taking place at Boston Public Library's Copley Square Branch or at a combination of difference branches, was a daylong session. Participants attended presentations, panel discussions with experts, group discussions, brainstorming, and so on. Its half-day session entailed discussions on the methods of enhancing institutional interests in the technology, collaboration between institutions and service providers, and so on. On the other hand, community workshops took place in at least ten different locations of the Boston Public Library. Daylong events consisted of sharing printed materials, presentations of videotapes describing how communities could use the cable technology to "organize, communicate and disseminate information," and panel discussions.[67]

The application included two reference letters that showed support of the city's application. One of them was the local television station, WGBH. The letter explained:

> Too often, cable companies do not take local origination and public access services seriously, and the facilities provided are unrelated to serving the needs of the community. the proposal for planning for the development of cable in Boston which the City is today submitting represents a thoughtful attempt to ensure the implementation of local cable services which are truly responsive to local needs.[68]

The other reference letter was from Boston Public Library. Assistant director of the library, Liam M. Kelly wrote:

> As a member of Mayor White's Cable Access Advisory Committee, I have become very aware of the pressing need for a major undertaking aimed at public education with respect to the planning for Cable TV in Boston. The imperatives for community needs assessment and public education were highlighted in the Committee's final report. I am delighted that the City Administration is now seriously working on implementing the Committee's recommendations.[69]

These letters showed the depth and seriousness of city's application and reflect the needs felt in the local community.

On January 7, 1981, in cooperation with the MIT Research Program on Communications Policy, the Mayor's Office of Cable Communications held a conference on cable television. More than one-hundred officials participated in the conference and discussed "planning and policy relative to the specific and limited subject of municipal government uses of cable television." The conference lasted for one day with an expert panel, four workshops, and an address via speakerphone from New York City. The divide between residents' attention to local involvement and city office's relative disinterest was obvious at the conference. Of the eight benefits from cable television seen by city officials, none directly addressed issues concerning local involvement or minorities. Those eight categories, including employee training, record keeping, sharing of data files, two-way conference facilities, and others, only reflected what departmental and commission position papers had already revealed, without necessarily proposing how local citizens could take advantage of the new technology.[70]

Issuing of Request for Proposals in Boston

On February 7, 1981, Mayor White issued the city's RFP to seek a franchisee for a fifteen-year nonexclusive citywide license. Although he had sought preliminary applications from cable providers previous summer, the RFP titled *Issuing Authority Report* signaled the end of establishing framework for the future cable system in Boston. The document gave applicants seventy-five days, until April 23, 1981, to file an amended application. It was a "crucial step in the process of licensing the operator" for Boston.[71] The more than two-hundred–page document consisted of five parts. Detailed instruction was provided after the introduction. The application section elaborated on the information that should be provided, such as service package, rates, and financial plans, and offered the basic outline of Boston's needs. The fourth part included Boston's Standards of Ethical Conduct and a copy of Construction Feasibility Study. The last section contained exhibits and supplementary forms.

Mayor White concisely outlined what an optimal cable provider was by stating that the city:

> [looks] forward to a partnership with an innovative and responsible cable system operator, one who is sensitive to the needs of

the various sectors of [the] City, who sees cable as a vehicle for increased interaction among a wide variety of people and institutions, and who shares [the city's] view of cable as a catalyst for a new era of communication.

Boston wants, and expects, cable television to play a meaningful role in every aspect of life in this community. At the same time, Boston feels it is a city where cable communication has an opportunity to explore and achieve every aspect of its potential. And we recognize that these goals will be realized only if the system is a successful economic venture for the licensee.[72]

The mayor's statement was significant in that it not only referred to the benefit that the city and its residents would gain from cable television but also to corporate revenue as a key factor to the positive impact for the city. To underline his point, White continued, "the City will continue to make every effort to maintain an attractive economic climate for the licensee."[73] From the mayor's perspective, economic benefit for the company was good for the provider, as well as for the service receiver.

Mayor White made two major requests to the cable applicants. First, he expected a wide variety of programming and cable services. Both the basic and premium services must include high-quality entertainment, cultural and educational content, public affairs programs, and sports programming. Two-way communication capacity also was required in the Boston system. Although Mayor White had only briefly discussed the launching of nonprofit access corporation in 1980, he wrote in his introductory statement that:

we expect the operator to support the non-profit access and programming corporation. . . . The needs of Boston's substantial Black, Chinese, Hispanic, Haitian, Polish, Lithuanian, Italian, and Greek communities, among others, have not been adequately met by broadcast television; cable can and must address these interests with the full participation of the groups themselves.[74]

In order to meet the diverse market needs and wants of Boston, the cable provider would need the capacity and resources to produce various kinds of high-quality programs.

White's second request concerned the sharing of economic benefits. Under the city's "resident-minority employment policy," which applied to all public construction projects, local residents had to fill at least 50 percent of all jobs. The rule also set numerical targets for female and minority employees. The franchisee had to adopt the same policy and meet the standard in its construction and operation. White explained that "Boston is the capital of a major industrial state and the principal city of the entire New England region." But he also recognized that the city is "also a mosaic of close-knit and diverse neighborhoods." For the cable system to truly reflect the interests of the city's residents, minority engagement was crucial.[75]

The winner of the application process was required to reach an agreement with a nonprofit corporation that specialized in local access. Although the need for the corporation had existed for almost one year, it had not been established yet at the time that the RFP was issued. But the corporation's responsibility was to ensure that everyone in the city had access to cable. The franchisee also had to provide at least five downstream channels and one upstream channel for local access. As for the subscriber network, seventy-two downstream and eight upstream usable channels was the minimum requirement. Although not required, the city recommended that the basic service should include all local off-air signals, locally originated programming, community and educational access channels, and children's services.[76]

The evaluation of the proposals was based on two major factors. The first was a quantitative criterion such as the corporate finance, technical system and channel capacity, construction plans, and conformity to the requirements as described in the RFP. The qualitative criteria included general qualifications and experience, the extent of benefits to Boston and its residents, quality of programming and services, local origination and special services, and community use. These were the issues that each applicant had to address in its proposal and convince the city officials that its plan offered the most beneficial service to Boston.[77]

The RFP underlined the importance of local employment. The document reiterated the city's intent to make sure that "all residents of Boston, minorities, and women will have an equal access to cable development in Boston." The city set its own target that the franchisee would be expected to satisfy. The goal was threefold. First, at least 50 percent of the employees had to be Boston residents. Second, 25 percent of employees were to be minorities. Construction had to involve at least 10 percent women. These figures were far from representative figures of the demographic composition of Boston where more than 50 percent of the population was female and about 20 percent African American. Applicants had to be more inclusive of these minority populations than the RFP required in order to be successful in bidding.[78]

In addition to employment, procurement was another key issue that was going to affect local communities in Boston. The RFP required at least

10 percent of the dollar volume of procured materials must be from qualified minority-owned business. Another important requirement was to prioritize Boston-based businesses over outside sources. In order to maximize the use of local and minority businesses, under the requirement of the city, the capitalist idea that the lower-the-cost-the-better would not be the case. The RFP stated, "Boston-based businesses must be used whenever their bids are within 5% of the lowest bid in a competitive bidding procedure, and the quality of the product is equal to or better than that of all other bids."[79]

In the RFP, the city officially explained that it planned to establish the Access and Programming Corporation, as recommended by the Boston Cable Access Committee. This decision aimed to ensure that the cable television's capacity to meet the "broadest communication needs of the people and institutions of Boston" would be fully used. The RFP explained that the company had "the responsibility for ensuring that all residents, businesses, and institutions in Boston have access to cable; for allocating channel time and facilities to users; and for stimulating the development of quality programming and new communications initiative."[80]

The corporation would engage representatives from various sectors of the communities in Boston to reflect "the City's unique cultural, ethnic, and social diversity." It also would nominate two individuals to serve on the licensee's Board of Directors. The responsibility and influence of the corporation over the franchisee and the attained system would be immense.[81] The winner of the bidding process had the responsibility to financially support the corporation. Because the type of programming that the corporation was in charge of, such as access and local origination, does not aim to produce capital gain, the corporation necessitated stable funds to continue its operation. For the first two years, the applicants would be expected to provide $250,000 to the corporation. After that, it needed to contribute 5 percent of its annual gross revenues to it.[82]

Although in agreement with the establishment of the corporation, Martin Kessel remained critical about Mayor White's motive. He argued that mayor had continued to "shut off opportunities for public involvement in the process." The mayoral decision meant that Boston might be successful in the efficient public access in which many other major cities had failed. But the lack of detailed information or thoughts on the corporation compromised the quality of proposals that applicants could produce. In other words, the organizational structure, for example, had not been finalized on the issuing of the RFP. Its institutional needs and resources had not been determined, either. As a result, applicants had a very difficult time calculating the cost relevant to the corporation. Additionally, with such uncertainties, they were not willing to contribute as much as 5 percent of their annual revenue to the corporation. Kessel argued that the exclusion of public as seen in the lack of White's interests in corporation might cause some groups to be neglected, and "creative input from the community" might also be overlooked.[83]

As a last-minute attempt to alleviate the sense of uncertainty about the nonprofit corporation introduced in the RFP, the Cable Coalition submitted the *Boston Community Television*, a draft public access system-design proposal, to the applicants on April 14. Although the document still lacked specificity about the characteristics and nature of public access services, it nonetheless redefined public access as "community television." It also emphasized that only community discussions would enable optimal access system design. Furthermore, the non-profit corporation that had been long discussed would be the Boston Community Television Corporation (BCTC). It would maintain the central support facility as well as fourteen neighborhood production centers.[84]

Submission of Amended Application

On April 23, 1981, when amended applications were due, only two applicants submitted their responses: Warner Amex and Cablevision. Seven others that had submitted preliminary proposals had withdrawn from the bidding.[85] Although the applications reiterated much of what the preliminary application already had mentioned, the updated documents specifically addressed the concerns and needs of the Boston community.

Amended Application by Warner Amex

Warner Amex submitted its amended proposal on April 22, 1981. Established in 1972 as Warner Cable Corporation, Warner Amex had nearly ten years of experience in the industry. The application included "the widest variety of programming in cable services available anywhere in the world." Warner also established "mechanisms for participation so that Boston residents and institutions may share the benefits of cable." Additionally, as the preliminary application emphasized, QUBE continued to be the "nation's only fully operational two-way interactive cable system" and Warner served major cities such as Houston, Dallas, Pittsburgh, and Cincinnati, as well as other cities in a total of twenty-nine states. Warner operated more than 140 systems in more than 350 franchise areas. Based on these experiences, the proposal promised to provide its first service to East Boston two months after the awarding of the franchise license, and to complete its construction requiring 682 miles of cable within sixty months.[86]

Warner's final proposal included the following:

- There would be eighty-six channels for subscribers and forty-seven for institutions;
- Subscribers would have two choices of services: Tier 1 for $5.95 per month and Tier 2 at $9.95.
- Installation would cost $14.95 for the aerial system and $24.95 for the underground system. For an extra $3.95, QUBE service would be available.
- Senior citizens would receive a free eleven-channel "Boston Elders Service" after paying the $24.95 installation fee. This program provided information and entertainment geared toward the elderly. Warner justified this plan by recognizing the city's senior citizens as "a special community resource with talents that can invigorate the City's communications efforts and contribute to programming that will benefit the entire City."
- There would be six optional pay television channels: Home Theater Network ($3.95 per month), HBO ($8.95 per month), Showtime ($8.95 per month), Galavision ($8.95 per month), The Movie Channel ($8.95 per month), and Cinemax ($8.95 per month), as well as pay-per-view channels ($0.50 and above).
- Two full video channels for data retrieval and a Home Security and Medical Alert System also would be available.
- Satellite networks would include twenty-five channels.
- Community communications would be improved through Warner's eighteen channels dedicated for this particular service.
- Parent control function, FM service, video games, home shopping and other commercial transactions, and college courses for credit also would be available.[87]

Warner also made various facility-related proposals. It planned to build a fully equipped local origination studio. Public access would be possible through seven community communications studios. Warner promised fifteen-year funding support for the nonprofit access corporation. City Hall also would have a municipal studio. Warner planned to invest $500,000 to support the development of the city channels. Three mobile field production vans could facilitate the production away from a studio. Seven sets of portable camera and audio recorder packages also would assist community use of the cable system. Viewers could visit twenty public viewing centers located throughout the city.[88]

In the final proposal, Warner continued to emphasize its technical advantage with QUBE in relation to local origination. It argued that "local programming should emphasize viewer participation, taking advantage of the advanced technology that has married the computer to the cable to produce interactive 'talk back' television." The corporation budgeted in excess of $1 million for local origination. Warner's Center for Interactive Programming, which would have a construction and equipment budget of $3.5 million, was slated to be the hub of this function. One of the programs where such local origination could take place was *Boston Alive*. The program reflected "the City's varied lifestyle and interests and making full use of Warner Amex's widely acclaimed interactive QUBE system which would permit viewers to 'talk back' and participate in the local programming. It would include opinion polling, local issues discussions, game shows, charity auction, local sports, news and other special events."[89]

Warner also was committed to access programming. It showed support for the city's decision to establish the Boston Access and Information Corporation (BAIC). The plan included seven community studios with full equipment dedicated to access production located throughout the city. Mobile production vehicles and portapacks would also be provided. As required in the RFP, Warner must provide the BAIC with $500,000.00 per year in funding support during the first two years, and contribute 5 percent of the gross revenue after year 3. If Warner Amex decided to change franchise fees, it had to offer 6 percent for access and franchise fees.[90]

Employment policy was another way the city strove to secure local and minority involvement in the development and operation of cable television. Warner responded to this requirement by claiming that "the percentages set by the City will be met and exceeded if Warner Amex is awarded the Boston franchise." In order to reach such a conclusion, Warner conducted a seminar to inform minority contractors about the opportunity to work with the cable system and met with representatives from Women in Construction. Thelma Cromwell Moss, president of the Contractor's Association of Boston Board of Directors expressed her appreciation about Warner's approach by writing "Your proposal to include minority contractors in the construction of Cable T.V. facilities represents the philosophy of the Contractor's Association of Boston, which is economic development and job opportunities for Boston's minority community." Meetings with the community development corporations and the Boston school system also helped Warner Amex learn about the training and recruiting possibilities of local minorities and females. When such employees receive education and career development, their job-related expenses could be fully reimbursed.[91]

In order to showcase the support Warner had attracted from minority communities in Boston, it enclosed a letter from Inquilinos Boricuas en Accion (IBA), a local Latino community group. The letter explained that there were approximately 150,000 Latinos in eastern Massachusetts. Considering the size of the market, IBA considered public access, local news, locally produced commercial and educational programs, Spanish programming, and training necessary. The organization showed its support by stating that "Warner/Amex is interested in and willing to support a Spanish-language service molded on the above. For this reason we will continue working with you until we can reach an agreement acceptable to both parties."[92]

Warner proposed an extensive collaboration with local academic institutions. One of the examples was an internship program for local students. The corporation explained that it "believes that an integral part of its service to the community is its commitment to provide the opportunity for local students to undertake and participate in a meaningful cable television internship." Instead of conventional internship opportunities restricted to production, Warner's plan included opportunities tailored to meet each student's individual major. Marketing, public relations, sales, and other positions would be available to local students. Warner Amex also planned to provide programming originating from academic institutions. By connecting Warner's production facility and universities and schools, superintendents could address the community from their school buildings. Similarly, high school sporting events would be put on air. Teachers would be able to teach on television for students who were unable to attend school. Warner saw various positives that it could bring to the local community by involving local educational communities.[93]

Warner's project included a minority/ethnic channel as one of the twelve downstream channels on the subscriber network that the RFP required. A women's channel, a neighborhood channel, A public access channel, three educational channels, as well as others, comprised the rest. The company explained that the minority channel would be a part of an effort "to address the minority and ethnic community needs of Boston." Being able to have a channel for such a specific demographical group was one of the largest benefits of cable's ability to have numerous channels. Warner stated that the channel would be "designed for and dedicated to programming produced by individuals, organizations and institutions which comprise the City's rich cultural heritage. . . . Warner Amex will ascertain programming interests, identify sources of programming and program funding and facilitate the creation of local productions appropriate for this channel." In addition to the Irish Social Club and the Polish American Citizens Association programs, the Haitian Festival at Franklin Field, the Black Nativity at the Elma Lewis School, and other events and organizations would be part of this channel production. Warner also expected that there would be times when not many local programs would be available. When this happened, Warner said it would supplement local production with the contents from BET, and many other ethnic and minority satellite programs.[94]

Although Warner's proposal tried to differentiate itself from its competing candidate through its technical experience and financial capital, it also emphasized the significance of its commitment to local services. Many of the services it planned to offer were similar to what Cablevision offered. Nevertheless, it underlined Boston's cultural heritage by claiming that its system could reflect the city's cultural richness.

Amended Application by Cablevision

The other cable candidate that submitted its final proposal was Cablevision. Although Cablevision was a young company with less than ten years of experience in the industry, its proposal emphasized management members' experience. Charles F. Dolan, chief executive officer and the founder of Cablevision, for example had been in the cable industry for twenty-five years. He established Manhattan Cable in 1961 and Home Box Office in 1971. Similarly, John Tatta, chief operating officer; Patrick Caruso, director of sales; Stuart F. Chuzmir, vice president of programming service; James W. Elmore, general manager; Wilton Hildenbrand, chief engineer; and many others had more than ten years of experience with cable television. The accumulation of such experience and expertise had already allowed Cablevision to operate approximately 4,500 miles of cable in greater New York, serving in excess of 450,000 homes, and 180,000 subscribers. In Chicago, its service encompassed more than 350 miles with 30,500 homes and 12,000 subscribers by May 1, 1981. By the end of 1981, Cablevision projected that it would serve 5,400 miles of cable with New York and Chicago combined.[95]

In the executive summary of the proposal, Cablevision listed two major objectives. It explained:

> Our plan rests first on the kinds of entertainment and communications services Bostonians perceive they need and want. Second, it incorporates an extensive analysis of presently existing and desired future information services for Boston's commercial and noncommercial institutions.[96]

Particularly, the first point reflected what Mel King envisioned cable technology to achieve: putting on air what his community members consider to be important rather than what others offered to them. In order to achieve this goal, Cablevision strove to offer "the most extensive program service offered anywhere" with a low monthly cost. Because the city expected that "the successful cable system in Boston must offer attractive low cost programming whose appeal will attract a majority of homes from the outset and generate sustained popular support," Cablevision responded with a universal basic service that would offer fifty-two channels of programming for $2 a month. Cablevision explained that the low-cost service would be particularly important because Bostonians had expressed concerns about cable television service rates being too expensive. Additionally, Cablevision also learned from a survey by Peter D. Hart Research Associates, Inc. that more than half of Bostonians were interested in subscribing to a cable service if it offered more diverse programming. Based on this information, Cablevision stated that its basic service will "carry access and local origination channels, all local over-the-air television signals and a wide selection of stations from outside Boston. No channels will sit blank." Cablevision made the point that the name "universal basic service" comes from the idea that its service must be available to everyone regardless of the income. The company elaborated:

> Our Universal Basic Service package for Boston developed out of Cablevision's particular marketing philosophy. All of our franchise areas, like Boston, have satisfactory over-the-air broadcast reception. In order to sell cable television, we must provide a diversity of quality programming as well as technically reliable service that makes subscription to cable television as trouble-free as over-the-air television.[97]

Cablevision's plan was to offer a low-cost service with diverse programming and high-quality contents.

Peter D. Hart Research Associates, Inc.'s survey findings underlined the interests that Bostonians had in cable television. The survey took place between March 11 and March 15, 1981, approximately one month prior to the deadline of the application. Twenty-minute telephone interviews were held with 635 adults from Boston. The study reported that 63 percent of those interviewed would consider subscribing to a basic cable service if the monthly fee was $2. This population consisted of 32 percent who said they would definitely subscribe and 31 percent who said they would probably subscribe. If the fee was $5 per month, 57 percent said they were inclined to do so. Bostonians also expressed their foremost interests in cable television because of its "ability to provide information about local community events and local government and to help young children and adolescents in the educational process."[98]

The research also found that many Bostonians felt that television could help make Boston a better city. Behind this hope existed dismal reality about existing television services. A very small number of respondents felt that conventional television was meeting their needs and expectations by providing sufficient local information. Of Bostonians interviewed, 39 percent said that the programming available on television appealed to them "just somewhat"; 30 percent said that it attracted them very little. Overall, more respondents were dissatisfied than satisfied with broadcast television. As a result, their expectation for cable television to cover more local matters and specific issues unique to Boston or important for its residents such as crime and race relations was significant.[99]

Cablevision expressed its willingness to create a partnership with the city in six ways:

1. It planned to make a $500,000 investment to finance a Boston-based broadcast-quality production studio.
2. The company also planned to invest $1 million per year to support the production studio with purchasing equipment.
3. A satellite uplink would be provided for Boston so that locally produced programming could be distributed nationally.
4. Staff and facilities would be used to "develop and produce a wide variety of local programming designed to meet the various tastes of the Boston population."
5. Technical and financial support would be available for the access corporation that had yet to be established at the time of the application submission.
6. The Public Institutional Network would have fifty channels operated by the access corporation.[100]

Although Cablevision emphasized its willingness to assist the local access corporation, it also stated that it would support local programming. The firm suggested a plan that included six news channels on which there would be a local newscast every evening. The newscasts would be anchored from a central studio, as well as from any of the neighborhood studios, the mobile studio, or news-gathering vans. The plan included four sports channels that would air local high school and amateur sporting events. Two youth channels had the "Extra Help" services every evening, through which students would be able to call in with homework questions. The program already had been successful in Long Island. Local teachers hired by Cablevision would discuss the question on the air. Similarly, Campus Boston was expected to air lectures, concerns, and students' production.[101]

Cablevision paid special attention to minority participation. It planned to accommodate their needs by securing sufficient channel space and providing facilities. The application stated that:

> Cablevision is sensitive to the needs of ethnic minorities, and hopes that this sensitivity is apparent in all of our programming plans. In addition, we have devoted 2 channels featuring programming with a special appeal to Boston-area minorities. All programming on the first channel will be in English; Cablevision has planned a second channel which will be telecast in Spanish and other foreign tongues.[102]

These channels would carry BET, the Spanish International Network, which would be aired twenty-four hours a day, and the Spanish News Network. These networks were expected to "promote interest and expertise in locally originated minority programming." Although they served a national audience, Cablevision foresaw that such programming could encourage Boston's ethnic minorities to produce their own media content. Cablevision also planned to have a studio at the Humphrey Occupational Resource Center. It planned to encourage its cable utilization coordinators to reach out to Boston's local ethnic communities, and particularly to the African American community to encourage programming production among this population.[103]

Cablevision listed several channels that would specifically deal with the interests of Boston residents. Boston City Channel on Channel 8, for example, would be its "major local origination effort [that] will preset alphanumeric, live and videotaped coverage of Boston's people, places and local events." Programs such as *Perspective, Cable Report, Dialogue, Best of Boston,* and *Boston Almanac* would keep Bostonians updated with the latest news and events in the city, as well as provide them with channel spaces in which their views and ideas could be represented. Channel 20 would be a place to fulfill some minority interests. The proposal explained, "our remote and studio resources will be utilized to develop series and special event programming of interest to Boston's Black and Hispanic communities, as well as the diversity of ethnic populations within the City." This was a special-interest channel "designed to produce and carry programming geared to the diverse needs and interests of Boston's population. Satellite services include 'Black Entertainment Television.' Also, local origination programming and access presentations [will reflect] the diversity of Boston inhabitants." In order to formulate their proposal, Cablevision had discussed programming ideas with minority groups such as Greater Roxbury Community Development Corporations, IBA, and South End Settlement House.[104]

Cablevision's application included a study issued by Kalba Bowen Associates, Inc., a research and consulting firm from Cambridge, Massachusetts, on the planning process for Boston's institutional network. The firm was requested to assess the needs of potential cable users and help identify key needs for the network design process. The study found that Boston's community organizations were interested in a cable system as "a transmission medium for locally produced community programming." Many have shown interest in obtaining access to production equipment and using video "as a tool for community organizing, documenting events, training, and creating programs on issues of interest." Cable, therefore, would be a way to distribute programming to other groups and the public.[105]

Cablevision included a letter from United South End Settlements, which expressed its interest in assisting the candidate in employing minority workers. Kenneth D. Wade, coordinator of youth services, wrote, "I would like to state, again, our willingness in assisting your organization in recruiting minority candidates for positions with your company should you receive the cable t.v. franchise for Boston."[106] Cablevision also had its own policy on equal employment and recruiting. It promised not only to hire and promote its employees regardless of race, color, religion, national origin, or sex, but it also had the policy to "place employment advertisements in media which have significant circulation among minority-group people and women in the recruiting area." It also would recruit through schools and colleges where a significant minority population attends.[107]

In many ways, Warner Amex and Cablevision's proposals were similarly impressive. In other respects, they were much different. In terms of public access, Warner outdid Cablevision. Warner Amex proposed to build seven community studios with the budget of $1.870 million. On the other hand, Cablevision planned to have a main studio and three neighborhood studios at a cost of $1.460 million. Although Warner proposed a $3.5 million Center for Interactive Programming, Cablevision's plan was more modest with its $1.102 million budget. The latter company planned to have a staff of fifty-four compared with thirty-two for Warner and more weekly hours of local programming with eighty hours compared with twenty-one. There was a large difference in the fee proposals of the two companies, as was discussed at a public hearing held at the end of June 1981. Cablevision proposed to have fifty-two channels and charge $2 a month. Warner on the other hand planned to charge $5.95 for thirty-six channels. As required by the RFP, both companies planned to pay 5 percent of their annual revenues to the access corporation.[108]

Although the drafting period emphasized so much on the financial and technical capabilities of the companies to ascertain successful implementation of the system in the city, Boston's officials and cable representatives, as well as community leaders and residents continuously expressed their interests in local origination and access channels. The fact that only two companies out of almost ten that had submitted a preliminary draft proposal submitted their final proposal proved that the city's demand was very high. Both Cablevision and Warner Amex submitted sophisticated applications that were to be reviewed during the following few months.

Drafting and Issuing Request for Proposals in Detroit

Detroit's Cable Communications Commission issued its *Request for Proposals for the Provision of Cable Communications Services in the City of Detroit, Michigan* (RFP) on August 16, 1982, in order to "seek qualified applicants to provide cable communications service to the citizens of the City of Detroit."[109] Almost twenty years after the city's initial interest in cable television and about a decade of serious talks and feasibility studies, the RFP was issued to explain the city's expectation with great detail. This document, therefore, was an immediate outcome of all the discussions, debates, and studies that had taken place prior to summer 1982. Drafted during the triangular power struggles among the mayor, the City Council, and the Cable Communications Commission, the RFP had a bitter history in Detroit. On the first page were seventeen names of those who played significant roles in the process. Along with Mayor Coleman A. Young, ten City Council members, Erma Henderson, Nicholas Hood, Clyde Cleveland, Barbara-Rose Collins, David Eberhard, Jack Kelley, Maryann Mahaffey, John

Peoples and Mel Ravitz; and seven Commission members, James Robinson, Alta Harrison, Otto Hetzel, Joseph Jordan, Jean Leatherman-Massey, Lester Morgan, and Myrna Webb, were listed as project members.

In the opening note addressed to Mayor Young and the City Council, James P. Robinson, chair of the Cable Commission reiterated that the RFP "reflects [Mayor and Council's] desires that the citizens of Detroit be provided with a state-of-the-art cable communications system." He then addressed concerns and interests of the local community by explaining, "the RFP requires a commitment to community, educational and municipal access. The RFP also insured that the system will reflect the City's commitment to equal employment and business opportunities."[110]

The RFP consisted of eights sections. The first was the body of the RFP that included instructions and service and application requirements. The second was the blank application forms that each applicant was required to fill out. The third was statistical data based on the citizen survey from 1980, Detroit Uniform Data Chart, and the subcommunity map from 1980. After that, Code of Conduct and regulatory ordinances followed, each comprising a chapter. The sixth section was a draft of the franchise agreement that the bid winner would sign with the city. The following chapter included draft franchise ordinances on franchising and rates. The document closed with addendum.

The RFP clearly showed the expected franchise schedule. The deadline for the application was set for 2 p.m. on Friday, December 10, 1982. March 15 was the deadline for the commission to finish reviewing each proposal and to issue a draft report. The report asked each bidder to make clarifications by the end of the month. Once those questions were answered, the commission had two weeks to review the clarifications, hold public hearings, and make a recommendation to the mayor regarding its proposed franchisee. By the end of April, the mayor had to evaluate the recommendation and forward the recommendation to the council. The council, as the beholder of the power to make the final decision, reviewed the mayor's recommendation, held its own hearings, and approved the ordinance to issue the franchise by June 1, 1983. The winner was expected to start its construction and complete the construction within five years after the awarding of the franchise.[111]

Explaining itself as a minimum requirement to give applicants more flexibility to realize more innovative cable system, the RFP outright identified that the new program must address prior exclusion of minorities. It argued:

> the City finds that there has been a historic exclusion of minorities from participation and ownership in the communications

field. The City has therefore included provisions in granting the franchise for the City that are intended to provide opportunities for minorities to remedy the discrimination that has prevented them from access to the communications media, their ownership, operations, management and policy boards.[112]

For example, the city required that at least 20 percent of the equity interest in the city must be offered to Detroit-based investors "specifically including minorities."[113] Because media ownership among African Americans had been a rarity in the television industry, such a promise, despite only 20 percent, could ascertain that local ownership of the system by the predominantly African American communities would happen. Ownership and fair representation were inseparable. Additionally, half of the locally allocated equity would be made available to local investors in Detroit in lots of $100. This also was a way to enhance local engagement and investment in the cable system.[114]

As discussions prior to the issuing of the RFP suggested, employment was a key factor in the introduction of the cable system. In 1983 Detroit, the African American unemployment rate was 30.6 percent for men and 29.9 percent for women. Whites' rates being 17.9 percent and 12.2 percent, African Americans were twice more likely to be out of job than whites.[115] To mend this condition, the RFP required that "the selected applicant will employ Detroit-based labor, specifically including minorities and women, to the maximum extent possible in the construction and management of the system." This meant that at least 20 percent of employees had to be minority at every level of employment and management, as well as in all phases of the construction and operation of the system. The RFP required applicants to elaborate on their commitment for equal opportunity for employment, on-the-job training, and other necessary measures.[116]

In order to ensure African American and other minority participation in the cable system, policies for individual level employment were not going to suffice. The city also required a special attention to be paid to minority owned businesses in Detroit to construct and operate the services. The introduction of a cable system meant a large business deal for any applicant. The types of Detroit-based businesses that could be involved included construction contractors, subcontractors, suppliers, consultants, banks, accounting and legal services, insurance, and other facility-related services. The RFP stated that at least 20 percent of the cost of construction services and supplies must come from minority firms during each year of the franchise.[117]

The RFP stipulated a set of minimum service requirements to make sure that minority and local resident participation would be realized. The successful candidate must wire a certain minimum number of the twenty-two lowest income communities each year, in addition to other affluent areas. By the end of year 5, all twenty-two communities had to be wired. It was a way to make sure that all areas of the city, regardless of the economic composition of the demography, rich and poor, were connected.[118] In addition to wiring communities, applicants were expected to have, out of seventy-two minimum channels, at least two public access channels, four "geographically dispersed, fully equipped and staffed" neighborhood access studios, and four leased channels reserved for minorities, among others. These were nothing but minimum requirements. The successful applicant would need to "demonstrate a firm commitment to public access, local government access, educational access and local origination programming."[119]

Stipulating more detailed requirements could hinder the construction of more innovative and unique cable system. It was a difficult balance for the city to maintain between giving applicants the widest possibilities and flexibilities and concurrently ensuring they would address the city's concerns. Along with financial concerns, minority-related issues such as participation, employment, training were a major part of the minimum requirements. Although the city encouraged applicants to be creative with their proposals, it was willing to make little compromise in terms of securing a certain level of African American and minority involvement at minimum. The city's recurring reference to the ethnic and racial gap and the cable's role to improve it serve as the evidence of Detroiters' awareness that the new communication technology might be able to help the city develop and help Black Detroiters be involved more.

The RFP demanded that the application covered information regarding the ownership, employment, and training for minorities in Detroit. A part of it was the certificate of nondiscrimination. The document stated:

> The applicant agrees not to discriminate against any employee or applicant for employment because of race, creed, color, sex, sexual preference, national origin, marital status, handicap, or ancestry. The applicant shall take affirmative action to insure that employees are treated without regard to their race, creed, color, sex, sexual preference, national origin, marital status, handicap, or ancestry. Such actions shall include, but not be limited to the following: employment, promotion, demotion or transfer,

recruiting or recruitment, advertising, lay-off, or termination, rates of pay or other forms of compensation and selection for training, including apprenticeship. The applicant and any subcontractor shall agree and applicants for employment, notices setting forth this provision.[120]

The certificate also required the same philosophy of nondiscrimination to be applied to all subcontractors. Although the clause did not aim to specifically eliminate race- and ethnicity-based discrimination, inclusion of such a certificate was to promise that the cable system was not just another form of white domination, as had been the case with broadcasting television.

The RFP also required very detailed data on prospective employment practices. First, it called for a description of "plans and policies for equal opportunity and affirmative action" for each year of the contract. This section would include the projection of minority and female participation in every level of management, as well as at every phase of operation. Therefore, it included positions such as the board of directors, marketing, programming, production and engineering, technical operations, installations, maintenance, and clerical and accounting. Similarly, training, education, job advancement, and other development programs also were subject to the same policy.[121]

Two forms were attached to the application. The first, Form M-1, described the number of anticipated hires for every year of the franchise. The total number of hires, the number of minorities, and the number of females had to be provided on the form for each of eleven categories: officials and managers, supervisors and foremen, sales, clerical, technical, craftsmen, operatives, laborers, services, apprentices, and trainees. The RFP also required that candidates state their goals for minority hiring by ratio. The second form, M-2, described the status quo, unlike the future projection illustrated in M-1. In this form, applicants needed to provide the total number of male and female employees, number of Black, Latino, Asian and Native Americans and the rest by sex for the same eleven categories. Applicants were expected to show their commitment in minority representation both at the time of application and in the future.[122]

In order to correctly and effectively reflect the voices of local Detroit community and organizations, the RFP included an addendum that compiled various proposals and requests about the Detroit cable system submitted by interested entities. This extra information was to allow applicants to meet not only the minimum requirement that the city had set, but also the needs of Detroit's residents and organizations. The addendum included nine requests. The organizations that had submitted such recommendations were the Office for Television Development of the Archdiocese of Detroit, the Center for Creative Communications, the Detroit Medical Center Corporation, the Detroit Public Schools, the Detroit Symphony, the United Community Services of Metropolitan Detroit, the University Cultural Center Association, the Wayne State University, and the Health Research Division of Wayne State University.

The Center for Creative Communications' proposal was articulate about the role of cable television for the retaining of ethnic identity among the minorities. Representing the view of the Alternative Arts Consortium (AAC), whose members included the Center for Creative Communications, as well as Metro Arts Complex, and Ann Eskridge public relations and media consultant, the recommendation argued the following:

> cable television and cable radio may offer the last opportunity to develop minority directors, producers, actors, technicians and other media professionals. We also believe that cable television and cable radio may service a minority audience that is being underserved by commercial and public radio and television.[123]

The AAC explained that it planned to provide a dedicated channel with programs by, for, and about minorities and their artistic work and talent. Additionally, the consortium considered cable television to be a talent development tool. It hoped to use the presence of cable television to help residents develop skills as producers, directors, operators, and engineers in the domains of music, filming, visual recording, and other types of artistic work. The technology also would serve as a tool to educate and instruct people about ethnic and minority music, techniques of acting, dancing, and the history of minority and ethnic arts history.[124]

The Center for Creative Communications listed eighteen concise but accurate recommendations. Their requests included the following: cable providers should designate "one cable television channel as an Ethnic Arts Channel to be programmed by the Alternative Arts Consortium," "[they should place] the Ethnic Arts Channel on both the subscriber and institutional networks with both upstream and downstream channels for carrying video, audio, and data and FM cable radio"; advanced editing suite and stations should be provided with a permanent professional staff; provide training opportunities; and so forth. These recommendations had the objective of ensuring that sufficient financial support, training opportunities, facility, and cable space would be reserved for ethnic and minority arts. The AAC's recommendation document was one of the earliest in the history of Detroit's cable television system that considered art via cable as a significant means for nurturing self-worth by acknowledging people's heritage. The center also listed forty-five minority arts organizations located in the city that would produce for and benefit from a channel dedicated for local ethnic artists.[125] Among racism and general disinterest, one of the reasons why minority art was not a popular theme of a broadcasting program was that it did not attract a large audience. But when it came to cable, targeting its audience narrowly was an advantage for a program. The aforementioned forty-five organizations would be able to produce programs on art of their specialized ethnic or national areas, and such programs would reach people and be appreciated.

The proposal, which the Detroit Public Schools Cable Television Advisory Committee submitted more than one year before the RFP was issued, also addressed the importance of cable television for ethnic minorities, particularly in the educational context. It identified the promising future of cable television, as well as a reason for care in planning the structure and the content of the system. The committee recognized that cable television's bidirectional system would make "a positive difference in the lives of [its] students." Because cable television can have multiple channels, programs could meet "the educational needs of the schools' diverse ethnic groups, of small as well as, large, numbers of students." However, it also was important to remember that the introduction of a cable system would not guarantee a positive impact. The success was dependent on "the interests and the ingenuity of the school system's 'in-house' planning as well as the arrangement that are made for cable television service to the schools and the educational community."[126] The important role that schools played by telling cable candidates about their needs was the rationale for the committee's proposal.

Applying to Wire Detroit: Barden Cablevision

In response to the RFP, three cable service providers submitted applications. The first was Barden Cablevision. Headquartered in Inkster, Michigan, this corporation was unique in a few ways. It was the only minority-owned cable company in the state. The founder, Don H. Barden was born in Detroit and resided in the city at the time of application submission. Although the company itself was not based in Detroit, there was a personal connection and reason for Barden to be particularly interested in establishing a system in the city. This personal interest became more apparent as the company applied for cable system implementation. Barden Cablevision had experience in the cable industry prior to its application to Detroit. It had five franchises: Van Buren Township, Royal Oak Township, and Romulus, as well as in Inkster where it established its first system in May 1982, wiring 12,000 homes with a budget of $3 million. Outside of the state, Barden served Toledo, Ohio. In December 1982, it was also in negotiation with Sacramento, California and Dearborn Heights, Michigan, to serve as its cable provider. With such a large network, Barden also was one of the largest minority-owned system operators in the nation.[127]

Not only was Barden Cablevision a young but experienced company in the cable industry, Don H. Barden himself also was "an aggressive, growth-oriented individual with a proven record of achievement." In addition to Barden Cablevision, he had established two more companies: The Don H. Barden Company and Urbanaction, Incorporated. The first was a real estate development and investment business; the second was a community development business. Barden also had eleven years of experience as a television news director for WUAB-TV in Lorain and Cleveland, Ohio. He also worked as a weekly talk show host for NBC in Cleveland. He had served as a city councilman in Lorain City for four years. In Lorain, he founded the *Lorain County Times,* where he worked as editor and publisher for five years. His community service in Lorain lasted for fifteen years, through various organizations including Fraternal Order of Police, Lorain Chamber of Commerce, NAACP, and others. He was no doubt an active citizen in Detroit.[128]

Barden committed $232 million to develop a cable system in Detroit. It included the $156 million construction cost. With such a financial investment, Barden promised to build the following:

> a state-of-the-art telecommunications system that will offer a full range of entertainment, information and education services to the City of Detroit and all of its citizens. Barden cablevision of Detroit will construct a system that the City can be proud of, one that will rival any in the country. The system is designed to greatly enhance cultural, educational and entertainment opportunities for all Detroiters and to revitalize communication between City residents, their neighbors and the government that serves them. Barden Cablevision is also committed to promoting the economic and social recovery of Detroit. It will create an integrated telecommunications infrastructure that is unequalled in any of the great cities of the United States. Barden Cablevision

is also committed to human and community redevelopment through its comprehensive manpower development, minority employment and community programming plans.[129]

The outcome would be a 2,143 mile-long subscriber network and a 412 mile-long institutional network. Six hubs located throughout the city would secure equal speed cable service. The construction would end within forty-eight months after the signing of the franchise agreement. Barden elaborated all of the aforementioned benefits it promises to bring in its seven-volume proposal, largest of all the proposals submitted to the city. Although the proposal included a great deal of technical and statistical data, it also dealt very closely with issues concerning minorities, local origination, community access, and others that would heavily influence the lives of African Americans in Detroit.[130]

In this executive summary, Barden Cablevision, as a minority-owned corporation, reiterated its emphasis on minority involvement through ownership, employment, and training. Its numerical goal was bold. It pledged to hire 75 percent minorities and 40 percent women before the end of the franchise term. It included a $1.186 million training budget to support this effort. It would fund more than 1,600 students who would take college courses and technical training. Additionally, access channels would be available for job training and placement to both minority and female Detroiters who were seeking to advance their career. On the contractors' side, at least 20 percent of the subcontractors' cost would be with minority contractors. It also promised that when no minority contractor was available for a particular service, a joint venture would be established with minority or female Detroit residents.[131]

Barden's local origination and community access ideas "[stressed] both community involvement and community development." Barden planned to have a local origination super station that would "beam local cultural and sports events and the good news of Detroit's Renaissance to cable systems across the country." Barden promised to support the effort by supplying more than $85 million in equipment and operation. The grants would help develop program endowments for arts, cultural, religious, and social service agencies, neighborhood news network, educational television network, scholarship and internship programs, social programs for the hearing impaired, antipoverty fund, and others. The financial support also would help establish five neighborhood studios, five FM audio studios, and five mobile vans. Organizations in Detroit also would be able to borrow fifty sets of video access modulators and twenty portable cameras to produce

Drafting of Democratic Communication Media 151

their own cable content. In order to secure viewership, Barden would locate forty-three public viewing and learning centers. Its main access studio, which would be located at Woodward and Kennilworth, would have a community theater that could accommodate five-hundred people.[132]

Barden Cablevision included letters of support from various contractors who had agreed to work for the company, should it win the bid. For example, the law office of Gregory J. Reed was willing to offer legal services regarding communication law, business, and tax paying.[133] Clarence J. Hall, vice president of Lewis & Thompson Agency located in Detroit, similarly agreed to provide local administration of Barden's insurance program.[134] For employee life insurance, Write Mutual Insurance Company expressed its interest one week prior to the deadline of the application.[135] For electrical construction and maintenance, numerous companies from within and outside the city wrote to Barden Cablevision between the end of November and early December. Just to name a few, Stribling Electric Company and Exclusive Building Company from Inkster, Sabbath Electric Corporation, Williams & Richardson Corporation, and Mello Consultants from Detroit showed their interest in cooperation.[136] Barden's application compiled letters of interests from more than forty additional companies. Those companies included American Office Products from Inkster, Center Lumber Company from Inkster, Jay Advertising Corporation from Detroit, Gold Star Exterminating Corporation from East Lansing, Bivins Management from Detroit, Welch Bros. Paper & Supply from Detroit, Huddleston Huddleston Incorporated from Inkster, Dukes' Service System Incorporated from Detroit, Urban Security Corporation from Detroit, Giant Plumbing & Heating Supply Company from Detroit, D & M Landscaping Corporation from Detroit, Mopkins International Services from Detroit, J. A. Hall Enterprises Incorporated from Detroit, Glenn E. Wash & Associates Incorporated from Inkster, Conyers Ford Incorporated from Detroit, Mobley Fabricating Corporation from Detroit, Alton Lewis Moving from Detroit, Ark Electrical Supply & Services Incorporated from Inkster, Satellite Communication Systems Incorporated from Inkster, Communication Enterprises from Inkster, and Pontchartrain Cablecomm Incorporated from Greenville.[137] Some of these companies were from outside of Michigan: BLR Electronic Distributors Incorporated from Elmsford, New York, Clias Brothers Incorporated from El Paso, Texas, TPI Construction Supplies from Atlanta, Georgia, Counties Contracting and Construction Company from Philadelphia, White House Electrical Supply Corporation from Philadelphia, Pennsylvania, Electro-Wire Incorporated from Skokie, Illinois, Innovative Technical Systems Incorporated from Denver Colorado, Rocky Mountain Electrical Supply

Incorporated from Denver, Colorado, Electro-Optics Corporation from Dallas, Texas, Raba Incorporated from Washington, D.C., Saiki Design from New York, Communiplex Services from Cincinnati, Ohio, International Telecommunications Projects Incorporated from Washington, D.C., Pukka Service from Columbus, Ohio, J. M. Worrell Communications Construction Company from Atlanta, Georgia, Urban Communications Corporation from Atlanta Georgia, Bianchi Enterprises from Baltimore, Maryland, U.S. Antenna Incorporated from Staten Island, New York, T. A. C. Cable Services Incorporated from Holbrook, New York, and Black Shield from Pittsburgh, Pennsylvania.[138] This comprehensive list of concerned parties showed the wide interests in Barden's plan for Detroit was shared across the nation from in-state corporations to those in Georgia, Colorado, and many other states. The list was substantial.

There were other nonbusiness-oriented supporters. Various organizations and individuals also endorsed Barden's bid to wire Detroit. Joseph E. Madison, director of the NAACP Voter Education Department expressed his support for Don Barden by writing, "During my on-site visit of your facilities, I was impressed with the professionalism of the staff and the first rate equipment you have. Your track record should make Barden Communication a viable competitor for the contract to service Detroit. You can count on my total support."[139]

Raymond A. Leporati's name appeared frequently on letters from interested companies. As vice president and chief engineer of Barden Cablevision, Leporati was one of the major figures in the history and development of the company. He was "a pioneer of television electronics systems development in the Midwest" and active for more than thirty years. He had been involved in cable systems in excess of twenty municipalities in Ohio, four in Michigan, four in Pennsylvania, one in West Virginia, two in Wisconsin, and one in Iowa. He had worked not only with Barden Cablevision, but also with Grafton Cable Communications, Matrix Cablevision, Clear Pictures, Times Mirror, American Cablevision, West Side TV Corporation, and many other corporations. He played a leading role in marketing and technical development in thirteen projects in particular. In those franchises, he had fifty-eight employees and laid almost 640 miles of cable.[140]

Many of Barden Cablevision's executive members had similarly extensive experience in the cable business, planning, urban development, and other relevant fields. Lawrence R. Baskerville, former executive staff member of the National Cable Television Association, served as vice president of new market development.[141] Another vice president, David M. Miller, had engaged in a scholarly exploration of the impact of cable television. While at Ohio State, he wrote his doctoral dissertation, *Governing Practices of Cable Television and Its Relationship to the Telephone Common Carriers*. Miller also had served as an advisor at various municipalities as they considered the introduction of cable.[142] Wade Briggs, executive vice president, was a radio disc jockey and host of an award-winning Detroit radio talk show. He was a past winner of the Outstanding Community Service Award, and had been selected as the Radio's Top Voice of 1980 by *Detroit News*. Briggs received the Community Leadership Award in 1981. He was named DJ of the Year in 1980, and was known as Butterball Jr., a nationally renowned character created and developed by Briggs. Barden's application elaborates on Brigg's serious commitment to community service. It explained that:

> While cultivating his community involvement over the past 14 years, Mr. Briggs served on numerous boards, committees, and task forces assisting community groups, churches, neighborhood organizations, politicians [sic], and individuals with the design and implementation of fund-raising activities and community awareness programs for civic and community projects in the areas of education, employment, youth, crime prevention, and political issues.[143]

Briggs' involvement in Barden's bid, therefore, was a way to connect the new media technology and social and community benefit.

As required in the RFP, Barden Cablevision planned to provide 20 percent of its ownership to Detroit-based investors, while Don H. Barden would hold 57 percent of the share. At least 10 percent of the investment would come from minorities. Within fifteen days after the franchise decision, Barden had to use the local media advertising to make the investment availability open to public. Additionally, Barden reached an agreement with various local organizations such as the Detroit Baptist Pastor's Association and the Minority Contractors' Association to distribute the information about the investment to their constituents.

Barden promised to exceed the minimum requirement imposed by the city in terms of contracting minority-owned and local firms. The application stated:

> Barden Cablevision has contracts that will run through from the initiation of the franchise process to utilize minority business enterprises and persons in a majority of all of its business dealings. This means that during the first year of operation

following the grant of franchise that Barden Cablevision will maintain contracts and subcontracts with substantially more than the City mandated minimum.[144]

The company would continue the same philosophy during the first four years of operation until the system was completed. During this period, minority contractors and subcontractors would make up for more than 20 percent of all contractors. Minority firms were not involved only in operation and construction but also in researching and representing Barden Cablevision. As stated earlier, it also promised that when no minority-owned companies were available for certain services, it would encourage minority groups to launch a venture.[145]

Local community involvement was an objective in institutional network, as well. Barden Cablevision responded to each of the four means through which the network would be able to benefit the local development of the city. First, institutional networks would "improve the productivity of the city and other public services." Second, the company also would "promote the economic development of Detroit." Barden's plan was expected to create "a modern telecommunications infrastructure capable of attracting and retaining firms in high technology fields." Third, the network would develop the human resources that the city possessed. Most significantly, the institutional network bridged the gap between the information-rich and the information-poor. Application documents explained that "the overall approach of Barden Cablevision's institutional network plan is to see to it that all citizens of Detroit, regardless of race, sex, creed or economic opportunity, have equal access to the information society."[146]

Barden suggested a three-tier structure service. The first tier, basic service, would include forty-two channels for $3.95 per month. It would offer four special leased access channels for minority use. Installation costs would be $29.95 for Tier 1, whereas it would be only $19.95 for subscribers to Tiers 2 and 3. Tier 2 would service thirty-eight additional channels for an additional $3. Tier 3, which would cost $10.95 per month, would have forty more channels.[147]

The proposal designated Channel 6 as an African American channel, available for subscribers of any tier structure. Using locally produced contents, it planned to offer eight hours of African American–oriented programs every day. Additionally, all the tiers had BET aired on the same channel for six hours every evening. The proposal described BET as:

> the nation's first and only specialized cable television satellite service to distribute quality Black-oriented programming. BET

offers feature films, Black classic films, Black college sports, gospel music, a video concert series, and programming of particular interest to Black women, young people, the Black business and professional communities, as well as coverage of the Black literary and performing arts scene.[148]

We now know that BET did not truly solve the problem in racial representation on television. But it was noteworthy that there was a written statement that BET would be available. It was not a correct assumption that it was the solution to the decade-long, if not century-long, problem. The proposal, however, at least reflects the cable provider's willingness to accommodate the needs of local African Americans.

Four channels would be reserved entirely via a special lease for minority use: channels 26, 27, 38 and 39. Tier 2 and 3 subscribers would also have access to Channel 47, the Nation of Islam channel, and Channel 81, Shrine of the Black Madonna. Barden pledged to "assist the Nation of Islam in the design and production of their local origination programming" as a part of the recognition of its responsibility to meet the diversity of religious beliefs and needs in Detroit. This plan secured fourteen hours of cable television contents geared solely to Detroit's African American residents. Additionally, not only would they be able to view African American–oriented programs, they also would have ample air space to produce their own cable programs.[149]

In order to strengthen the system's tie to the local Detroit community, Barden planned to establish *Barden's Detroit* on Channel 111 and a "flagship station" called The Renaissance Channel on Channel 12. *Barden's Detroit* would offer "the pot at the end of the cable rainbow." It would feature people and places in Detroit that made the city unique. Similarly, the Renaissance Channel would be the place where:

the local businessman will be able to showcase his or her product. BC-12 will bring a unique community perspective to news and sports decidedly different than that which is now available on broadcast television. Our Superstation facility with its professional high end equipment will be able to send programs from Detroit via satellite to the nation at large or utilizing our mobile units, bring a story "live" from the most remote corner of the city. BC-12 will spend millions of dollars over the duration of the franchise to solicit and engender quality local productions.[150]

Some of the programs to be aired on Channel 12 would include *Motown News* where "discussions of current events, weather, sports, social and religious activities that can assist Detroit residents with planning a more effective day" would be shown; *Children's Programming*, a morning program for children; *Sports Spotlight* airing "a complete summarization of high school, college, para-professional and professionals' sports activities [*sic*]." The NAACP planned to air an "NAACP Freedom Fund Telethon" for its annual fundraising. This channel would reflect the dynamism of the city by connecting different corners of Detroit and sharing information with its residents.[151]

Barden elaborated on its philosophy and on its plans for local origination and community access in its proposal. Barden recognized that television is important in a city like Detroit, because it was:

> a major media market and proudly possesses some of the best broadcast television stations in the United States. Detroit is also noted for superior, competitive radio stations and an outstanding record industry. Several major professional recording and broadcast careers were launched in the City of Detroit.[152]

Barden endeavored to carry on the city's rich tradition in media production. Its goal in public access was to provide "access to the cable television medium for the people of Detroit realizing that the talent of Detroit is contained in the people of Detroit." Its guiding philosophy had six components. First, Barden's commitment was in "providing high-quality and technically sophisticated local origination and public access facilities." Second, it also attempted to ensure Detroit residents would fully partake in public access programming. Minimizing rules and regulations was one of the methods to realize this second tenet. Third, by building community access centers, Barden would attempt to strengthen the idea of neighborhood and community engagement. Fourth, once such ideas were reinforced, Barden's system would become "conduits of communications for the people of Detroit." Fifth, Barden promised to provide training, especially for those who were afraid of or unfamiliar with new technologies. Finally, Barden offered, through its cable system, a means to battle hunger and poverty. Barden pledged to support charitable organizations in the city to counter the economic plights of people in Detroit. In these respects, Barden's involvement in Detroit was not just one of offering channels spaces and production facilities. It showed deeper commitment to the city's local communities.[153]

Barden believed in leaving the power of cable production in the hands of residents. It explained:

> The company believes that there are significant benefits in placing the power and tools of communication through the electronic media in the hands of local residents. . . . Access channels are an efficient means of meeting fully local communications needs. These access channels provide information that is often ignored the mass electronic delivery systems and enable a community to develop a vital exchange of ideas and services. . . . The various active neighborhood and community groups, extensive recreation programs, a strong religious base, excellent municipal services, an elaborate educational system, outstanding entertainment facilities, and the potential for a strong business recovery and growth are example of elements which can contribute to the cable television system for the City of Detroit.[154]

The application also continued:

> Barden Cablevision believes that the active local use of cable television can foster real dialog within the community, and meet the fullest range of community communications needs. the research and experience of Barden Cablevision shows that it is simply not large amount of dollars in equipment that make community programming successful, but the design of the community television operation, the accuracy of its response to the community's needs and resources and the ability of its staff to meaningful involve the public. Barden Cablevision sees its role as a facilitator of access in the City of Detroit. The goal is to encourage participatory television in every sense of the word by building into the plan for cable service an aggressive community programming component that will assist the economic development in the City of Detroit.[155]

Barden listed several ways to realize its objectives from traditionally recognized ways, such as public access, to more specific plans, such as job training and government services.

Through retraining and vocational education programs, Barden tried to address issues concerning jobs, employment, unemployment, and worker retraining. It analyzed the job condition in the early 1980s, and claimed "unemployment has become critical in Detroit in the course of recent recessions, and the 1981–1982 recession has been devastating for jobs and human resources. . . . It is now clear that the City will never return to levels of employment in the automobile industry experienced prior to 1981." Barden stated that as a minority-owned company, it understood the plight of Detroiters well. The company was "in a unique position to comprehend the requirements of persons who are unemployed from the automobile and related industries and those who have had difficulty or no success in securing employment or in learning job skills relevant to today's world of work." Barden's approach allowed both temporarily unemployed and chronically unemployed Detroiters to find and maintain the job through training.[156]

Barden's system planned to publicize job openings, give instructions as to how to find a job in which a viewer has some experience, connect job seekers and hiring companies, give advice on how to obtain skills, and so on. Barden called its own system "a quantum leap in the technology of employment information processing." In cooperation with the Michigan Employment Security Commission and private placement agencies, Barden would assign Channels 10 and 11 as the Job Opportunity Channel and the Job Retraining Channel. The former was to be a place where a subscriber would find a job. Once he or she began to work, Channel 11 would allow them to acquire better skills. because the lay-off and unemployment rates were high in the city, especially among African Americans, these services could bring positive improvements to their lives.[157]

Barden also made an effort to be engaged in the local community by providing funding opportunities with a $50 million budget. The company planned to allocate $10 million for minority-owned businesses and manufacturing entities. The fund would help minority residents gain work-related training and retraining. Similarly, an additional $10 million was earmarked for a Barden Communication Business Opportunity Trust for Minorities, Women, Handicapped and Socially Disadvantaged Persons. This financial assistance would help Detroiters form new communications ventures. Encouraging minority involvement in communications created future job and economic security in the neighborhood. Communications was particularly important in the post-auto industry period of Detroit.[158]

Starting with thirteen local originations and twenty-two access full-time and fifteen part-time workers for local origination and access, respectively, during year 1, Barden planned to expand its human resources to twenty-three local origination full-timers and forty-five access full-timers. Twenty-five hours of programming per week would be produced locally by Barden. An additional twenty hours of programs would be provided by other sources. In total of forty-five hours of local origination programming, exclusive of access programming, pay programming, satellite programming, and others also were part of the proposal.[159]

Barden saw two major ways to make its access programming successful. Its policy must "reflect the community through [its] local origination and access programming." It must also "meet community needs by making available professional equipment, training assistance, and management." In addition to local origination channels, Barden planned to provide at least thirty-two channels as access channels. Four of them were reserved for minority use. Barden explained that different segments of the local community, including minority groups, unemployed, those in arts and culture, and many others, must be reflected on public access channels. Minority groups included both ethnic minorities and linguistic minorities, in the case of Detroit. Therefore, some programs would address issues concerning people with different backgrounds. The others would be in different languages to allow people who do not understand English or other more common television languages.[160]

Various organizations in Detroit concurred with Barden's interests in minority groups. Horace Shetfield, president of Detroit Association of Black Organizations, explained that the organization expected cable television to make a broad range of programs available for Detroit's African American community. Similarly, the Detroit Urban League sought a community information channel and an access channel "reserved for the discussion of economic, social and political concerns unique to an urban community." The NAACP also saw the potential of Barden's system to "guarantee the Rights of Black Americans through legal and political action and educational programs."[161]

An "active and participatory approach" not only referred to Barden offering its facilities at no charge, but also to community education for local origination and access programming. It promised to offer support staff and facilities readily available to interested residents of Detroit. Its outreach programs would touch on all segments of Detroit's communities. Training was offered to people with no or little experience in television production. From lighting to microphone use and to editing, the training would include all the fundamentals of television production. The courses each would have different levels and foci to meet the specific needs and wants of interested parties. At the end of each training course, participants would receive a diploma. Advanced workshops would include television directing, advanced porta-pak, advanced editing, use of remote production van, camera skills improvement, make-up application, set construction, visual intercom system, and so on. The courses would provide education on how to produce media contents, as well as the concepts behind the idea of community television. As a result, when residents were ready to produce their access programming, they were not simply creating a show but participating in the process of community television activities.[162]

Offering information about employment was not the only way Barden saw its role in securing jobs for Detroiters. Barden also was willing to actively hire local Detroiters and minorities in local communities. Its application explained that "Barden Cablevision, a minority owned and operated company, is committed to utilizing local labor resources, including minorities and women." One of the most tangible signs of its seriousness in equal hiring was in its Equal Employment Policy Statement, which read:

> It is the policy of Barden Communications, Inc., including its subsidiaries to implement affirmatively equal opportunity to all qualified employees and applicants for employment without regard to race, creed, color, sex, national origin, age or handicap.
>
> Positive action shall be taken to ensure that promotion decisions are in accord with principles of equal employment opportunity by imposing only valid requirements for promotional opportunities.
>
> All decisions on employment will further our principle of equal employment opportunity. We will recruit, hire, train and promote personas in all job classifications without regard to race, creed, color, sex, national origin, age or handicap.
>
> All management personnel will continue to ensure that all personnel actions such as compensation, benefits, transfers, layoffs, return from layoff, corporation-sponsored training, education, tuition assistance, and social and recreation programs will be administered without regard to race, creed, color, sex, national origin, age or handicap.[163]

To nurture skilled workers in the field of communications, Barden planned to support one-hundred students annually at Cass Tech for their studies in electronics, technical, or computer fields. Similarly, ten students per year would be supported at Wayne County Community College in the field of electronics and computer programming. Barden planned to hire ten percent of these sponsored students upon their graduation. Barden spared at least $150,000 to provide such occupational opportunities. Additionally, in cooperation with Henkels & McCoy, Incorporated, a Pennsylvania-based training service provider, Barden planned to train 250 local residents for cable television–related jobs, such as linemen, splicers, installers, service technicians, or bench technicians. The program included both hands-on training and class activities.[164]

Barden planned to diversify its recruitment methods to promote equal employment. Traditional methods had been walk-ins, write-ins, or call-ins. Often newspaper ads and private employment agencies were the means for more mass-oriented recruitment. Barden promised to proactively approach minority groups such as the NAACP, the Urban League, minority churches, women's groups, and so on to request referrals of those who they consider qualified or available. If Barden were to put an advertisement in a newspaper, it would make sure that the same recruitment information would appear in a newspaper with a high circulation rate in minority communities.[165]

As required in the RFP, Barden listed its projection of minority hiring goals. At the time of application submission, Barden had forty-five employees, including twenty-seven minorities, only one of whom was not African American, but was a Latina clerical worker. Twenty-six minorities were African Americans, with eighteen males and eight females. There were three male and four female African American officials and managers. During year 1, Barden proposed to hire 141 employees. Seventy of these individuals would be minorities and forty-three would be women. In year 2, 158 of 284 employees would be minorities, with eighty-eight women. Year 3 had 248 minorities of 421 employees;138 of these would be women. This meant approximately 60 percent of the employed would be minorities and more than 30 percent would be women. Within ten years, minorities would constitute more than 70 percent of the employees (344 of 477); 40 percent would be women (188). Although 54 percent of the officials and managers during year 1 were minorities, by the end of the tenth year, that number would increase to 73 percent. Similarly, 72 percent of the supervisors and foremen would also be minorities.[166]

Applying to Wire Detroit: City Communications

The second company to submit a proposal was City Communications. Founded in 1981, this company had a strong local base in Detroit. Three local attorneys, Edward F. Bell, Lester D. Hudson, and A. Robert Zeff, renowned financer David Chase, and corporate developer Roger Freedman began a working group to seek the city's cable franchise. The group aimed to "provide solid cable television programming to all segments of the Detroit community." It was founded and staffed by Detroiters. Residents of Detroit owned 74 percent of the corporation.[167]

City Communications submitted its application to the Cable Communications Commission in early December 1982. In its cover letter dated December 3, 1982, Robert L. Green, an authority scholar in the study of American cities, and the president and chairman of the Board explained that the company's programming would "encourage unity, . . . [highlights] the exceptional artistic talent of [the] city; . . . [teach the city's] youth, and [direct] parents to motivate them." Although the letter did not explicitly address race- or ethnic-related concerns that were more apparent in Barden's plan, it nonetheless emphasized that the company's Board of Directors represented "a cross-section of men and women who are leaders in their respective fields who have made significant contributions to [the] city."[168]

The Board included Coretta Scott King, president of the Martin Luther King Center for Social Research. Frances Hooks, wife of Benjamin Hooks, the NAACP executive director, was also a member. From other domains, Arthur Jefferson, superintendent of Detroit Public Schools, Reginald P. Ayala from Southwest General Hospital, Rev. Charles Adams from Hartford Memorial Baptist Church, and Bishop David Ellis of Greater Grace Temple of the Apostolic Faith joined the Board to represent and reflect the concerns that Detroiters had about the city's finance, economy, culture, and other issues.[169]

The owners, partners, and officers of City Communications had vast experience in media and African American activism in Detroit. Thirty of the thirty-nine directors lived in the city of Detroit, as well as in surrounding communities. Edward F. Bell, one of the owners of the corporation and a local attorney in Detroit, had been involved in African American–related issues. Detroit's television anchor Carmen Harlan, actress Jayne Kennedy, and players from the National Basketball Association, all had been Bell's African American clients. He also was involved in the NAACP Freedom Fund dinner. Another local attorney and a graduate of Fisk University and Boston College, Lester D. Hudson also had been involved in African American–related issues, including working for Rev. Jesse Jackson. David Chase, real estate developer from Connecticut, had been involved in the Martin Luther King Center for Social Change through his financial support. Robert Green, former Dean of the Urban Affairs Programs and dean of the College of Urban Development at Michigan State University, was a "nationally known pioneer in the development of research and program strategies designed to assess and upgrade the status of the urban poor and racial minorities." He also had been an educational consultant to the national NAACP legal staff.[170]

William T. Johnson, operations manager, and Gilbert A. Maddox, vice president of programming, had the most experience in cable television of all the owners. At the time of the submission of the application, Johnson was the president of KBLE Ohio, Inc., the first African American–owned cable company in the United States. Although he was born in Ohio, he had an extensive background in Detroit. He grew up in the city and experienced its automotive and steel industries as a summer student worker. Johnson's consultation included that on African countries and telecommunication. Maddox had served as a host of *Profiles in Black*, *Detroit Exchange*, and other prime-time television shows on WDIV-TV for more than ten years. His commentary was supported by his firsthand experience of being raised in the city. Maddox wrote his doctoral dissertation on television programming based on research he conducted in local area communities. His work was particularly "suitable and appealing" to Detroit's African American communities. He had taught as a faculty member at Howard University, the University of Michigan, and Wayne State University. Johnson had been a communication consultant to the Congressional Black Caucus, the Detroit Urban League, and many other African American–oriented organizations.[171]

City Communications held interest in three cable systems. William Johnson owned the franchise in Columbus, Ohio, in its entirety since May 28, 1978. The construction had been completed in July 1982. Johnson also owned the 80 percent of the system to be constructed in Seattle, Washington. David Chase owned the system in Hillsboro County, Florida. City Communications emphasized that Detroit would be able to benefit from this past experience. The company also underlined that Johnson had founded the first Black-owned cable company and remained the "standard-bearer" for African American ownership of cable television systems.[172]

To realize effective local institutional involvement, City Communications suggested that twenty-seven buildings, including that of the *Detroit Free Press*, Renaissance Center, and General Motors World Headquarters, would have a local origination package of $50,000 for video production. Similarly, nineteen hospitals and health care institutions had a similar package. State buildings and educational institutions also were included in the plan. The enterprise foresaw that according to their study a lot of Detroiters were interested in the institutional production. It also was aware that there was "a significant lack of public awareness as to the meaning of and the potential in an institutional/commercial network." They demonstrated great enthusiasm about the technology. But such an excitement was not based on any specific knowledge about what cable could bring to the city. This was one of the reasons why City Communications promised to conduct educational and training meetings to inform residents of the new communication medium. By doing so, video transmission of educational programs, neighborhood cultural events, distribution of community requested materials, and other new ways of using a television would be materialized.[173]

City Communications proposed a two-tier structure for its service. In the Tier 1 service, which would cost $19.95 for installation and $3.95 per month for subscription, thirteen full-time local broadcast signals would be offered, seven of which would be from Detroit and one from Windsor, Canada. Five of thirteen would be independent companies. Three additional full-time signals would come from outside of the area. They would be independent channels from Atlanta, Chicago, and New York. Satellite programs would include C-SPAN, Cable Health Network, and fifteen other channels. With an extra fee of 75 cents per months, BET was available six hours per day. City Communications planned to provide seventeen access nonbroadcast channels, four of which, Channels 15, 37, 54, and 55, would be reserved for minority lease channels available twenty-four hours a day. In total, Tier 1 subscribers would be offered sixty channels. This meant that the plan had almost one third of the channels reserved for access channels.[174]

The City Communications plan emphasized audio programming. The proposal included a cultural and arts channel, a third world music channel, ten ethnic-language channels, three local sports channels featuring high school and college sporting events, four experimental channels for local entrepreneurs, one local news channel, four community bulletin board channels, and many others.[175]

At a cost of $8.95 per month, Tier 2 service would include twenty additional channels, which Tier 1 subscribers could choose to subscribe with per-channel fees. Many of the additional channels would be satellite channels such as ESPN, CNN Headline News, MTV, AP Newscable, and so on. Two major minority channels, BET and Community Channel, would be included as a part of the Tier 2 service. Those who wished to watch one or both of these channels would have to subscribe to Tier 2, or subscribe to those channels separately from the Tier 1 contract.[176]

City Communications planned to offer several channels especially useful for the residents of Detroit. Channel 21 would be for help wanted listings. It would aim to help audiences find and improve their employment condition by offering a listing of job openings twenty-four hours a day. Because a high unemployment rate among African American Detroiters was partially due to lack of information, African American subscribers would be able to benefit from the service. Channel 43, Detroit Lifestyle, would be a channel that connected Detroiters by providing locally originated programs featuring "the vitality, energy and style of Detroiters at work and play." It would aim to enhance residents' affinity to the city and strengthen ties with other residents. Regarding BET, City Communications explained:

> BET provides satellite services with high quality and selective programming focused towards Black audiences. This network features films, musical specials, video concerts, documentaries and sporting events, with specific emphasis on entertainment by and about Blacks. The programs are aimed at encompassing Black cultural lifestyles, experiences and environments. BET is the nation's first and currently the only available specialized cable television channel entirely oriented to Black America.[177]

Channel 75, Community Channel, would be a channel for minorities with more comprehensive interests without limiting itself in entertainment. It would include daily news programs dealing with issues concerning African Americans in the United States and around the world. It would also educate its audience about Black history, culture, religion, health, and so on.[178]

In order to realize local origination, which would "[enrich] and [improve] the lives of Detroit's citizens," City Communications proposed two studios on East Lafayette Street where the firm had its corporate business office. Additionally, it would offer six access studios. For the purpose of enabling outside filming, the company planned to provide a fully equipped mobile production van that would be for the priority of the access users. General Television Network, an Oak Park–based corporation, made proposals as to what equipment to prepare for optimal local origination and community access. These facilities would reflect the company's understanding that "Detroit is a city of many organizations, nationalities, neighborhoods and special interest/need groups that can be effectively served by the programmatic variety and sensitivity of a cable system." The system would enable "increased access to a wide variety of local, national and international events and performances." The application also made sure that studio and production facilities and services would be available free of charge for the residents of Detroit who were producing for noncommercial interests. Studio and remote production, supervision and instruction, leased channels, system playback, and editing would be free of charge for governmental, educational and community purposes.[179]

City Communications' plan promised that it would make an increasing amount of commitment to these two important services by securing sufficient operating budget and staff. The first year of production would have $833,000 for operation such as salaries, maintenance, tape stock, and other supplies. The following year, the budget would be almost doubled to $1.567 million. The budget saw an increase to $2.741 million, $3.859 million, and $4.383 million for each year to follow. From years 6 to 10, the budget would remain at $4.539 million. The next five years the budget would be at $4.560 million per year. Forty percent of the budget would be allocated to access channel production, and 60 percent for local origination programs that would be aired sixty-six hours per week. Similarly, City Communications planned to start its production with four local origination and two access employees. The numbers increased to eleven and eight, seventeen and eleven, and twenty and twelve for years 3, 5, and 10, respectively. These figures reflected the company's financial and human resource commitment.[180]

City Communications identified Detroit's African American community as being an important audience of cable programming, especially because it was a young community with likelihood to expand in the near future. The Detroit African American community was particularly interested in sports and entertainment. City Communications believed that basic services such as Alpha Repertory TV Service, the Black Entertainment Network, ESPN, and others, as well as Bravo, HBO, USA Network, and other pay channels, would satisfy such demographics. Additionally, channels and programs such as *Sights and Sounds of the City*, *At Play in the City*, *Life Style*, and other locally produced programs would meet the needs of African Americans to hear about local sporting and entertainment events. The potential of such shows was immense.[181]

Sights and Sounds of the City, for example, would feature the city's music scene such as jazz, folk, gospel, and classical music. It would also offer programming on theater and dance. Featuring third world music, the program included blues and spirituals of African Americans, Reggae and Calypso of the Caribbean, and drums and dance from Africa. Black Film Festival would recognize the history of African American film production since the early 1920s. The festival aimed to correct historical injustices in movies against African Americans by showing some of the old but accurate depictions of African Americans. By showing Black cowboys and gangsters, the event would counter some common misperceptions about Blacks in movies such as *Gone with the Wind*, *Tarzan*, or *The Birth of a Nation*. Musicals on Black life also would be a part of the film event. Such a program could attract more diverse audience by holding not only the festival for

African Americans, but also programs targeted toward the Arab-speaking population that had been on an increase.[182]

At Play in the City would feature sports. From baseball to football and hockey, Detroit was a city of sports. City Communications explained that despite the poor performance of Detroit-based teams, Detroiters remained loyal fans of their local teams. Even less-popularized sports such as soccer, biking, jogging, track and field had begun to attract more participation. Having sufficient channel space for sport programs would be vital to meet the expectation of the residents. The company also was aware that professional sports were not the only kind of sports Detroiters showed interests in. High school and college sports also were covered on this program.[183]

Life Style would be a more general program on the life in Detroit. It would discuss how to dress, what games people played, what to do with gardens and homes, and other day-to-day happenings. It would also talk about artistic and cultural events to realize "a cultural, social and psychological renaissance of the Detroit spirit." In addition, *Goin' Home* was a program through which many Detroiters could trace their heritage to the rural South. Because there was a large group of African Americans in Detroit that migrated to the city during the great migration right after the world wars, this program was particularly of their interest for introducing African Americans to some of the major southern cities and life in the south, and evoking a sense of nostalgia in the mind of the audience.[184]

With unemployment being a major social problem in Detroit, City Communications considered it be helpful to share job-related information on cable television. *Unemployment and Work* would be a program that attempted to inspire and encourage Detroiters. In the early 1980s, the non-white unemployment rate was close to 30 percent. When it came to minority youth, the figure was about 60 percent. Of Blacks with college degrees, 15 percent did not have jobs. This attested that education alone did not guarantee employment. Basing its production philosophy on late Vice President Hubert Humphrey's idea that unemployment, poverty, and illiteracy are interconnected, the program addressed exactly these three topics. For example, it informed its audience of job openings, different types of job training, the ideas deferred gratification and the inherent rewards, the social values, and attitudes that would make people more attractive in the job market, and other information. The program served as an opportunities for many Detroit residents, especially minorities, to learn how to prepare for an interview. *Careers of the 1980s* featured a documentary-like drama in which those, particularly minorities and women, who were successful in business and their walks of life, shared their experience with the audience.

City Communication hoped that the program educated both the youths and adults what kind of factors should affect their choice of careers, change their attitudes and thoughts on work, and prepare them to the world of work.[185]

Local news programs including *What's It All About* were the remedy for the lack of media coverage on the city during the preceding three decades. City Communications argued that broadcasting news programs had "distorted news coverage, failed to cover positive events, or day-to-day events, unfairly maligned Black leadership, and failed to identify the real leaders of the Detroit community." Therefore, cable news could correct such distortion and accurately depict Detroit's African American leadership. Local activities led by the New Detroit Urban League, the NAACP, Black Family Development, and other organizations also were covered, along with news on women or from women's perspectives.[186]

Access programs were to be as important as locally produced programs. As previously introduced statistics and figures showed, City Communications recognized the importance of community access and was ready to make a substantial commitment. It promised to offer training for technical and programming personnel during the first two years, develop programming schedules, and play other key roles in realizing access programs. Additionally, such training and assistance, as well as the use of equipment would be free of charge for Detroit residents who were producing access programs.[187]

City Communications emphasized that "it is the ultimate goal of this corporation to employ persons at all levels so that the total work force properly reflected the racial, gender, and ethnic composition of the City of Detroit." It promised to take "an aggressive posture" to hire Blacks, women, and other minorities. Its objective was to exceed the minimum levels of minority participation per job category as per Detroit's Human Rights Department, and eventually reflect the minority and female composition of the city. The company's Board of Directors already had begun to reflect the diversity of the city population little by little. Of the board members, 48 percent were African Americans, and 17 percent were women. Nearly all the members lived or had business in the city.[188]

Looking back in time, City Communications acknowledged that African Americans and other minority populations had their access to job opportunities denied or limited for a long time. As a result, they had difficulty acquiring skills and knowledge to make a progress in their career. To overturn such a historical trend, the company promised that both minorities and women would be actively recruited for all jobs in the company. Additionally, Psychological Supportive Services, Inc., a minority-owned firm specializing in psychological assessment and management training, provided testing for hiring and promotions. City Communications' decision to acquire such a support was to measure only skills required for the job the applicant was applying for and no bias would interfere with the hiring decision making.[189]

City Communications planned to implement the role of director of Equal Employment Opportunity Programs. The director would be responsible for the administration of the corporate policy and update the management on the progress the company was making. The director would assume both intracorporate roles and liaison roles. He or she would check the effectiveness of the current equal employment plans, suggest what changes to be made, and measure the extent of successful realization of the plans. With other officers, he or she would "work affirmatively in hiring Blacks and other minorities to ensure that possible problem areas do not emerge." They planned to reach minority groups through newspapers and publications such as the *Michigan Chronicle* with a large circulation among minorities and through organizations such as the NAACP, the Urban League, minority churches, community groups, and others. Additionally, they would hire African Americans and minorities and also would be required to engage in their organizations to encourage minority application for the job and promotion. The director would also serve as the point of contact for the Cable Commission and other agencies, as well as local organizations.[190]

The Urban Communications Institute of City Communications was one of the means for minorities with some skills to acquire more training. It provided job information both locally and nationally. It therefore served as "an employment pool" for the company. Such a training opportunity was part of a $250,000 annual budget reserved for training.[191]

City Communications also addressed concerns that urban and minority residents had shared about their difficulty conducting business. It suggested that the technical innovation that it planned to bring would benefit the economic welfare of the city. The cable industry in particular would provide Detroit-based businesses with various opportunities. Minority Purchasing Program procured office and production supplies from local vendors of quality. City Communications conducted interviews with vendor candidates to make sure that their services were sufficient. Once deemed as such, the vendor's name appeared as a "qualified participant" to the program. The company also made a partial front-end payment to facilitate purchasing of materials for the vendors. By conducting a paid management internship at the City Communications' office, it also attempted to improve executive and management skills. As some of the potential local businesses, the application listed fifteen accountants, seventeen consultants, six attorneys, nine insurance companies, two office suppliers, and many other agents, distributors, and vendors.[192]

Applying to Wire Detroit: Detroit Inner-Unity Bell Cable System

The third applicant to submit an application was Detroit Inner-Unity Bell Cable System (DIUB). Unlike Barden, the ownership of DIUB was heavily operating in New York. Its chairman of the Board was Wendell Cox, the first African American to be awarded an FCC license for a radio station in Detroit. He had applied to receive the franchise to wire the city with cable back in 1969 when most were unaware of the potential of the technology. Cox owned the nation's largest cable system in San Diego with more than 220,000 subscribers. He also had fifteen more systems, each of which served more than 25,000 customers, respectively. After working for the industry for nearly fifty years, he had operated sixty cable systems in twenty-three states. DIUB was the fourth largest cable company in the United States. Despite such a national presence, the firm included another local board member: Oscar W. King III, vice president of franchise development was a professor of economics at University of Detroit. Percy E. Sutton, Eugene D. Jackson, Sydney L. Small, and other board members also had many experiences in the industry in New York, Philadelphia, St. Louis, and other places through various corporations.[193]

In a cover letter attached to the application submitted on December 10, 1982, Wendell Cox explained its ties to Cox Cable Communications, Inc. The letter said that Cox would construct and operate the cable system in Detroit, while DIUB stayed in charge of establishing technical specifications and operational requirements. This structure considered DIUB as the designer of the system and Cox as the implementer. DIUB was confident that Cox would be successful in fulfilling its responsibility as "the fourth largest multiple cable system operator in the country [which] has substantial experience in building and operating cable television systems."[194]

Cox's experience in the cable industry had been extensive since its inception in 1962. Listing one to three systems per page, DIUB's list of systems under Cox was approximately eighty pages long. The annual report published in 1981 explained that the company operated sixty-two cable systems in twenty-three states. They were present in various locations in states such as Nebraska, Oklahoma, California, Florida, Washington, Illinois, Iowa, Indiana, Virginia, Massachusetts, and Georgia. It listed thirteen major cities where Cox operated or was developing systems: Spokane, Washington; Roanoke, Virginia; Tucson, Arizona; Norfolk, Virginia; Vancouver, Washington; Virginia Beach, Virginia; Pensacola, Florida; Portsmouth, Virginia; Oklahoma City, Oklahoma; Omaha, Nebraska; New Orleans, Louisiana; Fort Wayne, Indiana; and Jefferson Parish, Louisiana. the company had twenty-one systems in Michigan, including one in cities such as Saginaw, Zilwaukee, Marquette, Owosso, Corunna, Palmer. If DIUB were to win the franchise in Detroit, it would be adding another major market to its portfolio.[195]

Using the phrase "The Focus is on You" as its project motto, DIUB emphasized that it planned to offer a service that residents of Detroit truly needed. DIUB explained that it would "[offer] you the most modern, technologically advanced, efficient cable television system available today." It continued, "we offer personalized service and attention." DIUB argued that it knew the needs of residents well. As a part of the message to the residents, it said:

> Detroit Inner-Unity Bell Cable systems conducted a careful survey of the people of Detroit. We spoke with your friends and neighbors. And what you told us was that as residents of one of the world's most exacting, demanding cities, you know what you want, and don't want. Your way of life includes being involved. A cold, impersonal company that won't have the time or inclination to listen to you just won't make it in Detroit.
>
> At DIUB we understand that. With us, you'll have exactly what you want in cable television viewing because you'll be involved in its development and future growth.[196]

The cover letter elaborated on DIUB's commitment by introducing some budgetary involvement. It explained:

> The real strength of DIUB's proposal is its local commitment— DIUB will employ an estimated $89 million with identified Detroit business—DIUB will employ an estimated 385 individuals, hiring at least 60% minorities, with 80% from Detroit and DIUB will provide the Detroit Public Benefit Corporation $40 million—and much more.[197]

First and foremost, DIUB emphasized its role in the creation of job and business opportunities In its application, DIUB claimed that it understood that:

> its obligation to the citizens of Detroit goes beyond providing a state-of-the-art communications and entertainment system. . . . DIUB also recognizes there has been a historic exclusion of minorities and women from participation in the communications field. DIUB will pursue and implement an aggressive Affirmative Action Program, from the date the franchise is awarded, that it will encourage and ensure the further employment and participation of minorities and women.[198]

The company further elaborated on its affirmative action policies. The application explained:

> A successful Affirmative Action Plan consists of the methods and programs which will result in the Equal Employment Opportunity Policy and principles becoming organizational realities. It means recruiting, training, hiring, supervising, compensating, terminating and promoting qualified persons without regard to their race, color, religion, national origin, sex, or handicap. DIUB will implement its equal opportunity program in full compliance with all federal, state, and local laws.[199]

In order to implement such policies, DIUB planned to carry out five strategies. The first was to post information and notices in more than one language. It not only was going to produce pamphlets and brochures in English and Spanish. It was willing to translate documents into other languages depending on the needs at offices and studios. Additionally, it planned to ensure that there would be fair union representation, particularly for minorities, women, and people with disabilities. Another organizational effort was to involve local organizations such as the NAACP, the Urban League, Detroit Street Services, and Detroit Indian Center for Recruitment and Counseling. The foreseen outcome of these efforts would be increased minority representation. During the first year of the cable franchise, DIUB planned to hire twenty-one new employees. Fourteen of these were slated to be minorities and 50 percent would be women. The numbers for years 2 to 5 were 167 minority employees and 127 female employees. Between years 6 and 10, there would be 262 minorities to be hired and 196 women.[200]

The firm also underscored its commitment in minority involvement through its financial investment. It stated:

We'll employ Detroit residents who already possess special skills, but our search won't be limited to only those experienced in the communications industry. There are numerous jobs suitable for on-the-job training that will be provided by DIUB. We believe strongly in community involvement and people helping other people.[201]

DIUB promised to provide $1.950 million in scholarship and training funds to students and institutions in Detroit. The company would make $42.120 million available for the city's public and private schools, colleges, and universities as grants. It also planned to establish a $3 million loan fund for Detroit-based businesses owned and operated by minorities and women. DIUB explained that it had a belief that "often qualified and talented prospective business persons are excludes from conventional lending sources. DIUB's fund will provide qualified, cable-related businesses a part of their necessary capital, interest-free." DIUB's procurement policy also reflected its willingness to serve the local community. Of the total cost of contracts and subcontracts during construction and supplies, 20 percent would be purchased from minority and female-owned businesses. An estimated amount of such procurement was about $89 million dollars. Before submitting its application, DIUB had identified numerous firms owned and operated by women and minorities that were interested in providing services for the system. DIUB claimed that the list was continuously updated and expanded.[202]

Some subcontractors had contacted DIUB and Cox before the submission of the application. Such letters were attached to the application package submitted in early December. Many of these subcontractors underscored that they were minority-owned or were committed to affirmative action policies. Jack Martin & Co., for example, was a firm that offered to provide accounting and auditing services to DIUB. In its letter, it explained that it was "a minority-owned business that was established in Detroit in 1975." Gregory J. Reed and Associates, a law firm in Detroit, similarly explained itself as "a minority-owned business." Scales & Associates was an engineering services consultant. It wrote in its letter that it was "a minority owned business, 100% owned by a Black male, that has operated in Detroit for the twelve years of [its] existence." There were many other similar letters addressed to Oscar King that DIUB compiled in its application.[203]

DIUB's commitment in local origination and access programming through the development of human resources also was confirmed in its staff commitment. DIUB planned to have five full-time employees for local origination and two for access channels during the first year. Although the numbers were far from enough, DIUB promised to expand that number to fifteen and forty-seven, respectively, by year 3. The proposal anticipated thirty local origination part-time employees and forty access channel part-timers.[204]

Local training was one way to ensure such facilities were put to a good use. It was not only to create jobs and those who were properly trained for jobs. In addition to nurturing professionals, training was going to help amateur residents familiarize themselves with the new technology. DIUB said that any resident over the age of twelve, and anybody under adult supervision, would be able to partake in two different types of training at the local origination studio and neighborhood studios. The basic workshop was planned for first-time learners. It would last four days, with two hours of training per day. It would provide basic information on equipment and the technology and would train those attending in fundamental editing, scripting, producing, and other skills. After completing the training and passing a standardized proficiency test, a resident would be able to receive a certification that would allow him or her to use DIUB's equipment and facilities. DIUB expected that every five weeks it would be able to train and certify seventy-five people at its facilities. The advanced workshop for the graduates of the basic workshop would include location shooting and post-production skills in eight hours over four consecutive days.[205]

Another training program would be offered through an internship. With the $500,000 budget for training and internship programs, this would involve local educational institutions and pay $4.50 per hour to each intern. Each student would be able to acquire a semester worth of training in wide variety of relevant job activities. This program was going to engage a total of eleven schools and bodies: Wayne County Community College, Wayne State University, Center for Creative Studies, Detroit Public Television, Detroit Public Schools, Marygrove College, University Cultural Center Association, Center for Creative Communications, Archdiocese of Detroit, Detroit Producers' Association, and Mercy College. College students could also take advantage of a scholarship fund that amounted $1.450 million. Over the first two years, $800,000 was slated to be spent "for students graduating from Detroit public or private schools and continuing their education in communications-related fields, or to students attending any college or university located in Detroit majoring in communications."[206]

As Barden and City Communications did, DIUB also proposed funds and facilities for local origination and public access. Its Producer's Fund had a budget of $810,000. It would be available to any Detroit resident who had cable programming ideas and who was in need of funds to realize them. DIUB would invest $53 million in local and access programming to foster cablecasting. Additionally, it had a budget of $6.4 million for facility development and construction solely for public access and local origination. DIUB planned to have thirty viewing centers scattered all over the city at local city halls and other municipal offices for free community viewing of basic service. DIUB also planned to contribute 3 percent of its gross revenues in Detroit to the Detroit Public Benefit Corporation. Estimated contribution was $41 million to "promote, encourage and provide additional financial support for the development of access programming in Detroit." An additional $68 million would be paid directly to the city as franchise fees. DIUB's financial contribution to the city was immense.[207]

DIUB planned to construct its main access studio of approximately 3,000 square feet on Myrtle Avenue. The total budget for this studio was more than $300,000. The firm had plans for four neighborhood access studios, all of which had the approximate size of 3,000 square feet. The first one was slated for East Jefferson Avenue, the second for Van Dyke Avenue, the third between Schaefer and Keeler, and the fourth for West Seven Mile. Each had an allocated budget in excess of $275,000. DIUB's plan also included a cultural consortium studio. Although the location had not been determined before submission of the proposal, it was slated to be another local origination channel that would specifically focus on issues pertaining to culture. The educational consortium studio was another similar example.[208]

DIUB's plan included three different types of basic services. Basic 1, which would cost $2.95 for subscription, included twenty-eight channels with the home security service. Four of these channels were FM channels. Basic 2, with added channels and services, would cost $6.95. Super Basic was estimated at $11.95. This highest-level service would have programming from Atlanta, Chicago, and New York. DIUB proposed four different types of access channels to the city. It explained that residents of Detroit could present their own programming on public access channels. Residents would be able to obtain information on City government activities, voting, and community meetings on government access channels. An interfaith religious access channel where religious organizations and clergy in Detroit would be able to televise their services and messages, was planned. Using satellite, viewers would be able to stay home and partake in the gathering. Minority organizations in Detroit would be able to purchase time for their showing of commercial programming on minority-leased access channels. Additionally, DIUB planned to have employment guide channels, community information channels, and other local community-oriented channels and services.[209]

At the end of the application document, DIUB enumerated numerous organizations that were interested in the cable system and then explained how these organizations would be able to benefit from cable televsion. The Afro-American Museum wished to have some of its projects cablecast, especially those related to its Black History Month series, which included storytelling, lectures, and performing arts. Its banquet could also be cablecast and it featured a prominent Black person every year. It had also been active in producing films and videotapes featuring Paul Robeson. Once the cable system arrived in Detroit, the museum would be able to take advantage of its public access studios and other facilities to showcase its collections and air public announcements. The NAACP similarly hoped to air Black history programming, soap operas, and other programs that are of particular interest to the African American residents of Detroit. The ACLU also was interested in cable's access studios and technical help. It wished to have its speeches and debates cablecast so that a greater number of people would be able to learn from the experiences. The Detroit Urban League considered cable television to be an optimal means to disseminate social service and civil rights seminars and information. This organization even felt that cable television would enhance the popularity of it because, according to one of the representatives, not all residents were aware of its existence.[210]

Emphasis on Public Access and Local Origination in Detroit

By the end of 1982, all three firms had finished submitting their final proposals. They each similarly emphasized their commitment in public access and local origination services. They also underscored their respective willingness to employ local residents, use local subcontractors, offer job training, provide students with internship opportunities, and construct local origination facilities. It was now time for the city to compare and study each proposal and decide which franchisee would bring a system that was most useful for the residents of Detroit.

Chapter 5

Progress and Struggles in the Process of Franchise Decisions for Media Democracy

Once cable providers turned in their final proposals, the relevant offices and departments of Boston and Detroit closely examined the content of the submitted documents. Each candidate was expected to exceed the minimum requirements described in the RFPs in order to win the franchise. This chapter examines how city officials came to their final decisions to award franchise agreements to Cablevision in Boston and Barden Cable in Detroit. It investigates the interrelationship between African American racial and municipal politics.

For Boston, this period began in April 1981, when applications were due. Through December 1982 when the final franchise decision was made, the city and candidates had maintained frequent communication with each other to clarify and negotiate the details of the cable system in Boston. Once the interested cable companies submitted their applications, the city held four hearings to which not only city representatives and cable company representatives but also residents of Boston were invited. Various interested parties shared their recommendations with decision makers. Even after the franchisee was determined, city officials and Cablevision kept in close contact with each other to ensure that the promised service would be provided and improved, if appropriate and possible.

In Detroit, final applications were submitted in December 1982. Compared with Boston, Detroit had a longer review process, taking until 1985 to reach the conclusion that Barden Cablevision would provide the service to the residents of the city. Although the discussions during this period dealt with practically every aspect of developing the future cable system, a good deal of attention was paid on concerns by the city and candidates about African American involvement and participation. Both parties made efforts to evaluate the feasibility of public involvement as expressed in the applications received, willingness of the companies to realize their promises, and the appropriateness and sufficiency of the proposed facilities and equipment, as well as other resources. The commission also emphasized the importance of minimizing the subscription fee for the basic service, ensuring the financial burden would not be too great, thus making service more accessible to all.

Boston's Period of Application Review

Despite the large number of preliminary applications submitted in late 1980, only Warner Amex and Cablevision completed the official application process, which was due on April 23, 1981. During the twenty months prior to December 1982 when the final franchise agreement was signed between Boston and Cablevision, various evaluations of the submitted applications took place. In some cases, discussions took place internally. In others, citizens, applicants, and city officials sat together to discuss the issues, as was the case with the public hearings. At times, reviews dealt with applicants' technical specifications. For example, approximately one month after applications were submitted, John E. Ward from MIT wrote his review, focusing primarily on the technical system features of each proposal. He dealt little with access or the public use of cable television.[1]

In June 1981, Richard Borten, the city's cable television coordinator, made a site visit to examine the services that Cablevision and Warner Amex provided in their existing markets. On June 9, Borten and Dan Jones visited Woodbury, New York to examine Cablevision's cable service there. Charles Dolan, president; John Tatta, vice president; Sheila Mahony, and other senior staff met them. Unlike Warner, whose service area Borten and Jones would visit about a week later, Cablevision kept and shared its complaint logs. Borten expressed his satisfaction with how Cablevision responded to some of the complaints that the company had received. Both Borten and Jones were impressed with the facility and commitment by the staff. Jones for example wrote, "I was impressed by the motivation of the staff and in general their excitement about the particular roles they were playing in the development of the company."[2]

On June 17, Borten and Jones visited Columbus, Ohio to study Warner's operation. John Clark, visitor coordinator; Larry Wangberg, division vice president; John Petrie, program manager; Peter Meade; Alan Austin; and David Davidson joined them during the tour. In his visit report, Borten explained that Warner representatives underlined the large number of locally produced programs in the Columbus system. Jones, however, realized that not many programming efforts had been directed toward series programming. Although there were series shows like *Columbus Alive*, Jones was concerned about the lack of serialized programming through which continuous community involvement would be assured. During his visit, Borten also met Jerry Hammond, a city councilman. Hammond was an African American community leader who also served as vice president of the Columbus Area Electric Utility. Hammond explained to Borten that there had been very limited involvement by the African American community in cable programming. He also felt that Warner Amex had not made substantial efforts to reach out to the African American community.[3]

Public Hearings in Early Summer 1981

One of the highlight moments in the decision-making process in 1981 and 1982 came two months after the city received applications from two companies. Four hearings took place in June 1981. The first hearing was held on June 12 and the second on June 16 with Richard Borten, cable television coordinator, chairing the meeting. Charles Beard; Micho Spring, deputy mayor of the city; and Harold Carroll, the city's counsel from Fole, Hoag, & Eliot, joined the meeting as panel members. Bob Albee, Carmen Bono, Dan Jones, Marylou Batt, Margie Cohen, Peter Epstein, and Jay Cowles also represented the city. Four employees from Warner Amex attended the hearing, including Richard Aureolio, vice president of franchising. Eight Cablevision employees attended, including Charles Dolan, president of Cablevision Systems Boston Corporation. The purpose of the hearing was to clarify some of the points made in the final application. Many of the questions asked pertained to financial and technical issues.[4]

The third and fourth hearings focused more on issues concerning subscribers and local origination. Unlike the first two hearings, the latter two were open to public. The first hearing took place on June 23 at the Boston Public Library. Chaired by Richard Borten, Micho Spring; Charles Beard; Harold Carroll, corporation council for the city; and Joseph Casazza of the Public Works Commission and chairman of the Public Improvement Commission, lead the meeting. The procedure of the hearing had been determined approximately two weeks prior, on June 12 and 16 at another set of hearings where the two applicants offered clarifications on their amended applications. During these hearings, Cablevision and Warner Amex agreed that Cablevision would make its presentation and answer questions first at the June 23 meeting. Warner Amex had its turn two days later. Borten explained that the purpose of the hearings was "to arrive at the clearest possible understanding of the applicants' proposals for building cable systems in Boston."[5]

The hearings consisted of presentations, questions, and answers. Cablevision first made a thirty-minute uninterrupted presentation. An hour-long question-and-answer session followed. Only city representatives were allowed to ask questions at this time. After a fifteen-minute recess, Warner Amex gave its thirty-minute presentation, followed by a question-and-answer session. Then each applicant was allowed to make an additional fifteen-minute presentation to supplement information provided earlier. Applicants were not allowed to suggest any additional plans beyond those that had already been presented in the proposal.[6]

Sheila Mahony, vice president and director of Cablevision Systems Boston Corporation, was in charge of Cablevision's presentation. Nine of her colleagues also were present. They were Charles Dolan, president of Cablevision Systems Boston Corporation and the founder of Home Box Office; John Tatta, chief operating officer; Bill Bell, chief financial officer; Bill Quinn, director of engineering who would play a key role in building the system in three and a half years; Ned Lamont, franchise coordinator; Donna Garafano, Boston product director; Pat Felese, producer and director; Charles Ferris, former chairman of the FCC and communication counsel to Cablevision Systems Development Company; and David O. Wicks Jr., managing director of an investment bank active in the industry.[7]

Mahony's presentation included two issues from Cablevision's proposal that would be especially beneficial to African American and other minority communities in Boston. She emphasized that one of the benefits of Cablevision's plan consisted of lowering the access barriers to communication technology. In other words, regardless of citizens' buying power, they would be able to take advantage of cable technology. Setting its basic rate at $2 for fifty-two channels and offering channels for Boston's diverse population including CNN, BET, Nickelodeon, and others, Cablevision promised to make a price breakthrough. Mahony said that "Cablevision has shown that no one need ever again go without full participation in the information age because of its cost."[8]

Mahony also explained that Cablevision's commitment to Boston "is a commitment to a public access system [and] a local origination program that will reflect the rich diversity of this city fitting together its neighborhoods but allowing each its own identity." For example, Cablevision was committed to making a capital investment of $500,000 in the production center and budget $1 million annually for production services. Such financial assistance was expected to help Bostonians have modern production facilities, including three news-gathering vans and a fifty-person staff dedicated to local production. By doing so, Cablevision envisioned that it would "prove to the rest of the country that cable television can actually be an original medium of local expressions in Boston [and will] reveal a city of cultural riches and vitality."[9]

Once Mahony and Wicks made their brief presentations, Charles Beard opened the question-and-answer session. He explained that both Cablevision and Warner Amex were national companies that had formed partnerships to obtain the Boston franchise. As a result, he was concerned that many decisions would be made outside of the city, which could mean that some vital decisions would not reflect the interests of regular Bostonians. Beard reiterated the city's desire that "residents and institutions here in Boston play a meaningful role in the policy-making decisions of the cable operator in the governance of the system here in Boston, a system which we believe will have a profound impact on the lives of the residents of the city." Hence, he requested that Cablevision representatives outline "how the City of Boston residents and institutions will play a role in the decision making for the proposed Cablevision system in Boston."[10] The very fact that the question-and-answer session began with this type of question investigating the potential for local involvement in the decision-making process strongly reflected the city's interest in ensuring that cable television was leveraged beneficially for the city.

Charles Dolan's answer to the question was twofold. He first assured that the chief operating officer would remain in the community of Boston and would be responsible for all operational decisions. Cablevision's headquarters in New York would only provide policy-related direction. More importantly, Dolan explained, the nature of the system would leave no choice for Cablevision but to respond to the needs of the local community. Cablevision planned to make "a considerable capital investment in the city and . . . mount a very heavy operating cost." In order for the investment to pay off for the company, the system must develop well. In order for it to progress, the support of the community was indispensable. Dolan continued, "in order to have the support of the community, it is really very necessary for the system at all times to remain sensitive to what the community wants, not only in entertainment programming but in all aspects of its operation."[11]

Although Dolan said that the company would welcome the support and participation of an advisory board, he also showed some reservation regarding the balance between Cablevision's autonomy and the advisory board's involvement. He explained:

> [The advisory board's] input will be very helpful, but we don't think that we should come to you depending on advisory boards for the management of the system that we're not proposing here. We think we should be able to do that ourselves. You should be able to rely on us to do whatever is necessary to make good on the commitments that are offered in the application.[12]

This answer raised a flag for Micho Spring. She explained that the city wished to have two representatives of the Access Corporation serve on the Board of Operators. She was concerned whether Cablevision's corporate structure and philosophy allowed the city's wish to be reflected and materialized in Cablevision's decisions and actions. To this question, Mahony answered that although the proposal did not intend for the management of Cablevision of Boston to rely on an external resource, Cable Systems Boston Corporation would include two representatives from the city.[13]

Ned Lamont spoke about financial and facility contribution that Cablevision planned to make to the city. The company offered a $300,000 access studio that would be under the exclusive control of the Access Corporation. The proposed package included three neighborhood studios, mobile vans, and electronic news-gathering vans. Bill Bell added that of fifty-two channels in the basic service, many were access channels. He promised that "there is an active real programming on those access channels and for that reason, we have looked forward to contributing to those access channels in any way that we can."[14]

Cablevision's idea of increased local involvement was reflected in its fee scale. Dolan repeated what had been explained in the proposal. Establishing an appealing and affordable basic system for the resident was the only way to obtain support from the local community. Cablevision projected an offering of fifty-two channels that included "just about every desirable service" for $2 a month. If the benefit of the system was properly explained to each household by trained staff, the subscription rate would include at least 75 percent of the city's households.[15]

Dolan also talked about the future of cable television. His response to Spring's question regarding what might happen to the industry during the next ten years resonated with what Mel King felt about having media content created by and for individual audiences. Dolan expected that there were be:

> Programs that are addressed specifically to the interest of individuals so that cable will become in the home, will make television in the home much like a library function for the individual today, that you go to television to pursue your own interest to find something that you want, and that you can identify with and that is valuable to you rather than being a passive presence in front of the television set.[16]

This comment reflected King's vision of a new communication technology through which individuals could produce and consume media images that truly matter to them, rather than those offered by other people.

After the fifteen-minute recess, it was Warner's turn to make its presentation. From the city of Boston, Dan Jones in charge of the access issue, Peter Epstein in charge of legal issue, Jay Gold on construction, Bob Albien from the Public Works Department, John Ward who was the technical consultant from MIT, and Chris Brown from Foley, Hoag, & Eliot, attended the second half of the session, along with Borten. Warner had nine representatives: Richard Aurelio, senior vice president; John Fletcher, vice president of the New England system; Peter Meade, assistant vice president of the New England system; Eileen Connell, director of the access programming; Richard Berman, senior vice president and general counsel; John Dowling, vice president of financial analysis; David Davidson, vice president of engineering design; Bruce Byarkman, New England design engineer; and Gus Hauser, chief executive officer.[17]

Hauser started his presentation by emphasizing that Warner Amex offered extensive experience in the New England area, including Massachusetts. He said "We're not the largest cable TV operator in the state of Massachusetts. We have been operating in the Boston area since 1974 and in fact some of our systems in Massachusetts go back to the beginning of the industry." After serving in nine neighborhoods in the Boston area for years, Warner had identified Boston as "a very hospitable environment for advanced services." Hauser stated that:

> with its renowned cultural and its academic and its medical and its business resources, there's a music center here for new communications ideas, for new projections, new programs that can be expanded to the rest of the United States. In effect, [it's] a

very good place for us as Columbus was one to innovate and to apply cable, the most advances services.[18]

Whereas Cablevision's presentations raised concern as to whether the company had sufficient presence in Boston as a New York-based company, Warner outright stated that it had a long history in Boston and its surrounding areas.[19]

Aurelio highlighted some of the major differences between Cablevision and Warner's applications. First, he explained that Cablevision's financial plan was "shaky." Indeed, although Warner had secured financial sources before the submission of the application, Cablevision only had promises from investors to support the company if the franchise agreement was signed between the city and Cablevision. For example, David O. Wicks Jr., managing director of Warburg Parisbas Becker, a leading investment banking firm serving the cable and communications industry, wrote to Mayor White on June 23, 1981 advising him of the firm's interest in working with Cablevision. The letter only stated that "should Cablevision Systems Boston Corporation be awarded the franchise for the City of Boston, we would be delighted to be chosen to arrange the financing." Wicks, one of the participants at the hearing, did not specify the kind of assistance that could be made available for Cablevision. Additionally, Aurelio was critical of Cablevision's fifty-two channel for $2 plan. He called it the "difference . . . between the real world and an imaginary world."[20] He also said that the difference between two applicants existed in the difference between "reality versus fantasy."[21] He speculated that 55 percent of Boston's households, not around 75 percent as Cablevision expected, would subscribe to cable services. Aurelio also mentioned that in terms of access channels, Warner had more focus on the elderly who constituted about 30 percent of the population in the city. He then concluded his presentation stating "in the past two years, Warner Amex has won 13 such awards [for excellence in programming], the highest number of any cable company in the nation. Our competition in Boston has won none."[22]

The first question for Warner Amex was identical to the one posed to Cablevision. Beard asked how "residents and the institutions of Boston [would play] a role in the decision making for the Boston system."[23] Despite Hauser's emphasis on the company's long history in the New England area, the fact was that Warner Amex was a joint venture between two New York-based corporations. After reiterating Hauser's claim, Beaman also added that there were at least four specific ways in which local decision making could be achieved. He explained:

The first is that . . . the local community groups could assist us, very possibly with job training and job improvement and that's significantly so within the minority communities. Second . . . they could be of assistance and would enjoy the role of helping us to identify and recruit minority women's business organizations, to participate with us in constructing our system and in supplying us goods and services. The third [is] that they will have a meaningful role in helping us complete our construction commitment to particularly within minorities and women. Forth, we have identified a role as helping us develop our customer relations procedures and then finally, it's more generic, but it's equally applicable, we think that such a community organization would help us play as community leaders across the board.[24]

These examples suggested that minority group members, especially women and their businesses would be able to play a vital role not only in production and consumption, but also in the construction, procurement, and even decision making of the cable system in Boston.

Toward the end of the question-and answer-session, Hauser elaborated his views on public access. He stated, "the concept of public access was that people should come and view what they please with this news video medium. So, we welcome the access, we support it. . . . I assure you that the people engaged in public access are free to do the maximum extent of whatever they please on those channels."[25] This statement reflected not only Warner's view on access channels, but also that of Cablevision and the city. As all parties agreed, leaving programming on access channels up to residents included the risk that inappropriate shows containing obscene and violent materials might be aired. Consequently, questions about control and censorship, as well as independent production by residents, became imminent. Hauser explained that there was no consensual definition for censorship, obscene materials, and other key ideas. The discussion at the hearing, however, was geared toward ensuring that residents' right to produce and consume would not be compromised in favor of unnecessary legal concerns.[26]

Aurelio claimed that Warner Amex showed stronger support for access than Cablevision. Warner had more stable initial funding and more access studios. It promised to offer free service to the elderly. Within two months of signing the franchise agreement, Warner promised to wire East Boston, which the company had identified as most needy of the cable system. It also planned to establish "an aggressive, resident minority program, employment policy and procurement policy."[27] These were the reasons Warner used to position itself as superior to its competitor, Cablevision.

The second hearing took place two days later, on June 25. Borten once again served as the presiding officer for the meeting at the Rabb lecture hall at the Boston Public Library. As was the case with the first hearing, Micho Spring, Joseph Casazza, Charles Beard, and Harold Carroll attended the meeting representing the city. Unlike the first hearing, the second hearing allowed the public to ask questions. Those who had questions had been asked to be registered on a first-come, first-served basis thirty minutes prior to the hearing. Each resident was limited to three minutes for their question or comment. At the beginning of the hearing, Borten showed the public a stack of thirty cards that contained the names of the first thirty people who could ask questions. Whereas the first hearing gave city officials the power to clarify uncertainties in the applicants' proposals and residents were only able to listen to the dialogue, at the second hearing, residents were able to play a more vital role.[28]

The first to make a comment about the applications was Thomas J. Butler, South Bostonian president of the South Boston Citizens Association. The association was 103 years old, with approximately 15,000 members. Incidentally, South Boston and Dorchester both had a large African American population. Butler shared his own and association members' excitement about the future arrival of cable television. But he added that their concern was about public access. Butler hoped to underline the importance community members felt regarding access to cable television not merely as audience but also as producers. From this perspective, he stated that "we are very excited about the fact that one of the companies, Warner-Amex, has decided to locate a public access studio in the confines of the South Boston and Dorchester area. . . . [W]ith a public access studio in the community, we will therefore be guaranteed some type of neighborhood participation."[29]

The second person to speak was Dawson V. Johnson, representing the Cable Television Access Coalition. He raised a few concerns about Cablevision's commitment to local ownership and control of production facilities. He asked how Cablevision planned to balance its control of the cable system and ownership of the Access and Programming Corporation. The coalition cast doubts that Cablevision's plan would be sufficient to meet the needs of users in the city in terms of the use of three neighborhood origination studios. Cablevision stated in its application that the corporation would only have down time use of the facilities.[30]

Responding for Cablevision, Garofano explained that the company would control the main local origination studio while the Access Corporation would control the main access studio. She added that the coalition should not be disappointed by the fact that Cablevision offered fewer studios than Warner Amex did, because "putting lots of studios in without any indication of what their use would be would result in excessive costs that would only reflect in the consumers rate." On this matter, Meade spoke for Warner Amex: "Warner Amex not only would be willing to, but it is our proposal" to offer more studios. Warner Amex seemed to offer a proposal that was closer to what was desired by the city and its residents.[31]

Nancy Perkins clarified that Boston's residents who wish to produce programming on their nonprofit cultural activities would not have to pay production costs.[32] Tim Clegg from the Boston Community Cable Association said that he considered Warner to have made more commitment in the "full range of objective that BCAA [Boston Community Cable Association] has proposed." From local employment to maximization of community participation, Clegg explained that Warner's proposal seemed more likely to fulfill the requirement made by the city.[33]

Polly Robinowitz from Cultural Educational addressed issues about local art and culture. In order to make sure that the Access and Programming Corporation would not be the only entity fully invested in public access, she inquired how much support each candidate would be willing to "give to local museums, performing groups or arts and science centers for the production of programs for specific local audiences." She additionally asked the applicants to provide specific examples of such support from other municipalities in which they had been operating. Connell, speaking for Warner, said the company would work jointly with cultural and educational institutions rather than working separately. Meade added that Warner planned to give a $250,000 grant to the Repertory Theater. In Columbus, Warner had raised money for the Boys Club, the Girls Club, the YMCA, and other organizations. Warner foresaw similar efforts being made for Boston's cultural and educational institutions. Cablevision's Garofano also referred to the importance of maintaining a close relationship with local organizations. She tried to be more specific about art and culture than Connell and Meade. Garofano said that because Cablevision planned to offer more channels than Warner, there was sure to be more channel space for arts and culture.[34]

Robert McCausland, a founding member of the Cable Television Access Coalition, asked when the access service would become available.

He wondered if it would be available as soon as the first set of houses were wired or if residents had to wait. Neither Warner Amex nor Cablevision had a clear answer. Connell answered negatively. Because the first houses to be wired in its plan were in East Boston, technically speaking Warner would not be able to offer the service on day 1. Cablevision also did not provide McCausland with a specific timeline while explaining that it would be available some time in the first year, although if required by the city, Mahony said Cablevision would be able to offer it on the first day of its service.[35]

WGBH's Carol Obertublesing reminded both the city and the applicants of the importance of local interests. She stated,

> above all, the City and the operator should remain responsive to local interests, aware of national developments in cable and other technologies, and alert to the pitfalls in viewing cable TV as a cure-all for the City's and society's problems. While cable TV can overcome the isolation of some citizens, it can create isolation for others. Only by being aware of the danger as well as the potential of this new medium can we control it. The technology is here, the challenge is to make the medium work for us.[36]

Additionally, she explained that WGBH would air a TV show on cable television on July 15, at 9:30 p.m. The channel also supported community workshops related to cable television. To this statement, Mahony agreed with Obertublesing, claiming more forums would be beneficial.[37]

There were many others who stated their ideas, posed their questions, and shared their enthusiasm at the meeting. From the hearing-impaired community, Patti Wilson, Jim Sullivan, and Lois McDonald asked questions to make sure that they would be able to receive the services they needed and they would be able to enjoy cable television programming just like other residents of Boston could.[38] Cynthia Patton from the Gay Community News and Wade Nichols from Community Media Group expressed their desires to be fully represented in cable programming and not just on a token basis. Patton argued that there were 75,000 "gay people" in Boston. She said, "we want more than token representation on other people's shows." Nichols made sure that this population would not be excluded from cable employment as well.[39] This resonated with the demands of ethnic and racial minority groups. Jorge Hernandez introduced Latino community-based programs that he wished to put on air. He also sought to make sure that the city's Latino population would have access to the service, because "if we don't have access or control of management of our facility, we're going to be second class citizens."[40] From the Jamaica Plain Neighborhood Development Corporation where many African Americans lived, Tomas Riviera spoke about the importance of local ownership and fee structure, to make sure that the service would remain accessible to all residents of Boston.[41]

Analyzing the Final Applications

By the end of June 1981, the city of Boston received the document titled *Analysis of Cable Television License Proposals* from Malarkey, Taylor & Associates, Inc. The city had requested the firm provide an analysis of the final applications submitted by Cablevision and Warner Amex. Consisting of three sections, the document submitted on June 30 compared the two applications in detail. The firm explained that both applicants submitted highly sophisticated applications especially because of the high standard set by the city's RFP. The report was to "[constitute] only one of several inputs into the City's decision-making process." The firm also avoided making direct comparisons between the two applicants. Instead, the analysis reviewed some of the latest industry standards, status quos, foreseeable future development of the technology, and so forth, to provide a foothold that city officials could use to further evaluate the applications. In other words, the outcome of hearings, internal discussions, and other studies would affect the city's final decision of which candidate would win the franchise.

The analysis was diverse in content. It examined technical issues such as specifications, infrastructure, economic and financial feasibility, and other aspects, while it also examined issues that potential subscribers were more likely to be interested in such as the tier structure, rates, public access and community programming, and other service concerns. For the latter category of issues, Malarkey, Taylor & Associates, Inc. made both positive and negative comments on the submitted applications. This part of the applications revealed most about the impact of cable television on the lives of African American and minority Bostonians.

The analysis made positive comments about Cablevision's rate and service tier structure. Being able to subscribe to a fifty-two channel service for only $2 a month made Cablevision's plan "a very attractive package which should receive a reasonable acceptance level in Boston." Although Warner Amex was critical of the plan at the first hearing on June 23, the firm believed that Tier 1 service captivated many Bostonians. The document stated that "for only $2.00/month, subscribers would have access to a variety of locally oriented and national programming available in most state-of-the-art cable systems." Although Cablevision's Tier 1 service was attractive, the evaluation recognized Cablevision's strategy for higher tiers and pay services as too optimistic. Subscribers might not be able to afford some of the pay services, at least to the extent that Cablevision foresaw. It suggested that one of the two things might happen. The first is that the service achieved a high penetration with a low total monthly price. Or, it might attain a modest penetration with a high total monthly price. Cablevision attempted to achieve a high penetration with a high total monthly price. This plan, according to Malarkey, Taylor & Associates, Inc. was "questionable."[42]

Regarding community channels, Malarkey, Taylor & Associates, Inc. correctly explained that instead of designating an access channel to a particular interest group, Cablevision left it to the Access Corporation to "ensure that the access programming available in Boston will be relevant to the broad cross section of Bostonian's tastes and interests." Consequently, the evaluation stated that Cablevision's plan as to allocation of channel spaces remained unclear in the proposal.[43] Although comments made at the hearings suggested that Cablevision envisioned its system would carry diverse programming to meet the needs of city residents, the evaluation was correct in that the activation period and allocation information was lacking in the application.

Uncertainty existed with community programming. Although Cablevision's promise to make a capital investment of $500,000 in the Boston Production Center and annually purchase $1 million worth of programming by the center was "commendable," it was short of detail on how the company would determine the "competitive value of locally produced fare." Additionally, access coordinators' roles remained unclear.[44]

On special services, the review recognized that Cablevision plans to have services for the elderly, women, hearing impaired, and other underserved and underrepresented populations. The candidate also planned to have two ways of catering to minority groups. The first was a separate channel in foreign languages and for those who spoke English as their second language. This mainly targeted the linguistic minority population. The second was more cultural. Cablevision serviced separate minority channels that offered BET, Spanish News Network, and others. Additionally, local origination programming, with a special emphasis on African American programming, would be provided on this channel. Equipment that Cablevision proposed for local origination impressed Malarkey, Taylor & Associates, Inc.[45]

In the third section, the review characterized Warner Amex as a large cable company that was formed in 1980 and was constructing and operating 143 cable systems with 760,000 subscribers in twenty-seven states. This was in sharp contrast to Cablevision, which only had two operating systems at the time its application was being reviewed. At the hearings, Cablevision argued that due to its small business size, the Boston system mattered more to Cablevision than to Warner Amex. For Warner Amex, the Boston project would only be one of many. Its success in Boston heavily affected the entire corporation.[46]

Warner Amex's emphasis on senior citizens was "a commendable idea" that distinguished the candidate from its competition. Warner proposed an eleven-channel service for the elderly without monthly fees, after the $24.95 installation fee was paid. The evaluation listed seven community communications channels for public access, health, neighborhood, religion, adult learning, and other foci. The concern that Malarkey, Taylor & Associates, Inc. expressed was the fact that much of the service was dependent on community communications channels. It stated:

> the quality and the quantity of programming that will actually be available on this tier is largely dependent on the continuing commitment of local institutions and organizations. Whereas the community programming opportunities are impressive, the entertainment needs of seniors are not meaningfully addressed in this service package.[47]

The evaluation therefore required more specific examples and strategies to truly meet the needs of the elderly.

The review positively evaluated Warner Amex's Tier 1 service just as it did that of Cablevision's. Although Warner's plan had only thirty-six channels and its monthly subscription fee was $3 more, the evaluation explained that the service "provides all local off-air broadcast stations as well as a panoply of channels appealing to a broad cross section of interest groups." Twelve more channels were available to the access corporation on a priority basis. Seven more channels dealt with information services such as a community bulletin board, City Service, and others. Summarizing this plan, the review stated that "the overall makeup of Warner Amex' Tier I service is clearly community oriented." However, the document also pointed out that "it is noteworthy that the community communications channels designated specifically for women and health professions . . . are not included in this tier."[48]

The review document also identified uncertainty about Warner Amex's communications channels. In order to facilitate municipal use of the system, there was going to be a fully equipped studio in City Hall. Ten portable modulators were available. The problem the application evaluation pointed out was that there was no clear explanation of the type of equipment that would be available for this purpose. Similarly, despite the company's pledge to commit $500,000 and technical staff and assistance for the first three years of the franchise to assist program development, it was unclear from the application if such resources would be exclusively dedicated to local origination.[49]

Warner Amex promised more financial commitment. It stated that it would invest up to $3.5 million to build and equip the local origination center to be located in the Theater District. The annual budget for local origination was going to be in excess of $1 million. The company planned for twenty-one hours of locally produced material and fifty hours of external programming on air per week. Warner Amex's QUBE option also characterized its local origination effort. The system allowed subscribers to use the interactive nature of cable television. There were fourteen channels for QUBE, at $3.95 a month. Such channels dealt with special events, self-improvement and self-help programs, continuing education courses, foreign-language programming, and ethnic movies. This meant that all local origination channels were available on the basic tier. BET and other minority channels were available under Tier 1. Additional "multicultural features" such as foreign-language programming were available with subscription to the QUBE service.[50]

On July 15, 1981, Sheila Mahony sent a letter to Richard A. Borten, Boston's CATV coordinator. While she expressed her general satisfaction about Malarkey, Taylor's analysis especially in its affirmation of the viability of Cablevision's construction plan of three and a half years and the feasibility in its financing, she found the evaluation of the Omnibus service was inappropriately critical of the proposal. The analysis indeed claimed that Cablevision expected its customers to be able to afford more than Malarkey, Taylor believed. Mahony, however, offered her perspective, without answering to the affordability of the service, that the purpose of the Omnibus tier was to offer flexibility to its subscribers.[51]

To reemphasize the financial soundness of Cablevision's plan, Charles F. Dolan sent a letter to the city. Although the analysis issued at the end of June confirmed that the company's financial strategy was sound, Warner Amex had attacked Cablevision's unstable finance sourcing plan extensively during hearings. The letter stated:

> Cablevision has a $90 million package in place for the construction of the Boston cable system. Unlike Warner Amex, which

has a single source of funds to construct at least four major urban systems (Dallas, Cincinnati, Pittsburgh & Houston), Cablevision's financing package is for the City of Boston exclusively. . . . Cablevision has never failed to build a franchise for lack of financial backing. Not only do we have the financial capability to build Boston, we have the financial and management resources to build the cable communications system we believe will lead the way into the 21st century.[52]

Warner Amex also sent a letter to Borten regarding the Malarkey, Taylor report. Richard Aurelio wrote in the cover letter that while reviewing the draft, they "have found some errors and some misunderstandings" of their proposal. Much of their response was related to financial and technical specifications. Warner's comments and corrections, however, also reiterated its emphasis on the elderly subscriber service, which differentiated it very much from Cablevision's proposal. The response said:

> our ascertainment convinced us that the elderly were a significant population to serve, and that the predominant interest among the elderly were local issues having to do with their neighborhoods, city services, consumer issues, health, religion, etc. The Elders Service Tier was designed in response to those expressed interests.[53]

It also emphasized that the city would be able to spend $500,000 for the development of the City Channel as it saw fit. Also referring to leased access policies, the correction underscored Warner's willingness to allow the city to reserve the right of decision making.[54]

The city announced its cable franchise decision in summer 1981. On August 12, Mayor White awarded the franchise to Cablevision. He delivered *Issuing Authority Final Decision and Statement of Reasons* in which he explained:

> As the Cable Television License Issuing Authority for the City of Boston, after careful analysis and evaluation of the final amended applications filed by Cablevision Systems Boston Corporation and Warner-Amex Cable Communications Company, of the record of the public meetings and hearings conducted in June 1981 as a part of the cable licensing process, and of other materials which are part of the public record of the licensing procedure,

I hereby award the Provisional Cable License to Cablevision Systems Boston Corporation.[55]

Mayor White stated that once hearings were conducted, the final state of the review process included his "personal and detailed analysis" of the final applications. He reviewed information included in the analyses by Cable Television Office staff, the Public Improvement Commission's cable committee and the outside consultants' comments, videotapes of the public hearings, applicants' videotaped presentations from the public hearings, and reports of site visits.[56]

Choosing Cablevision Over Warner Amex

As reasons to choose Cablevision rather than Warner Amex, Mayor White wrote that "Cablevision has demonstrated such imagination and innovation in its entire approach to the Boston franchise. And it has shown, more clearly perhaps than anyone in the cable industry, a creativity, flexibility and cooperativeness that is essential to the partnership we seek."[57] *The Boston Business Journal* explained that Cablevision's offer was "the most far-reaching cable television system the country had ever seen." The article continued to praise it by saying that "Boston's system will not be equaled in this decade—possible even this century."[58] Mayor White in particular listed five areas in which Cablevision's application excelled:

1. Quality of service at lowest possible price.
2. Unprecedented channel capacity.
3. Commitment to nonprofit corporation's independence, flexibility, and financial stability.
4. Opportunity for Bostonians to share in the system's success.
5. Innovative leadership.

Providing fifty-two channels for the $2 monthly subscription fee put Cablevision ahead of Warner Amex. Cablevision also guaranteed that the subscription fee would remain the same for the first five years of the franchise. Additionally, special interest channels, sports programming, twenty-four-hour news programs, and other wide variations were available through Cablevision's plan. Not only did Cablevision offer 5 percent of its annual gross revenues to the access corporation, it promised independence from the city and from the operator while supplying adequate resources such as facilities and funds. Bostonians had the chance to benefit from the success of cable by purchasing bonds yielding at least 16 percent of interest, if they chose.[59]

On the other hand, Mayor White listed three reasons why Warner's application did not fully meet the city's requirement: basic service level, local investment opportunities, and philosophy of control. In Warner's application, the basic package offered thirty-four channels at $5.95 a month. White explained that "both the content of the package and the price are fair but unremarkable; it is doubtful that the package will attract the high level of acceptance that the City feels is important." Local investment opportunities were limited in Warner's scope. The city hoped to see as many residents as possible benefit from the cable system. White stated that "the private offering plan proposed by Warner-Amex, although involving local groups, does not offer genuine economic participation in the system to a significant number of individual residents." In general, Mayor White added that "Warner-Amex's commitment in [channel capacity, funding, and independence of the Access Corporation] appeared weaker and less responsive than that of Cablevision." He did acknowledge, however, that Warner's tendency to retain control over the system was reflective of the operator's "concern for the integrity of its system, and its strong sense of responsibility for the management of its channels, this philosophy is a more restrictive one than what Boston considers ideal."[60]

Cable Access News from November 1981 focused on the power of the public over access governance. As had been the case for the previous years, Mayor White was not attentive to the structure and governance of the nonprofit Access Corporation. By November, he had failed to decide who exactly would serve on the board, how long their terms would be, how future board members would be selected, and what exactly the corporation would do. Borten made a comment that "community groups still have a chance to influence [Mayor White's] decision." Although Borten explained that he would continue to address his concerns to the mayor, "the most effective way to get the message across would be for those community groups with political influence to contact the Mayor directly." He suggested to "draw on the existing recognition that your individual groups have."[61]

Although Borten expected the corporation to be established by the end of November 1981, it still had not been accomplished by January 1982.[62] In December, Mayor White announced that he had sent a letter of invitation to four people to be a part of the fifteen-member board: Kenneth G.

Ryder, president of Northeastern University; Daniel J. Finn, vice president of Boston University; John G. McElwee; John Hancock, president and chief operating officer; and Martin Kessel, co-chairman of Boston's Cable Television Access Coalition. Finn was expected to be the first president of the corporation. Eleven other members' names had not been disclosed. The terms remained undetermined as well. With the Provisional License with Cablevision slated to come up within a few months, the city had many uncertainties with its Access Corporation.[63]

On March 24, 1982, the corporation was finally formed, under the name of the Boston Community Access and Programming Foundation. Upon launching, there were only three trustees, previously called board members, listed: Daniel J. Finn, Martin Kessel, and Bill Chin. Finn, rather than Mayor White, would appoint the other fifteen trustees. Additionally, Deputy Mayor Spring had announced in February that there would be a fifty-member Board of Overseers. It was a response to the fear that the corporation might not sufficiently reflect public opinion. Finn also appointed those fifty overseers. Unlike trustees who continued to be appointed by the president of the corporation, future overseers were to be elected by their constituencies. In this respect, although the access foundation did not reflect the voice of Boston residents as much as it could have, lobbying efforts were successful to some extent.[64]

The day after the Boston Community Access and Programming Foundation was launched, the city issued *Provisional Cable Television License: Granted to Cablevision Systems Boston Corporation*. This more than one-hundred–page document consisting of sixteen sections, dealt with issues concerning construction and installation, maintenance, service and programming, rates, financing, subscriber and user rights, employment training, and others. The agreement between Cablevision and the city read:

> This Provisional License agreement entered into this 25th day of March, 1982, by and between Cablevision Systems Boston Corporation, a corporation organized under G. L. c. 156B, and Kevin H. White, Mayor of the City of Boston and Issuing Authority for the Award of cable television licenses under G. L. c. 166A.[65]

In the section titled "Grant of License," the document explains, in a highly legal term, that:

> The Mayor of the City of Boston, as the ISSUING AUHORITY of the City hereby grants a Provisional and non-exclusive license

to Cablevision Systems Boston Corporation, a Massachusetts corporation established for such purpose, AUTHORIZING and permitting Licensee to proceed to qualify for a Final License which shall authorize and permit Licensee, acting pursuant to the rules and regulations of the Public Improvement Commission to construct, install, operate and maintain a Cable Television System in, under, over, along, across or upon the public ways and places within the City of Boston for the purpose of reception, transmission, collection, amplification, origination, distribution or redistribution of audio, video or other signals and for the development of broadband telecommunication services in accordance with the laws of the United States of America and the Commonwealth of Massachusetts.[66]

This Provisional License was the beginning of a new phase in Boston's cable history. Prior to the issuing of this Provisional License, the city's Issuing Authority solicited preliminary applications on August 13, 1980 and more concrete final applications on February 7, 1981.[67] Although the Final License did not come out until December 1982, the Provisional License signaled that the soliciting and discussion period were near their end. Once all the proof and detailed documents required in the Provisional License were submitted, the Final License was provided to Cablevision.

Although much of the Provisional License focused on technical, financial, and structural details of the system, it also specified the types of services provided to the citizens of Boston. During the first five years of the contract, the basic service carried fifty-two channels. If any of the programming was removed from the list of basic service channels, the Provisional License required the licensee to "use its best efforts to assure that substantially the same mix of programming shall be retained as those that existed in Licensee's Application."[68] In general, the services offered to the subscribers must reflect the licensee's "endeavor to provide a wide range and assortment of programming services serving a variety of needs and interests."[69] The license also guaranteed five access channels for the residents. As for local origination programming, the documents stipulated that by the end of the first year after the activation of the system, sixty hours of local origination programming was offered weekly. By the end of year 5, the number was slated to go up to eighty hours. By the end of year 10, the city could expect one-hundred hours of local origination programming per week.[70]

Section 13 of the Provisional License dealt with employment training and procurement, two fundamental ways to ascertain local community involvement in the project. Not only did the license require the licensee to adhere to regulations regarding equal opportunity and affirmative action, it explained that at least 25 percent of the employees had to be minorities, 10 percent women, and 50 percent Boston residents.[71] The licensee also was required to provide its employees with a vocational training program. It developed a joint curriculum with the Occupational Resource Center, Boston Public Schools, or other institutions when possible.[72] Additionally, at least 10 percent by dollar volume of the materials, services, or equipment purchased for the construction, operation, or maintenance of the system must come from qualified minority-owned Boston businesses. When prices quoted by a local business and a non-local business were equal, the licensee was required to procure the service or product from the local business. Through the minority hiring policy, job training, and local procurement, the Provisional License promised to utilize Boston's local resources.

Although March 25, 1982 marked a significant moment in the development of cable television in Boston because of the signing of the Provisional License between the city and Cablevision, the day also commemorated the culmination of the eighteen-month planning for the Boston Community Access and Programming Foundation. Right before the mayor and Mr. Dolan signed the Provisional License, the foundation and Cablevision signed the contract to officially form the community organization. Daniel J. Finn, president of the foundation, promised an "absolute guarantee that every voice in the city, at any level, at any time, can speak to their fellow citizens." Despite the power play between the mayor's office and the foundation, Finn, not the mayor, was going to announce some of the trustee members whose names had not been finalized as of March 25. The foundation was built not only to "ensure public access to the cable system" but also "to provide an independent entity with the resources to develop unique and innovative programming designed specifically to meet the needs and interests of Boston residents."[73] The purpose of the foundation was fivefold. It was "to ensure access to channels and facilities for all Boston residents, groups and institutions and to provide public education and training regarding the use of access facilities and channels"; "to provide a mechanism through which the City's institutions and organizations can effectively share their educational, health care and cultural resources with the community"; "to foster and generate experimental uses of cable's unique communications capabilities"; "to protect the interests of Boston residents in receiving a wide variety and diversity of quality programming which addresses their special needs and unique potential through the production, acquisition and distribution of such programming"; and "to operate channels or programming space on any cable television system operating in the

City of Boston." The establishment of the foundation was a way to make sure Boston residents' interests would be protected against the corporation's potential pursuit of its business profitability.[74]

The formation of the foundation, however, did not go without questions. The Cable Coalition claimed that the foundation had a lot of work to do to meet the challenges it faced. The coalition said, "it requires the Foundation's becoming partners with the community by providing the outreach, training, and guidance that will truly foster quality community television." More accountability in the foundation's decision making and public involvement also were going to be necessary. The Committee for Community Access was more skeptical, questioning the legitimacy of the foundation itself.[75] Even in May, the foundation only had a temporary office space in the Boston Public Library without office staff and with only three trustees. *Cable Access News* further introduced a curious error that was made in the Articles of Organization. Although the contract signed with Cablevision had the correct name, the Articles omitted the word "Community." With the foundation's work handled by the mayor's Cable Office, the omission ironically reflected the very slow and struggling start that the foundation was experiencing.[76]

Two months after the Provisional License was signed, the City began to discuss its plan for the public institutional network (PIN) in detail. Although Boston's individual residents would benefit from Cablevision's emphasis on narrowcasting and local origination features, the city's various institutions would benefit from PIN. The city's introductory document explained the cable system with PIN as a structure of:

> four cable, 208 channel, two-way system [that] will provide more than one hundred channels of home entertainment and informational programming as well as unprecedented one hundred channel network for meeting the bi-directional communications needs of Boston's business, medical, cultural, educational and other world renowned institutions. All of Boston's public institutions will have the opportunity to participate in this new public telecommunications system through both the use of the Subscriber Network as well as the innovative Public Institutional Network.[77]

Although Cablevision's plan would offer state-of-the-art services to subscribing individuals, it also offered to provide a separate institutional cable application for public, municipal and nonprofit institutions.[78]

Institutions that benefited from PIN ranged from academic facilities such as colleges and universities, conservatories, libraries, and research institutions, to medical institutions and cultural organizations such as the Museum of Fine Arts and the Boston Symphony Orchestra, as well as community orchestras, repertory theater groups, dance companies, and other community-based groups. These organizations and groups represented the diverse interests of Boston's ethnic, racial, and occupational communities among others and allowed for institutional participation through cable television.[79]

Although the foundation was going to play the central role in local programming efforts, Cablevision also was willing to undertake initiatives. Patte Falese, Cablevision's franchise coordinator for programming, repeatedly explained that there were twelve local origination channels, a Boston City Channel, and a variety of special interest channels covering topics from local sports, to public affairs and Boston arts. She also anticipated that the local origination staff would be placed within a few to several months prior to the Final License approval. The staff members worked closely with community groups especially in production training for the community.[80]

The updated list of ethnic interest groups submitted to the mayor's Office of Cable Communications on November 4, 1982, included organizations that would receive notifications of employment positions available as the cable system expanded in the city. As was the case with Asian, Hispanic, and Latino communities, numerous African American groups and organizations appeared on the list. Newspapers that would carry advertisement and notification included *The Bay State Banner*, *The South End News*, and other media that had high African American readership rates. Special interest groups that received notification included the Boston NAACP, Roxbury Community College, and other corporations and groups in South Boston, Jamaica Plain, and other neighborhoods that had a high African American demographic concentration.[81]

Discussion With Cablevision

The Office of Cable Communications and Cablevision continuously kept contact with each other to avoid potential confusions and misunderstandings. One of the clarifications regarded employee training and Cablevision's affirmative action policy. The addendum to the Provisional License read:

> Cablevision has already made contact with many agencies, representing women and minorities, in order to establish a liaison which will help assure their constituencies to receive information

about job availability. Since we intend to make every effort to surpass the employment requirement to which Cablevision and the City of Boston agreed, position availability will be published in the daily newspapers, neighborhood-oriented weeklies, as well as sending the same notice to community and special interest groups.[82]

Cablevision planned to hire several hundred people when the construction of the system was complete. In order to allow Boston residents to "qualify for entry into the cable television industry," the corporation was willing to fund and conduct vocational training programs. Its budget for this purpose was between $200,000 and $300,000. Training programs focused on installation, customer service, sales, converter repair, as well as more technical and advanced skills needed for promotion within the company.[83]

The training program began with an installation training session including twenty-five Boston residents. The target of the training would be women, minorities, economically disadvantaged residents, and unemployed residents. The training was repeated through the end of the first quarter of 1984. On-the-job training was offered from the second to fourth quarters of 1984. Just as installation would be the first step necessary for residents to gain access to cable services, it would be the first step for them to gain access to the industry.[84] Other training programs for sales representatives, customer service representatives, and others also started in 1983 and continued in 1984. Some were longer than others and had more trainees, depending on the nature of the job responsibility.[85] Cablevision summarized in its entry-level training program agenda that it only trained the number of residents for whom positions were available. In other words, after successful completion of the training, the trainee was guaranteed to work for Cablevision and guaranteed to become of a member of the cable system industry. Furthermore, Cablevision expressed its goal to "provide necessary skills which will allow otherwise untrained Boston residents to become cable professionals."[86]

On December 7, 1982, Cablevision provided the city with an updated construction schedule. Its original projection was to finish the construction within three and a half years. In the new projection, the construction would be completed within two years.[87] Similarly, the information on subscription rates for residential customers in the Provisional License needed correction. In order to correct the misunderstanding, Laurence M. Bloom wrote a letter to Richard Borten with accurate data. According to the corrected pay schedule, customers paid $2 per month for Universal Basic Service. The installation fee was $25 and could be paid out over five months. The charge for moving or reconnecting an outlet was $10. If a customer wished to have Home Box Office service, the subscription fee was $7. The $7 basic security system, $25 video-text service, and $6 omnibus service also would be available, along with other additional services.[88]

Another clarification letter from Laurence Bloom on December 7 touched on a few issues relevant to Boston's minority communities. Bloom explained that most of Boston's narrowcast channels would feature programming for special interest groups. The combination of satellite-distributed programming and locally originated programming would jointly meet diverse interests and needs in Boston. Cablevision also had started its effort to make sure that there was sufficient local production. Bloom wrote:

> by establishing contacts with many of the community organizations and institutions that will provide raw material for production, we have already begun the necessary groundwork for local programming in Boston. In fact, Cablevision took this first step in the development of local programming two years ago during the franchising process and has maintained contacts with Boston groups and institutions ever since. Recently we began a more aggressive campaign to reacquaint ourselves with organizations we have already met, and introduce ourselves to others. . . . We have also subscribed to virtually every neighborhood newspaper in Boston in order to keep abreast of each area's particular concerns and activities. It is this kind of contact with groups, institutions, and opinion makers that forms the base of any local programming operation.[89]

Bloom saw that Cablevision's more than two-year–long effort to be involved in local communities in Boston would be paid off once the amicable relationship brought both quality and quantity of locally produced programs.[90] Late in November, for example, Kathleen O'Keeffe, access coordinator of Cablevision, sent a letter to the East Boston Social Center. In the letter she provided a list of institutions with which Cablevision had been in close contact. O'Keeffe requested that the list be updated so that the company would know what residents of the area wanted and the residents could learn what was happening in the city through Cablevision's system in the future.[91]

Various production efforts would take place as soon as the contract was signed. Starting in the first quarter of the franchise, subscribers would have access to local video production. Similarly, neighborhood studios would be built in a timely manner. Roxbury and West Roxbury, where many African Americans, lived each would have a studio by the end of 1983 and 1984 respectively.[92]

Granting the License to Cablevision

Only a few days before the Final License was signed between the city and Cablevision, John Tatta, president of Cablevision, and Charles F. Dolan, general manager, signed the *Assignment of Provisional Cable Television License*. The document from December 10, 1982, read that Cablevision:

> accepts the assignment [described in the Provisional License]; agrees to be bound by, and to perform each and every provision of the Provisional License; [and] agrees to indemnify and hold harmless Cablevision Systems Boston Corporation from any loss, liability, damage of expense arising out of any failure or alleged failure on the part of Cablevision of Boston to perform all obligations under the Provisional License.[93]

On December 15, 1982, Mayor White granted the Final Cable Television License to Cablevision of Boston. Its content remained very similar to that of the Provisional License. While technical, financial, and mechanical issues accounted for the great majority of the content of the license as was the case with the Provisional License, many of the issues described in the document issued in mid-December in relation to local origination, community access, and involvement of African American and other ethnic and racial groups as specified in the Final License had unprecedented significance since the development of the system was going to be based on the agreement signed on that day. The Final License was the testament that the city of Boston was going to be wired by Cablevision.

The Final License listed several key dates and meetings that formed the outline of the agreement. It helped summarize concisely how the discussion on the development of the cable system had taken place in Boston. The Issuing Authority's solicitation for application came out on August 13, 1980. It then announced more detailed goals and specifications for the construction, installation, operation, and maintenance of the system on February 7 of the following year. Responding to the call, Cablevision Systems Boston Corporation submitted its amended and final application to the city on April 23, 1981. Two clarification meetings took place between the city and

Cablevision, on June 12 and 16. Public hearings took place at the end of the month, on June 23 and 25.[94] In the summer, the city decided to award its franchise license to Cablevision. In early 1982, the Provisional License was signed, based on which the Final License was drafted.

Sections 2 and 3 of the Final License help explain the general outline of the project. Boston's commitment to establish a successful cable system in the city is clearly visible in the terms and conditions of the license. On the one hand, the term of the Final License was for fifteen years. By the end of the sixth year after the signing of the license, Cablevision was required to complete its system construction. The license reflected the city's serious investment in its efforts with Cablevision. On the other hand, the grant was a nonexclusive one. In other words, the Final License did not prevent the city from offerings rights to use the system for any purpose. As seen fit by the city, the cable system was going to be used for the advantage of the residents of Boston.[95]

The Final License specified some of the programming and service content. Through its Universal Basic Service, Cablevision would have at least one channel for each of the following demographic groups: senior citizens; children; non-English speakers; and ethnic, racial, and cultural minorities.[96] As the many years of discussions between city officials, residents, and cable companies had emphasized, a minimum of five access channels would be offered as a part of the Universal Basic Service. As for local origination programming, sixty hours would be offered weekly. By the end of year 5, the number would go up to eighty hours and to one hundred by the end of year 10. To help develop local programming, at least one local programming coordinator would work at each neighborhood studio, which would be located throughout the city.

Cablevision's employment policy reflected the ideals of equal opportunity and affirmative action. During the construction of the cable system, at least 25 percent of the employees would be minorities, 10 percent women, and 50 percent Boston residents. To guarantee that the policy was followed appropriately, Cablevision and its subcontractors were required to submit the Issuing Authority information regarding the name of all people who applied for jobs and who were minorities, residents, and women. When a new hire did not belong to at least one of the aforementioned category or were hired for fewer hours, the licensee and its subcontractors had to explain why they were not hired.[97] To guarantee that various minority residents including African Americans, Latinos, and other ethnicity and race-based underrepresented groups along with others, the non-discrimination policy of the license stated:

> Licensee shall not discriminate against any person in its solicitation, service or access activities on the basis of race, color, creed, religion, ancestry, national origin, geographical locations within the City, sex, affectional preference, disability, age, marital status, or status with regard to public assistance. Licensee shall be subject to all other requirements of federal, state or existing local laws, regulations and all executive and administrative orders relating to nondiscrimination through the term of this Final License.[98]

There had been too much and too long of a history of American misrepresentation and deprived representation of numerous groups of people. Women more than men, Christian churches over Jewish temples, whites more than Blacks, and people without disabilities more than those with have been omnipresent both behind and in front of television cameras. This policy was one of the ways for Cablevision to reverse such a trend by the use of local community-based medium.

The period of application review ended with the signing of this Final License. The license reflected what Bostonians discussed not only after the issuing of the *Authority Report*, but also before the introduction of a cable system was originally announced in 1980. It was finally official that Cablevision would wire the city.

Detroit's Period of Application Review

Detroit's application review period began early in December 1982 when all the applicants aspiring to wire Detroit submitted their final applications. Unlike in Boston where this period consisted simply of a series of review meetings and discussions with limited hardships especially with the establishment of the Access Foundation, Detroit experienced a longer, slower, and much more controversial review period.

The letter from Mayor Young to the City Council stated the plan for the following several months. By early December, Barden Cablevision, City Communications, and DIUB had submitted their proposals. The Cable Commission was then in charge of reviewing the proposal documents and was slated to release its draft report on March 15, 1983. Each applicant had approximately two weeks to respond to the preliminary report to provide clarifications to the commission. On April 16, the commission made a recommendation to the mayor as to which candidate should receive the franchise contract. On April 30, the mayor forwarded his report and recommendation to the City Council. The month of May was reserved for review and public hearings on whose basis the council would issue the approval of the franchise ordinance.[99]

Tim Kiska wrote about the magnitude of the system by examining the amount of investment Barden made during the application process. He explained that Barden Cablevision had spent $350,000 on an eight-volume report that weighed fifty pounds. Wiring the city could cost Barden between $150 and $200 million. With 450,000 homes in the area to be wired—120,000 more than the largest system in the nation at the time, in San Diego, Kiska speculated that the system could become the largest single cable television franchise in the country.[100]

While the commission reviewed the applications, endorsement letters continued to arrive in city offices. Joseph E. Madison from the NAACP's Voter Education Department, for example, sent a letter of endorsement to Mayor Young, and a copy to more than ten city officers, including Erma Henderson, Barbara-Rose Collins, Maryan Mahaffey, and Don Barden. The letter stated, "[d]uring my on-site visit of Barden facilities, I was impressed with the professionalism of the staff and the first-rate equipment." Although the letter did not describe what exactly caught the NAACP's attention, it was a testament to the fact that race-based organizations saw the introduction of an appropriate cable system as beneficial for racial communities.[101]

While Detroit city officials reviewed applications submitted by three corporations, Maryan Mahaffey listed several questions she wished to ask each applicant. Her notes are extremely honest. She even put, "DIUB—a real loser! Vote is between Barden and CCI [City Communications]." In addition, DIUB had "bad news." Not only was it a non-Detroit–based company, its "franchise in St. Claire Shore [had] been poorly managed and residents [had] filed lawsuits already." Despite such a negative review for DIUB, Mahaffey's note revealed that in general candidates had good proposals in terms of minority involvement. She wrote, "I am not concerned with compliance for minorities/women, etc. All do well there." But some questions were more relevant for the topic of local involvement than others. For example, a few questions for Barden dealt with issues concerning local origination and Detroit Public Access Corporation. Mahaffey wrote, "[i]t appears that studios and equipment designated for local origination and public access are to be shared? Is this correct? If so, don't you see a problem with competition for use by profit making groups and non-profit groups?" Regarding the Access Corporation, Mahaffey tries to truly understand the purpose of it by writing, "[w]hat is the Detroit Public Access Corporation [and] its purpose? . . . What is their business arrangement with Barden?"

She elaborated on her own question by stating that she wanted to find out if the corporation was formed for programming purposes and what commitment Barden was willing to make. Similarly, Mahaffey questioned City Communications on its plan for the Detroit Urban Communication Institute was. From the proposal, it was not clear who would be involved and what its purpose was going to be. These questions suggested that as important as financial and technical questions were, Mahaffey was as much, if not more, concerned about the entity that was going to ascertain community involvement in the cable project.[102]

After the proposals were submitted in early December 1982, the Detroit Cable Communications Commission closely reviewed and evaluated each proposal and asked each applicant for clarifications through the first half of 1983, with the cooperation of Cable Television Information Center Associates (CTIC), a consulting firm to local governments on cable television franchising. On March 9, 1983, the commission released its Preliminary Report entitled, *Evaluation of Proposals for the Provision of Cable Communications Services in the City of Detroit, Michigan*. Through multifaceted analyses such as financial status, technical features, and more detailed examination, the report attempted to systematically evaluate which one of the three would be the best franchisee.

As an outcome of the reviewing process by the Detroit Cable Communications Commission, it issued its preliminary report that assessed the proposals submitted by each candidate. It covered issues from finance, technical capabilities, access channel, minority hiring policies, and almost any topic that would influence the life of cable subscribing Detroiters. Albeit not the final report, this preliminary report was to summarize what each applicant was planning to offer to the city, how the city viewed service content, and identify some of the unclear issues regarding the proposals.

One section of the analysis was dedicated to community services. Under this category, the report examined three firms' schemes on local origination and public access. Arguing that local origination was "one of the principal ways a cable system may increase program diversity," the commission evaluated each company's proposal in terms of expense budgets, local origination staffing, locally produced programming hours, and equipment, facilities, and capital available for local origination programs. The report summarized that all applicants proposed five local origination channels. Barden had two job opportunity channels and another local origination channel in the first tier of service. For the second and third tiers, Barden planned to include multicultural children's programming and a Jewish television network. City Communications offered a lifestyle channel, local news channel, and other local origination channels. Similarly, DIUB planned to include two local origination channels in the first tier, and an institutional network guide and access clips in the second and third.[103]

In terms of equipment and facilities, Barden's proposal included a 3,000-square-foot studio with nine color cameras, fifteen videotape recorders, and other equipment. The total equipment cost was estimated to be $975,000, in addition to a van worth $300,000, an electronic news-gathering van worth $100,000, and microwave equipment worth $30,000. City Communications proposed two 1,200-square-foot local origination studios with two color cameras, three videotape recorders, edit control, lights and audio equipment. Each studio cost was estimated to be $100,000. The cost for City Communications' van and onboard equipment was $250,000. As for DIUB, it planned to have a 7,500-square-foot studio with a $73,507 master control, $67,720 in light equipment, a van, and other equipment. The report concluded, in relation to these equipment and facilities, Barden's plan, with the total cost of $1.4 million, was "very well equipped" and almost double the cost of the next applicant, DIUB with its estimated expense at $751,864. It even argued, "it may be too much." On the other hand, the report claimed that DIUB's plan is more "reasonable" both in terms of cost and equipment. City Communications, to the contrary of Barden, did not seem to have sufficient equipment or facilities.[104]

As for public access and channel allocation, DIUB was slightly ahead of Barden and City Communications. Barden, however, was the only applicant firm "to address a policy for triggering access channels." This is why the report did not show a clear preference for one over the others.[105] As for educational and municipal access facilities and equipment, access staff and budget, the report did not state any preference. Although it argued that Barden seemed to provide the best equipped facilities not only for local origination but also for other access facilities, the commission hesitated to make any statement in the preliminary report.[106]

The report argued that all three companies had positive policies regarding employment practices, a key factor in minority labor force representation. They all included plans for the continued education of employees, minority workforce targets beyond the level required by the RFP, as well as scholarship and internship programs. Barden and City Communications had a formal corporate policy regarding equal employment and affirmative action. DIUB stated, in its proposal, that it was committed to equal opportunity and affirmative action. All three corporations proposed recruitment through minority and women's organizations to facilitate the hiring process. They also planned to recruit through local academic institutions with high minority and female concentrations. Barden also proposed to work with the Michigan Employment Security Commission and use radio, television commercials, and freeway signs to publicize employment opportunities.[107] George J. Dunmore, in his article, explained as well that Barden planned to provide careers for minorities through his cable television project.[108]

Although these analyses of three proposals described in the preliminary report showed that the firms expressed their commitment to fully use cable television's empowering capabilities in the city, the Cable Commission was not the only entity interested in and closely examining the proposals. Many national and local organizations sent recommendation and endorsement letters to Mayor Young, the City Council, and the Cable Communications Commission to show their support for their firm of choice. The aforementioned letter from the NAACP for example reflects such interests. Similarly, Clarence Welcome, a former colleague of Maryanne Mahaffey, sent a personal note to Mahaffey saying, "My request from you is that you vote for *Barden Cable TV*."[109]

The preliminary evaluation, however, raised some concerns at City Communications. Although the report was simply a tentative evaluation document, City Communications claimed that "press reports and self-serving pronouncements of another applicant have given undue weight" to the conclusions drawn in the report. Robert Green, president of the company, expressed great concern that the report included numerous errors and omissions that resulted in "a preconceived and inaccurate opinion" of the consultant.[110]

After the release of the preliminary report, the Cable Communications Commission required more time than originally planned to reach its final decision and to release its final report. The review process became an intense one. At a meeting in early April, all three corporations made a presentation in front of the city officers. Barden brought a fifteen-minute–long film featuring his plan for Detroit. City Communications brought Coretta Scott King, Jayne Kennedy, and other celebrities. DIUB made an elaborate presentation on its extensive experience in the industry.[111] As a result, the commission had to revisit its franchising schedule on April 28, extending the release day of the final report to the week of May 9, 1983.[112] About two weeks later, the revised schedule was, once again, amended, extending the report release day to the end of May 1983.[113]

While the review was under way, Barden sent a letter to the Detroit Board of Education to underscore what the firm could bring to the city should it be awarded the franchise. In his letter, Lawrence Baskerville, vice president of new market development explained that as the first African

American-owned cable corporation, it would "greatly enhance cultural, educational and entertainment opportunities for all Detroiter's and to revitalize communication between city residents, their neighbors and the government." It planned to offer scholarship endowments totaling $1.37 million, at the Cass Institute of Technology, Wayne County Community College, and Wayne State University. One-hundred students from Cass would receive $250 to acquire basic training opportunities. Ten from Wayne Community College would receive $1,400 a year as they pursued their associate degrees. Six students from Wayne State University would receive $4,000 as they acquired their bachelor's degrees in science. Barden also proposed to invest $1.5 million for vocational training, $1.5 million for scholarship and internship programs, and other educational purposes.[114]

While Barden tried to appeal to Detroit's Board of Education, City Communications also made a presentation before the same entity on the same day. The company promised to provide the residents with facility, equipment, and financial resources to make the new technology most useful for educational purposes. City Communications wrote that $2.25 million would be committed to training for all levels of employment. There also would be scholarship and internship opportunities. The firm said that $26 million would be dedicated to access and local program origination. Additionally, $25,000 would be available as grants to produce such programs. It also promised to offer preschool activity programs, enrichment courses, bilingual education, homework alert services, televised study hall, university-level courses, and other educational opportunities via its cable system.[115]

DIUB also made a presentation in front of the Detroit Board of Education. For local educational institutions, it made nine promises. Some of these promises included connecting all schools and administrative facilities via cable in Detroit; establishing a fully equipped studio production and editing facility worth $170,000; providing the city with a $77,000 production van; reserving four educational access channels; and others. DIUB also made financial commitments. The firm planned to offer more than $2 million in cash grants. There also would be $1.45 million in scholarship funds for students in Detroit to attend vocational and educational institutions. DIUB planned to make $800,000 available to develop programs to air. Along with other financial contributions that DIUB was willing to make in Detroit, it attempted to emphasize its willingness to serve the local academic and educational institutions.[116]

Because of the rigor of the selection process, the actual release of the final report did not take place until June 1983. After the release of the preliminary report in March 1983, the commission conducted a two-session, seven-hour public hearing in April. It was an opportunity for the three applicants to make presentations not only to the commission but also to the public and to be engaged in direct dialogue with the people the winning franchisee was going to serve. More than five-hundred people attended the session.[117] Although the official evaluation did not come out until June, local newspapers issued their own evaluations. The Michigan Chronicle, for example, had a positive opinion of Barden's plan to offer career opportunities to minorities. George J. Dunmore explained that under Barden's plan roughly about five-hundred new permanent jobs were expected to become available. This would help alleviate the unemployment rate, which was higher than 20 percent.[118]

Issuing the Final Report

The Final Report, entitled *Evaluation of Proposals for the Provision of Cable Communications Services in the City of Detroit, Michigan: FINAL REPORT, June, 1983*, came out on June 17, 1983. At the time of publication, the City Council included nine members: Erma Henderson, Nicholas Hood, Clyde Cleveland, Barbara-Rose Collins, David Eberhard, Jack Kelley, Maryann Mahaffey, John Peoples, and Mel Ravitz. Henderson continued to serve as president. The Detroit Cable Communications Commission had seven members: James Robinson, Alta Harrison, Otto Hetzel, Joseph Jordan, Jean Leatherman-Massey, Lester Morgan, and Myrna Webb. The commission appointed Michael E. Turner as executive director on March 28, 1983. Kathryn A. Bryant was to serve as deputy director.[119]

James Robinson, chair of the Detroit Cable Communications Commission, underscored three points of the report in his cover letter. First, the document marked the final phase of the commission's evaluation process. It started with its release of the RFP in August 1982, reviewing of proposals submitted by three candidates, and release of the preliminary evaluation report in March 1983. In April, the commission held a public hearing. Second, the document reflected the assistance of the CTIC. Robinson explained that throughout the review process the commission and CTIC were in close contact with each other to accurately evaluate submitted proposals. Third, the evaluation presented the consultant's analysis of each application. The outcome of the extensive evaluation process that lasted approximately six months was the final evaluation document.[120]

The purpose of the final report was to analyze the responses from applicants regarding questions posed in the preliminary report. Although the preliminary report issued in March stated clearly some of the advantages that Barden's application had over others, there were several uncertainties that needed clarification by each applicant. The final report discussed financial integrity, engineering and technical capabilities, services and rates, and other pertinent issues. The body of the report started with the summary of proposals reprinted from the preliminary report with a few revisions.[121]

Because each applicant had amended details in its applications, it was important to understand the fundamental service plans of each candidate at the moment of final review. The final report summarized that 57 percent ownership of Barden Cablevision of Detroit was with Barden Communications, of which Don H. Barden was the sole owner. Twenty-three percent of the ownership came from six local residents and the Detroit Public Access Corporation. The six residents were Esther Edwards, Yvonne Murray, Abraham Cherry, Karl Gregory, Raymond Leporati, and Wade Briggs. The access corporation was owned by four individuals: Donald Hunly, Nanci Rowe, Dennis Silber, and Harold Varner. Twenty percent was from minority investors. Barden's system proposed to have capacity for 124 channels with a separate institutional network. Tier 1 had thirty-six basic channels for the monthly fee of $3.95. Tier 2 had sixty-one channels for $6.95. For $10.95 a month, Tier 3 offered eighty-eight basic channels, as well as ten pay options, three pay-per-view channels, and interactive services. A forty-eight month-long process of construction would wire the city. Barden proposed to have a local origination and access facilities, five neighborhood access studios, viewing centers, and production vans. Their plan also included five local origination channels and twenty-seven access channels.[122]

City Communications, Inc. found 80 percent of its ownership in the hands of five residents of Detroit, one resident of Hartford, Connecticut, and one resident of Columbus, Ohio. Minority investors owned the rest of the corporation. City Communications planned to have a 120-channel network with a separate institutional system. The construction was slated to last forty-eight months. City Communications' plan had two tiers. The first would include fifty-six basic channels for $3.95 per month. Nineteen additional services would be available for prices ranging from 25 cents to $2. Tier 2 would cost $8.95 per month offering seventy-five channels. City Communications' plan offered two local origination studios, a production van, four access studios, and five local origination and twenty-two access channels.[123]

DIUB was a partnership between six New York corporations, a New York resident, and a Detroit-based corporation. Twenty percent of its ownership came from minority investors. DIUB's plan was to offer a 112-channel system with a separate institutional network. The construction period was sixty months. DIUB planned to offer three tiers in its service plan. The first tier cost $2.95 per month and would offer twenty-five channels. Tier 2, provided fifty-nine basic channels and optional pay-per-view channels for an extra $4 per month. Tier 3 cost $11.95 and has seventy-nine channels. It planned to offer a local origination studio, a van, and five local origination channels. Thirty-six access channels also were part of its proposal.[124]

In many ways, Barden's superiority to the other applicants was apparent in its service plan. The evaluation ranked Barden and City Communications over DIUB for broadcast signal carriage. As for automated programming, Barden was ranked highest, with City Communications and DIUB as second and third, respectively. Similarly, Barden received the best evaluation for satellite service, pay-per-view, and FM and audio services.[125] Under community services, the category of great consequences for African American residents of Detroit, Barden's application and responses received the highest review. The evaluation explained, "Barden makes the strongest local origination proposal based largely on the facilities and budget proposed." Although City Communications' offer included two studios, the evaluation concluded by listing Barden's proposal as the most attractive, followed by City Communications and DIUB.[126] In access services, Barden also led the race. The evaluation explained:

> The applicants' proposals for access all met the City's RFP requirements. Overall, Barden and DIUB made proposals which were more complete than CCI's. DIUB proposed more access channels, and larger staff than did the other applicants. Barden's equipment and facilities proposal for public access is seen as more complete than the other applicants, offering a fixed main studio and five neighborhood studios. All propose educational municipal facilities, and Barden and DIUB also propose equipment for the library and cultural consortium, respectively; Barden's proposal includes staff. In conclusion, Barden's proposal is the most extensive, followed by DIUB. CCI offers less, particularly in staff, budget commitment, and public access facilities.[127]

The final evaluation ranked the three applicants in thirteen categories. It was only in two categories, pay cable programming and tier structures, that Barden was not rated the best. From engineering to local origination and access, as well as enhanced services, Barden's application outperformed that of City Communications and DIUB.[128] Assuming the same organizational format of the preliminary report but with added analyses from the applicants' answers to the commission's questions and clarifications requested by the commission, the final report concluded that Barden had submitted the best application. Although the commission showed some concerns about the financial feasibility of Barden's plan, it made a strong recommendation for Barden, particularly, in the area of services.[129] The report argued that Barden showed "the greatest commitment" in community programming. The firm also "showed the most strength" in supporting institutional use of cable. Its proposal for enhanced services demonstrated a sign of commitment, according to the final report. While the commission recognized that "CCI [City Communications] was given a strong second ranking in a variety of sections in the [service] chapter," Barden outperformed in every aspect of their service proposal.[130]

In his official memorandum addressed to Coleman Young, James Robinson, the chairperson of the Detroit Cable Communications Commission, explained extensively why the commission had chosen Barden over other applicants. He stated that after various types of study sessions, it became clear that Barden offered "the superior cable communications system to the institutions, residents and businesses of the city of Detroit." Along with its technical and financial superiority, Barden also "demonstrated the strongest commitment in support of public, municipal and educational access in Detroit." The commission believed that Barden's proposal offered "important benefits" to Detroit in minority business, investment, and employment opportunities, commitment to the Public Benefit Corporation, training opportunities for local residents, and a serious commitment to charge reasonable rates for its services.[131] The commission was very much impressed with Barden's financial plan, technical capacity, and experience in the cable television industry. Barden's commitment to public services both in the final report and the commission's report to the mayor reinforced cable television's perceived power to be a community's self-esteem raising vehicle. Concurring with the memorandum, the recommendation resolution claimed:

> that Detroit Cable Communications Commission, and its meeting of June 13, 1983, for the reasons set forth in the . . . memorandum and based upon the proposals submitted, by the applicants, as reviewed by Cable Television Information Center Associates in its Final Report to the Detroit Cable Communications Commission, recommends to the Honorable Coleman A Young, Mayor of the City of Detroit, that the cable franchise be awarded to Barden Cablevision.[132]

One month later, on July 12, 1983, Mayor Young officially expressed his agreement with the commission to award Barden the Detroit cable communication franchise. As reasons for his decision, he reiterated Barden's promising proposal emphasizing its minority employment, commitment to training and education for Detroit residents, and reasonable fee scheme. In his letter, Mayor Young urged the City Council to complete its review process within thirty days so that the city could move forward with the implementation of the system.[133] City Communications and DIUB offered a plan to participate in a joint venture with Barden after the announcement of Barden's franchise. Gilbert Maddox from DIUB also explained that he was pursuing the possibility of having the city split the contract among three firms.[134] This offer made by these two corporations and Don Barden's refusal to conduct a joint venture, eventually caused a major problem that would further slow down the process of franchise agreement signing.

After Mayor Young showed his support for the commission's decision, there were two more important steps left. First, the City Council scheduled a public hearing on the Cable Franchise Ordinance on Monday, July 25, 1983. Although the commission and the mayor agreed that Barden should be awarded the franchise, a public hearing was necessary to discuss the ordinance. On July 27, 1983, based on the outcome of the hearing, the franchise ordinance was going to be voted on. Josephine A. Powell, director of the City Council, explained that the schedule was especially tight because the council was on a recess on August 3. A letter addressed to the council seemed to anticipate a delayed introduction of cable to the city.[135]

On July 15, 1983, Erma Henderson, president of the City Council, requested a special meeting on the ordinance. She felt that substantive amendments were needed in the ordinance. Most of the six amendments that Henderson requested were related to finance. She also mentioned that she wished to fulfill her responsibility of creating "new jobs for Detroit residents." The special meeting was set for July 18. According to Henderson's plan, there were four steps that needed to be taken by the end of the month. The meeting took place at 10 a.m. On July 25, at 10:30 a.m., a public hearing took place. Discussions continued the following day between 1 and 3 p.m. The final consideration remained to take place on July 27.[136]

On July 20, Barden provided Michael Turner with an additional document to clarify a few questions. The purpose of the Detroit Public Access Corporation was made clear in its pages. Barden explained that the name of the corporation was first filed on April 21, 1981 by Nansi I. Rowe, and later incorporated on December 22, 1982. There were four purposes of the corporation.

1. "Operate and act as a consultant and agent to business associates, corporations, or other organizations, whether private or public necessary for carrying out the purposes of this corporation."
2. "Provide communications, management, and resources."
3. "Invest in communications oriented properties such as, but not exclusively, cable television."
4. "Engage in all other activities permissible under the Business Corporations Act of Michigan."[137]

Politics Delay Media Democracy

A major obstacle arose on July 25, 1983, a date between when the final decision was made and December 1985 when the agreement was signed. The Detroit City Council received a letter from City Communications regarding its concerns surrounding the evaluation process. The company was clearly dissatisfied with the process and considered it to be unfair and faulty. It argued in its letter:

> The Cable Communications Commission (Commission) has recommended to the Mayor, and the Mayor has recommended to you, a grossly unfair modification of the Request for Proposal (RFP) and proposed Franchise Ordinance, and the award of the franchise to Barden Cablevision of Detroit, Inc. pursuant to such improper modification. City Communications, Inc. paid the City of Detroit $10,000 at the time that we submitted out proposal, and have expended about $440,000 to prepare and submit our proposal, all in reliance on the declarations and promises of the City and in the RFP that the bidding process would be conducted fairly and impartially.[138]

After listing several reasons why the process was unfair, the letter implied that City Communications would sue the city. The letter continued:

> We feel that such conduct, if followed by the requested action of your Honorable Body, would not only be improper but would be unlawful. We are accordingly advising you of our intent to commence appropriate proceedings in the event that the Council

should permit this impropriety and seriously consider awarding the franchise to Barden.[139]

Mahaffey wrote in the margin of the letter, "they will sue—ramifications important."[140]

City Communications was not the only company to file a lawsuit against the city. DIUB did so also. The two companies argued that the city had delayed the selection process on purpose to allow Barden to raise the capital needed to win the bidding. As a result, a Wayne County Circuit judge signed two separate orders to prevent the City Council from awarding a contract. Although the commission "categorically" denied such claims, the Cable Commission was no longer able to vote to award the franchise to Barden on July 27, as recommended by Mayor Young. Michael Turner concluded his news release by stating, "It is extremely regrettable that these firms, having lost that fair and open competition they voluntarily entered, would resort to reckless allegations and groundless lawsuits in a blatant attempt to subvert what has been an orderly, methodical and fair process."[141]

While the city faced lawsuits, many business and political organizations agreed with Young's decision. Similar to the public endorsements that came in for Barden during the review process, the City Council continued to hear from Barden supporters. For example, the Association of Minority Contractors argued that Barden would create precious construction opportunities that would "perpetuate the growth and experience of Detroit-based minority contractors."[142] Dorothy Brown, advisory board chairperson of the Detroit Repertory Theater, showed her agreement by writing that "[t]he Repertory has been impressed with the fact that, being a minority owned company, Barden Cablevision shows a unique degree of sensitivity to the high percentage of minorities in the metro Detroit area." She continued, "Barden Cablevision will provide the kind of service that will be most valuable to Detroit."[143] The UAW, International Union, United Automobile, Aerospace & Agricultural Implement Workers of America, offered a similar endorsement.[144] These examples suggested the important role that the commitment and attention paid to local minorities in Barden's cable television franchise application. Not only the Cable Commission and mayor, but the NAACP, UAW, and other politically powerful organizations also supported Barden particularly for its sense of community engagement.

The day after Michael Turner issued his news release, Mayor Young also issued his press release. On the voting of the franchise award, he claimed that it was "the final step in what has been an exhaustive, highly professional selection process designed to assure the best possible service for the people of Detroit." He continued that "the Commission, in my opinion, made the correct choice in recommending that the franchise be awarded to Barden, which clearly had the superior proposal." Although he only referred to the lawsuits at the very end of the statement, language of fairness and professionalism dominated his statement.[145]

Mel Ravitz, a council member, was not as optimistic as Young. He wrote his honest reaction to the status quo of the city in his statement. He said:

> Although I continue to hold serious reservations about the financial capability of any of the three potential franchisers to do the immense job of wiring all of Detroit for Cable TV, I have nevertheless voted for Barden Cablevision. I have done so in order not to delay the implementation of Cable TV in Detroit and to give Barden the opportunity both the consultants and the Cable Commission said it needs to achieve full, solid financing.
>
> It is my hope that Barden Cable will succeed in its financing and in subsequently implementing its cable system. However, I believe that Barden cannot be allowed an indefinite time during which to try to achieve its financial. Six months or so should be the maximum and I would hope for a much earlier date. I am not prepared to have the citizens of Detroit wait indefinitely for any franchiser to prove a financial capability it really ought to have demonstrated in its initial proposal. If such capability is not firmly secured in a reasonable time period I shall be among the first to call for termination of the Barden franchiser and prompt selection of a more viable cable TV agent.[146]

This letter reflected not only ongoing concerns about the financial conditions of Barden and the other two firms but also the degree of attention that council members paid to the benefit of city residents.

Another reason to delay the signing of the agreement arose in late 1983 when Barden and its affiliate companies signed agreements. On December 13, 1983, MacLean-Hunter Cable TV, Inc., Barden Cablevision of Detroit, Inc., Barden Communications, Inc., and Don Barden signed a financing agreement. MacLean-Hunter was a Canadian firm located in Toronto. A loan agreement also was signed among MacLean Hunter, Barden Communications, and Don Barden. Additional agreements also were established among them.[147]

Although MacLean Hunter was willing to help Barden finance the

$160 million franchise and was going to suggest an improvement in Barden's procurement of funds, this decision raised questions among both city officials and unsuccessful applicants, City Communications and DIUB. On the one hand, by selling limited partnerships of the franchise to MacLean, Barden would raise approximately $60 million. On the other hand, the arrangement appeared as if "Barden had turned the franchise over to the Canadian firm." First, the Canadian company was originally interested in bidding for Detroit's cable franchise. In March 1982, however, the firm announced that the city's economy was too stagnant and the unemployment rate too high, making its potential business in Detroit too risky. Some referred to this history to question if allowing MacLean Hunter to partake in the project would be fair to other companies.[148]

Others expressed their concerns that the Toronto-based company was white-owned. Both the commission and Mayor Young argued that along with City Communications and DIUB, Barden was an African American–owned company with a strong commitment to living up to its affirmative action policy and providing local minorities with job and training opportunities. As had been expressed since the early 1970s, the cable television system for Detroit was a new tool that could be used to solve some of the social issues and enhance the self-esteem of Detriot's African American population. Additionally, those involved had worked hard to ensure that the cable system would reflect the racial composition of the city. Barden's decision to collaborate with MacLean Hunter worried Detroit residents and officials who thought that the system would no longer achieve its original objectives.[149]

Mel Ravitz, who had shown some concern about Barden in July of the previous year, paid special attention to this structural change at Barden in early 1984. He was concerned that such alterations would affect the cable system that the city would eventually have. He posed eight questions particularly at Barden.

1. Has there been any transfer of ownership or control of the proposed cable TV system? What has been transferred?

2. If there has been a transfer of ownership or control of the system, is this transfer permitted in the ordinance passed by the city? Have the procedures . . . be followed?

3. Does Barden have the financing necessary to carry out the franchise?

4. Why did Barden Cablevision of Detroit choose a firm that

had chosen not to bid for the franchise when other companies that had bid appeared willing to form a cooperative venture?

5. What role did the Cable Commission play in the reported alteration of Barden Cablevision?
6. How will these changes in the franchisee's structure affect the rate that will be charged to the citizens of this city?
7. Has the Cable Commission been meeting since the City Council passed the cable ordinance?
8. Has the City Council's intent been changed from awarding the franchise to a minority firm.[150]

Based on the report issued by the law firm of Honingman, Miller, Schwartz and Cohn, which reviewed the numerous agreements and transactions among parties involved in the structural change, M. D. Farrell-Donaldson, auditor general of the City of Detroit submitted a letter to the City Council outlining three major procedural recommendations. He suggested that Michael Turner should appear in front of the City Council to answer any questions on the report by the law firm, and to submit a follow-up report within sixty days to explain what the commission would do in response to concerns raised in the report. Additionally, the legal staff of the City Council Research Division would review and provide feedback on the submitted report.[151]

Michael Turner, however, denied appearing in front of the City Council. He explained that "because we are still in negotiations, and the Franchise Agreement is not ready for consideration by the Council at this time, I am unable to respond to the subpoena." Although the final evaluation had come out and Mayor Young had expressed his agreement with the commission to award the franchise to Barden approximately one year earlier, the negotiation had not been finalized by early summer 1984.[152]

On June 19, 1984, Farrell-Donaldson wrote that he understood that the council hoped to approve the yet-to-be-signed cable agreement before August 1, 1984. In his letter, he did not clearly say whether he felt it was a feasible plan. Instead, he stated that his staff had not been attending the commission meetings and had not been familiar with the agreement that had been developed. The Office of the Auditor General, therefore, requested at least four weeks to review all the relevant documents.[153]

By summer 1984, it became clear that the agreement was not going to be signed by August 31, as originally scheduled due to the financial difficulties that Barden began to have. It argued that money lenders were hesitant about investing in a $160 million cable system and wanted the project to be downsized. Barden claimed that market conditions had shifted against their favor nationally, regionally, and locally. The news was disclosed in the *Detroit News* on July 31.[154]

On the following day, the commission released a memorandum on the status of discussions between the city and Barden. The document reiterated and was sympathetic to Barden's claim. The cable company had a difficult time meeting the time requirements and finance-related promises made in the ordinance. It explained that the economic factors and the FCC's increased involvement in the business had caused the delay. As a result, Barden began to urge the city to accept a smaller system with 72 channels, instead of 120 as originally planned. This change would cut the cost of the system to $100 million, rather than $160 or $180 million. Michael Turner argued that at this point, the commission was unable to accept or reject Barden's proposals. Instead, it requested Barden submit official documentation assessing the status for further review.[155]

While Barden delayed its process and the city had little to do to improve the situation, frustration clearly built among the City Council members, residents, and commission members. Ravitz wrote in his letter to his colleagues, "[m]y reaction [to the slow cable development] is a mixture of anger, disbelief and despair that Detroiters will never be able to enjoy cable television." The anger was not simply directed at Barden. Ravitz expressed his frustration at the fact that the cable status report dated August 1 was not delivered until after the council's summer recess. He stated that "this is another example of the arrogance with which the Cable Commission has treated the council." To complicate the already intricate situation, Ravitz correctly speculated that if Barden was deemed incapable of implementing a system and a new firm was chosen to wire the city, a new lawsuit might follow. Ravitz suggested a radical change. He stated, "it may be time for the Council to call for new Cable Commissioners, a new director and new consultants who will be able together to develop a workable, quality system."[156]

Far from alleviating any of Ravitz's frustration, Barden submitted the document set requested by the commission on August 21, 1984, while remaining unable to secure enough financing to build the $173 million system. It argued that while the company had "worked vigorously to bring cable television to [Detroit], . . . [the proposal] must undergo modifications in order to make the cable system for Detroit an economically viable entity . . . [because of] [s]ignificant economic and regulatory changes." Calling the decreased revenue projections and increased cost of production a "double whammy," it claimed that it was "impossible to run a profitable cable television company" without adjustments. Although the company continued to show its interest in wiring the city, it was apparent that the city would not be wired soon. Barden wrote:

> We are committed to the City of Detroit and we're proud to hold the franchise for this community. We have started to build out staff, we have done extensive preliminary engineering and other work towards bringing a state-of-the-art cable television system to the residents of Detroit.[157]

Such optimism was tempered by the actual climate of business and economic conditions. As Barden argued in its document, while modifying the plan, the firm attempted to minimize the impact on Detroit's viewers and potential employees. Without wiring the city, such lofty goals would never materialize.

The corporations that did not win the franchise were not content with the fact the original conditions that the city and Barden agreed on were no longer valid. DIUB sent a letter to both Mayor Young and Erma Henderson, president of the City Council, saying that because Barden was no longer required to meet the original terms and conditions, it was going to cause "injury" to DIUB and other bidders. Even Don Barden himself knew that the company or the city could be sued for this modification. On the one hand, the *Detroit Free Press* was understanding of the proposed changes by stating "it boggles the mind to imagine what Detroiters could get on 124 channels that they couldn't find on 78." On the other hand, it foresaw lawsuits if the city decided to adopt the change. As Barden tried to dissipate the criticism and the threat of litigation, frustration with the company steadily accumulated. The *Free Press*'s sympathetic writing was not as the result of its sympathy for Barden, but because it understood that Detroit's residents wanted a cable system. The paper was cognizant of the fact that "the city invites lawsuits, charges ad suspicions that could tie Detroit's cable system in knots for years." The letter by DIUB continued to claim that they were entitled to partake in any discussions and negotiations pertaining to a cable television franchise.[158]

Analyzing the situation, the commission recommended with a 6–2 vote and the mayor approved a plan extending the deadline for decision making from August 31, 1984 to September 30. Although organizations such as the Detroit Association of Black Organizations, Inc., continued its support of Barden, it was a rare case.[159] By then, some council members had already begun to feel that the extension was "a waste of time." Ravitz commented, "[w]e won't be any better off in 30 or 60 days from now than we are now." He continued to offer radical plans, including seeking bids again from any corporation interested in developing a smaller system in the city. He believed that hearings and the council procedures would produce nothing but "rhetoric and loss of time."[160]

Ravitz kept proving himself correct. Even in October, Barden had not secured adequate debt and equity capital to wire the city. Farrell-Donaldson sent a letter to James Robinson requesting documents and information so that his office could start reviewing materials and move the process forward as swiftly as possible. In his letter back to the auditor general, Robinson explained that he had not received any materials or presentations from Barden regarding this bottleneck process, either. Although his language was calm and diplomatic, the letter showed a sense of frustration in the party of Robinson that progress was slow, if at all existent.[161]

By mid-October 1984, there were only two choices left for the City Council: terminate the franchise with Barden and seek new bidders, or allow further modification of its proposal by Barden. Josephine A. Powell and Phillip S. Brown from the Division of Research and Analysis explained to the City Council that regarding the former option, the mayor and the commission must agree with each other. The City Council had nothing to do with determining if the city should terminate the franchise or not. The same could be said about the second option. The report showed that nothing prevented further extension legally speaking. It also identified the risk for more lawsuits by the unsuccessful bidders.[162]

Without much progress made, the year was coming to an end. In late 1984, despite talk of discarding the franchise ordinance with Barden, it became more likely that Barden would retain its franchise. As the city would lose its control over rates and programming if rebidding was to take place. For the city, it was more beneficial to work with Barden to maintain its control over the system.[163] John E. Boddy, deputy auditor general, wrote that "Barden Cablevision has not yet presented evidence of its ability to finance cable TV in Detroit." The auditor general required access to such information as soon as it became available. Barden had not submitted any documents or information. Barden continued to make it impossible for the auditor's office to review any of the materials.[164]

The lack of development or progress continued into 1985. Mel Ravitz sent a letter to his colleagues on January 10, 1985, with a clear message of discontentment. When the council decided to extend the franchise ordinance for thirty days, the expectation was that progress was soon to be made. Although Michael Turner explained that the commission and Barden had been working on the language of the ordinance "to avoid further delay and to facilitate a positive conclusion," day by day, the delaying of the process became apparent. Consequently, Ravitz explained that "no franchise agreement is in place and work on a cable system has not begun." He not only questioned the validity of the ongoing process but also wondered if the commission's negotiations with Barden were illegal.[165]

By mid-February, the tension between the City Council and the Cable Communications Commission had become apparent. The council sent a letter to Michael Turner with an opening statement, "the Detroit City Council is very concerned about the Detroit Cable Commission's refusal to comply with specific mandates" of the ordinance. It had failed to submit franchise agreements and had not shown adequate financing on the part of Barden despite the repeated extensions of the deadline. As a result, the council explained that "[the commission's] continuing negotiations with Barden and evaluation of Barden's proposed modified cable system . . . are without authority." The council requested immediate action and "candid discussion about the status of cable television in Detroit" It also warned the commission that it had been operating outside of its authority.[166]

The city's law department clarified that the commission still had the authority to develop an agreement with Barden. According to the Franchise Ordinance that became effective on August 31, 1983, it remained valid for fifteen years. In other words, in an ultimate case, the commission would be legitimately able to work with Barden until August 31, 1998. Such a legal clarification, however, did not satisfy Detroiters.[167] On April 2, the Division of Research and Analysis submitted to the council a list of cable consultants that might be able to help the cable plan advance more quickly. The three firms and five individuals listed had great experience in technical, financial, and legal aspects of cable television.[168] Carlton T. Stanton, for example, was a Detroit resident whose consulting firm had more than five years of experience in the cable television industry. It had consulted for the Comcast Cablevision system in Flint, Omnicom Cablevision system in Hamtramck, and others.[169] Similarly, Peter J. Christiano explained that his corporation, Century Telecommunications Services, worked for Barden Cable in Inkster, Michigan in 1983.[170] Although these corporations showed their interests in helping launch the system in Detroit, Barden submitted its modification on February 5, and the council was expected to revise it by June 7, 1985.[171]

The process was further delayed in the fall. In early 1985, Ravitz questioned the "possible illegality of the Commission's continued negotiations with Barden" in the "interminable, unproductive cable process." He argued that citizens continuously asked about the progress while he had no legitimate answer.[172] As the relationship between the City Council and the Cable Commission—the former urging the latter to comply with the ordinance and the latter refusing it—became increasingly tense, the City Council began to question if the commission still maintained the authority to develop a franchise agreement with Barden.[173] As the law department confirmed, against the council's hopes, the commission still held sole authority for the decision, the project continued to be delayed.[174] By the end of April 1985, the earliest deadline possible was June 7, 1985. By early May, the deadline was set for December 31.[175]

Signing the Final Agreement With Barden Cablevision of Detroit

It was in December 1985 that the city finally signed its agreement with Barden Cablevision of Detroit. The agreement, entitled *City of Detroit Cable Communications Service Franchise Agreement as Amended and Restated* came out almost eighteen months after the city agreed to franchise Barden. At the time of signing, the Detroit City Council included nine members: Erma Henderson, Nicholas Hood, Clyde Cleveland, Barbara-Rose Collins, David Eberhard, Jack Kelley, Maryann Mahaffey, John Peoples, and Mel Ravitz. The Detroit Cable Communications Commission had seven members: James P. Robinson, Alta Harrison, Otto Hetzel, Joseph Jordan, Jean Leatherman, Charlie Primas, and Myrna Webb. Michael E. Turner continued to serve as executive director, assisted by Kathryn A. Bryant, deputy director.[176]

The agreement outlined the basic features of the franchise. It stated that the franchise was valid until May 30, 2000. Construction would start within 180 days of the commencement day. Barden was responsible for paying $500,000 per year as the franchise fee. Because one of the major concerns for the city was Barden's lack of capital to realize the system that the corporation had originally proposed, the agreement stipulated that, by March 1, 1986, Barden had to provide the city with sufficient evidence that it had secured enough financing.[177] After revision of the proposed system, Detroit would have a maximum of seventy-eight downstream channels, instead of 120. It would have nine hubs, one of which would serve as a master control center.[178] As for local origination studios that would be important for residents, Barden was the only entity that funded for construction and equipment. It also was responsible for providing a production van and equipment necessary for it.[179] As for public access studios, there were two. The first one opened before the end of the first year of construction. The second was built no later than six months after subscriber penetration was above 25 percent in the coverage area. Under the same condition,

Barden had to provide the city with an access mobile van. Three additional studios were constructed once the penetration rate went beyond 55 percent. These studios were open to the public between 6 a.m. and 12 a.m., five days a week.[180]

The agreement established a clear framework for Barden regarding its responsibility for local origination and community access programming. It had the obligation to produce at least seven hours of local origination programming, excluding repeats, per week, as well as obtain three hours of local origination programming from other sources, for a total of ten hours per week. Although the channel space might be used for access programming and other types of programming, the agreement clearly stipulated that the residents of Detroit deserved at least ten hours of local origination programming every week.[181] The agreement also stated:

> Any Detroit-Based person, group, organization, or other entity who has adequately fulfilled the requirements of the Company's established operating rules governing public access channels shall have the right, on a first-come, first-served, non-discriminatory basis, to use public access studio facilities and equipment for the express purpose of developing access studio facilities and equipment for the express purpose of developing access programming for use over the System's public access channels.[182]

As promised by Barden and other applicants, and as the city required, this outline guaranteed that any Detroit resident was able to have access to its access facilities. It was free of charge to use any of the facilities for public access channel users. There were full-time personnel on site who would hold production workshops for free, and provide assistance to users.[183] Ten channels were reserved for access services, three of which were activated at the launch of service. One was for public access, one for educational access, and the other for governmental access.[184]

Barden and the city also expressed a commitment to equal employment and procurement in the agreement. The document elaborated on the equal employment policy in detail. To summarize its policy, it read:

> The Company agrees that it will not discriminate against any person, employee, consultant or applicant for employment with respect to the individual's hire, tenure, terms, conditions of privileges of employment or hire because of the individual's religion, race, color, ancestry, national origin, age, sex, affectional

preference, height, weight, marital status, or handicap that is unrelated to the individual's ability to perform the duties of the particular job or position.[185]

Additionally, Barden's Equal Employment Policy Statement read that it was the company's policy to "implement affirmatively equal opportunity to all qualified employees and applicants for employment" without prejudice. Such a policy aimed to impose "only valid requirements for promotional opportunities." The agreement listed thirteen ways to ascertain that the policy be disseminated and adhered to both internally and externally. Managers were required to follow this philosophy. The statement explained:

> All management personnel will continue to ensure that all personnel actions such as compensation, benefits, transfers, layoffs, return from layoff, company-sponsored training, education, tuition assistance, and social and recreation programs, will be administered without regard to race, color, religion, sex, creed, national origin, age or handicap.[186]

Additionally, Barden and the city prepared a staffing plan with an emphasis on equal employment for each of the twelve job categories: officials and managers, supervisors and foremen, professional, technical, sales, clerical, craftsmen, operatives, laborers, service, apprentice, and trainees. For each category, the document specified the number of anticipated hires for both minorities and females, as well as their goals for each category, from years 1 to 15.[187]

Barden's policy dealt not only with employment, but also with procurement. The agreement continued by stating that Barden:

> will utilize Detroit-Based businesses, specifically including businesses owned by minorities, to the maximum extent possible in the installation, construction, maintenance, management and operation of the System. The Company shall submit to the Commission a program describing in detail its plans and policies for each year of the Term of the Franchise for utilizing and contracting with Detroit-Based minority construction contractors, subcontractors, suppliers, vendors and other minority business enterprises and persons or the provision of services, supplies, equipment, consultation, banking, financial, accounting and legal services, insurance, and other necessary facilities

and services that will be used in installing, managing, operating, marketing, programming, and maintaining the Cable Communications System.[188]

Based on this policy of using local businesses, the agreement required that of all the contractors and subcontractors, at least 20 percent be minority owned, and 5 percent owned by women. Similarly, unless terms and conditions quoted by a Detroit-based company were inferior to that of a non-Detroit corporation, the preference would be given to local businesses.[189] This policy included contracts for services, supplies, equipment, consultation, banking, financial, accounting and legal services. It also included all the relevant services needed for installing, managing, operating, marketing, programming and maintaining the cable system. The agreement elaborated on how exactly Barden was required to use minority owned firms during the first fifteen years of the contract. In order to ascertain that the policy was adhered to, Barden promised to keep updated information on business contracts, legal structure, type of work and services provided, and more details in its record.[190]

On the one hand, the agreement symbolized and reflected the need of Detroit residents and the city to have a system that guaranteed citizen participation through community access and local origination. The agreement protected the public's right to access those services. Additionally, it clearly stated the importance of engaging local businesses and minority firms during all processes relevant to cable television in the city. It was a way to make sure that the system was not taken away from residents. On the other hand, the issuance of the agreement came very late. It took the city more than a year after the mayor made the final decision to award the franchise to Barden. Although the delay was partly to ensure that the system remained in Detroit and was not taken to Canada with Barden's business alliances, a handful of lawsuits also caused the delay.

Chapter 6

From Agreement to Production

Period of Struggling

Despite the signing of franchise agreements in both cities, residents did not get a cable system right away. In Boston, the Final License issued on December 15, 1982 gave Cablevision six years to complete the system.[1] As for Detroit, the amended version of the franchise agreement signed in December 1985, stated that the completion of the construction had to come before the end of the sixth year after the approval of construction was given.[2] Due to the size of the systems, both in terms of the number of expected subscribers and areas to be covered, it was only natural that the construction was going to be a time-consuming process. In Detroit, especially, the post-agreement phase faced process delays. Although some residents began to have access to the service relatively soon, it took almost ten years until many customers were able to enjoy the system after the initial study of cable television introduction took place.

This chapter focuses on the period between the issuing of the final licenses and roughly the end of the decade. Although the struggles of residents and the development of cable systems continued well into the 1990s, this study ends with the closure of the decade that witnessed an immense progress of cable television systems. What Bostonians and Detroiters produced through cable television was important. The premise of fair representation, however, existed in formulating the structure of the local cable production system in such a way that minority values and ideas would be reflected. The white-dominated mass media industry had ignored the interests of various ethnic and racial minorities. The same was true with gender, economic, and other minority groups.

The process of ascertaining such a media system was not an easy task. Detroit had already seen a major delay, as shown in the previous chapter. Boston, similarly, had suffered from political power plays among interested parties. This chapter investigates the types of issues that prevented the smooth development of the cable systems in both cities that affected residents, including African Americans. It also examines how race and racial politics affected some of these reasons for the delay. A close study of Detroit, for example, reveals that there was a fear among residents that they would lose control of the system if Barden collaborated with a Canadian firm. They voiced their concerns by claiming that the system needed to be rooted in Detroit. Because the city was heavily African American, such requests also implied that race was not a separate issue. It considers how Black leaders and residents perceived the potential of the cable system during the post-agreement phase, worked to quicken the building process, and voiced their discontent with the status quo.

Boston and Its Post-Agreement Phase

In 1983, Boston's cable system expanded rapidly. One month after the signing of the Final License, the city had thirty subscribers. That number increased sixfold in March. By the end of May, there were two-thousand subscribers. By the end of year 1, there were already approximately 23,000 subscribers in Boston. During this early phase of operation, the city and Cablevision continued to keep in close contact. Although the city considered the construction to be moving forward smoothly, it also felt that the project was not without problems.[3] Cablevision made revisions to some of its employment and other policies to provide better service to the local community. The city made sure that the residents' interests were met. Bostonians continued to play an active role in observing the implementation of the system and communication between the two parties to ascertain an appropriate development of cable television.

On October 12, 1983, for example, Cablevision announced a new policy for employing Boston residents that had taken effect on September 1 of the same year. As the final agreement had stated, the service provider considered the cable business to be an opportunity to provide the city with jobs. Because the system belonged to the city, the jobs generated by the project belonged to the people who lived in the city. Based on this idea, Cablevision stated that, "Upon the date of hire, each of our contractors and subcontractors will be required to verify the address of each employee who lists him or herself as a Boston resident." The verification was going to be done with a Massachusetts driver's license, a utility bill, or a rent receipt.

This change in the corporate policy reflected Cablevision's commitment in vitalizing the local employment and economy through its system.[4]

In addition to hiring Boston residents in general, Cablevision was active in hiring minorities and women. On December 9, 1983, Cablevision issued a memo describing the composition of its employees. In management, 17 percent were minorities, 88 percent Boston residents, and 32 percent women. As for technicians, numbers were 21 percent, 64 percent, and 16 percent, respectively. In sales, minorities composed 27 percent, Bostonians 67 percent, and women 20 percent of the employees. For clerical jobs, 46 percent were minorities, 80 percent residents, and 84 percent women. In operation, minorities were 21 percent, residents 91 percent, and women 8 percent. One third of laborers were minorities, 91 percent residents, and 16 percent females. Of sixteen part-time employees, 75 percent were Boston residents, 43 percent were women, and 37 percent were minorities. This set of statistics shows that minorities and women were more present in some categories than others. For example, only 8 percent of operation staff were women. Similarly, minority representation in management remained low.[5]

Local procurement was another way to enhance local involvement in the system. The license stipulated that a minimum of 10 percent of all services, materials, and equipment should come from Boston's minority businesses. Cablevision, however, failed to meet the number. It had less than 1 percent of all services and equipment from minority-owned Boston businesses. The city urged Cablevision "to increase its advertising and outreach to identify potential minority-owned businesses in Boston."[6]

Revising the employment policy was not the only way to ascertain protection of subscribers' interests. Cablevision's general manager, Arthur Thompson, elaborated on the subscriber complaint procedures in his letter to Margie Cohen Stanzler, a member of the Boston Cable Commission. As the system expanded and the subscriber base grew, Thompson claimed that it had become increasingly difficult to process complaints in a timely manner. In order to remedy the problem, he articulated that the company had installed an upgraded system and planned to include thirty more lines within a few weeks. The Customer Service Department would have thirty-six representatives to handle complaints and other requests. The training provided to new staff also was extensive. Service technicians' training lasted for two weeks in the field. Cablevision's service crew only included those with knowledge and experience in cable television. These efforts made sure that customers' voice was properly heard and processed as promised in the Final License.[7]

Public hearings were another way to assess the operation of Cablevision. In order to review its first year, a public hearing was scheduled for

December 13, 1983 from 11 a.m. to 3 p.m. at City Hall. At the hearing, the public had a chance to make comments or ask Cablevision representatives questions. The public included any Boston residents and anyone who represented an institution located in Boston. They were also allowed to submit written testimonies.[8]

Martin Kessel, a resident of Boston and a trustee of the Boston Community Access and Programming Foundation submitted his review on the Universal Basic Service to the city. Although he highly valued Cablevision's five-year rate-freeze policy as well as the content that it had been providing its subscribers, he identified some areas of improvement in local programming and access. Citing the Final License, Kessel explained that the promise of Cablevision was to offer sixty hours of local origination programming weekly by the end of the first year following system activation. It also proposed to offer a wide range of local video programming. The license included a list of forty-eight full-time employees and forty interns who would produce such programming.[9]

In reality, Cablevision had no studio and no program director by mid-December 1983. It only offered four hours of local origination, including repeats. From this perspective, Cablevision was far from meeting its promise. Kessel wrote, "Cablevision must . . . be held responsible for the minimal amount of programming on channels operated by the Boston Community Access and Programming Foundation. By continually delaying delivery of studios and equipment, Cablevision has made it impossible for the Foundation to produce significant amount of programming."[10]

The city shared the same concern about the lack of facilities. A city document read, "we are concerned that Cablevision has not built a studio after one full year of operation. Without studio capacity, the Foundation cannot fulfill its mandate to ensure public access to the cable system." It urged the firm to honor its commitment in a timely fashion so that the infrastructural restriction for public access would disappear.[11]

Jacob Bernstein evaluated Cablevision's development during the first year. He submitted a written document that expanded on his oral testimony at the public hearing on December 13. Bernstein represented the Committee for Community Access, a nonprofit communications policy organization based in Boston. At the beginning of his written testimony, he touched on a major structural problem in Boston. He wrote:

> For whatever reason (still unarticulated), Boston—the state's largest population center—unlike most other Massachusetts cities

and towns had no formal citizens advisory board (CAB) to focus public interest and participation throughout the franchise process and beyond. This cannot be, however, a mere oversight on the Mayor's part since the very first cable task force chaired by Vice Mayor Sullivan in 1979 recommended that such a permanent CAB be established upon its demise.[12]

What Bernstein perceived as the lack of attention to citizens' needs by the part of the city was also a reflection of over control by the city. Although he recognized that the $2 basic fee idea was beneficial for subscribers, Boston's system had become an example of "franchising excesses." The review criticized that too much involvement by the city, including high tax rates, compromised Cablevision's ability to function as a free enterprise in providing services that the local community needed.[13]

Global Village Associates, Inc. was another cable watch-dog group that questioned Cablevision's practices. The organization had a review prepared by Mackie McLeod, coordinator of the Roxbury Task Force on Cable TV with thirty-five representatives from predominantly African American neighborhoods such as Roxbury, Dorchester, and Mattapan. It explained that the organization was most concerned about the lack of affirmative action compliances. An issued statement stated that there was "the absence of people of color in senior-level, policy-making positions" in the Boston system. It was not the first time for the organization to point out such problematic hiring schemes. On April 11, 1983, the task force had held a press conference in Roxbury "to protest and condemn Cablevision of Boston's apparent policy of deploying all-white, all-male construction crews, recruited from Long Island, NY, and New Jersey, in predominantly Black, predominantly unemployed Roxbury."[14]

According to the task force, Cablevision also manipulated the number of minorities hired by keeping the number of fired minority employees counted in its statistics for the compliance purposes. The organization also argued that Roxbury had been redlined "by changing the Roxbury wiring schedule and public access studio construction to an undefined date in the future." McLeod recommended that the city hire an affirmative action team to find "qualified and authentic Boston residents of color." The team also would be responsible for minority contractor review along with the affirmative action team. From the study, it became apparent that the city lacked personnel and groups to make sure Cablevision truly abided by what had been described in the Final License.[15]

Accountability on the part of Cablevision was not always existent. Despite the promise to increase the number of customer service employees to handle subscribers' complaints efficiently, it had become obvious that Cablevision was not dealing properly with questions and concerns shared by its customers. As a result, many subscribers began to call City Hall or the state Cable Television Commission. Thomas P. Cohan, director of the Mayor's Office of Cable Communications explained that each office had been receiving more than fifty calls per day regarding Cablevision's services in Boston. He elaborated further that:

> [he] found it arrogant that Cablevision makes so little effort to reinforce an obviously inadequate customer service staff, yet wastes no time in dunning subscribers whose major offense, in many cases, is that they couldn't reach Cablevision to question their bills or have them adjusted.[16]

Cohan warned that Cablevision's practice could constitute "an unfair or deceptive act of practice" under the provisions of Massachusetts General Laws. By the end of the month, the Department of the Attorney General began its investigation into Cablevision's business practices.[17]

The amount of available training and employment opportunities for Boston residents, minorities, and women continued to be insufficient. To address this concern, along with a few others, Mayor Flynn sent a letter to Charles Dolan to reach an agreement on how to amend the situation. The mayor wrote:

> While I understand the myriad problems of staffing a build such as yours in Boston, I am nonetheless committed to insuring that Boston residents, minorities and women are adequately represented at all levels of Cablevision's workforce and that these groups receive a fair share of the economic benefits to be accrued from the growth of the cable communications industry in Boston.[18]

The letter to Dolan was not a simple one. On the bottom of it was a space for Dolan to sign. Flynn expected Dolan to sign the letter and to return a copy to make sure that Cablevision abided by the agreement of the Final License.

Mayor Flynn requested that Dolan agree on eight issues. Cablevision was not living up to the original expectations.

1. Cablevision had to agree to adhere to the provisions relevant to employment and minority business participation described in the Final License.
2. Cablevision was to meet with the city's cable office to discuss how to implement the company's Resident Minority Job Policy.
3. The increase promised in the policy must be achieved by May 15, 1984.
4. Cablevision had to fund additional training programs for train Boston residents, especially "the unskilled and semi-skilled, for subsequent positions in the cable communications field." The budget for this training program was $250,000, to be equally distributed over five consecutive years.
5. Cablevision had to initiate a school–work internship program for high school students in the Boston Public Schools.
6. Half of new permanent jobs had to be slated for Boston residents, 30 percent for minorities, and 30 percent for women. Although there was no penalty if Cablevision was unable to meet the goals, it was required to make its best efforts. There was a system through which Cablevision would receive referrals regarding qualified Boston residents for all job openings.
7. Cablevision also would have to hire minority- and female-owned firms for construction and operations.
8. All the reports pertaining to these agreements would be issued and submitted to the Neighborhood Development and Employment Agency.[19]

As a result of this letter, Cablevision and the city reached an agreement that all the provisions would be respected. Cablevision agreed to "provide more responsive customer service to Boston's neighborhoods." Mayor Flynn explained that he would monitor the access process very closely with Tom Cohan, cable commissioner.[20] Although Dolan agreed to abide by the Final License, the city continued to be cautious of the company's practice. Cohan was particularly critical of the firm. Although he acknowledged that Cablevision had been able to speed up the construction of its system, he also argued that it was mostly thanks to its bankers. They were able to come up with positive "financial benchmarks" to fund and speed up Cablevision's construction. Therefore, he explained:

> although we view this agreement [between Mayor and Dolan] as a first step toward developing a positive working relationship with Cablevision, the City will not hesitate in the least to publicly initiate breach proceedings if they fail to honor any portion of this agreement or fail to comply in full with any other contractual agreement with the City.[21]

With the general attorney's office conducting its investigation into Cablevision's business practice, Cohan was highly critical of the cable company.[22]

Cohan believed that the city needed four assurances from Dolan:

1. Cablevision had to increase the number of telephone lines and customer service representatives to handle customer calls.

2. Cablevision had to first solve the phone situation before it cut off services from their subscribers who had questions and issues.

3. The firm must set a certain number of employees to clear up the backlog of complaints.

4. The company's general manager was required to meet with the director of the city's cable office and the general manager of the access foundation on a monthly basis to discuss concerns and to deal with problems.[23]

As a response to the city's expressed concerns and the attorney general's investigation, Cablevision issued a letter to Mayor Flynn regarding its employment and procurement policies. Arthur Thompson explained that two minority contracting firms had been given pre-bid conferences for the renovations of the Allston offices, as had been agreed on in the license. Similarly, three minority cleaning services had been contracted for custodial maintenance of the Allston office. In order to procure appliances for the Allston office cafeteria, Claudette Bailey and Cablevision had begun to work together to find a minority business firm. Thompson also promised to "make a good faith effort" to hire minority and female residents and businesses during the upcoming phases of the system construction.[24]

By the end of 1984, Cablevision had undergone various changes to better meet the city's requirements. However, not all Boston residents were satisfied with the services. On the one hand, as Tom Cohan explained, Cablevision had improved its business practices substantially. Cohan wrote that it had become a "very commendable employer in Boston" with its minority employees comprising more than 32 percent of the overall employees, more than 7 percent women, and in excess of 45 percent Bostonians.

On the other hand, in October 1984, some citizens decided to boycott Cablevision. Cohan received a leaflet on October 23, although he was unable to identify the reason for such boycott leaflets. He found out that they had been distributed door to door in Fenway and other neighborhoods where Cablevision was working at the time. Many of the claims listed on the back of the leaflet, however, were far from true according to Cohan. Unlike what the pamphlet claimed, Cablevision was not using foreign-made materials, was not using tax money to pay for the cable-franchising process, and was not hiring out-of-state workers illegally. Although the headlines of the leaflet, "Boston's Second Massacre: By Cablevisions of Boston," was sensational, the claims made by International Brotherhood of Electrical Workers were not substantiated.[25]

In November 1984, both the city and Cablevision spent some time discussing access facilities. On November 1, Mayor Flynn wrote a letter to James Kofalt, executive vice president of Cablevision, to clarify the use of access studios and to establish guidelines. Mayor Flynn reached an agreement with Cablevision to increase the accessibility to such facilities. The main access studio would be available a full year earlier than originally planned and would be considerably larger in size. The studio was available for public access on a full-time basis. The neighborhood access coordinators also had storage space for equipment and office space at each studio, making their access work much easier. Cablevision agreed to offer production assistance on municipal programming and studio and equipment maintenance service. All these agreements finalized between the two parties in very early November facilitated access programming and improved the accessibility of the system for Boston residents, including minorities. Residents of Roxbury had a neighborhood studio that was open to them in the evening five days a week and all day on Saturdays. The African American presence in the process was significant.[26]

At the end of 1984, Cablevision issued its evaluation report. Although it had made substantial progress in system construction over the year, it also was suffering from financial difficulties. At the national level, Cablevision had a $15 million to $20 million deficit in the total capital available to the Boston system. A confidential memo submitted to Mayor Flynn analyzed that Cablevision's purchase of the Brookline system for about $10 million at the end of December 1983 had hurt the company's finances. Cablevision was not receiving as much cooperation from local landlords as was needed to meet the projected number for multiple dwelling unit hookups. It turned out that Cablevision had based its business projection on a faulty number of households in Boston. Whereas a service candidate, Warner, based its projections on 234,000 households, Cablevision's proposal was on 250,000.[27]

Cablevision failed to fulfill other requirements as well. For example, although it had an obligation to produce sixty hours of local origination programming weekly, it only aired a fraction of the required time. Although the members of the Office of Cable Communications acknowledged that produced programming appeared to be "quite responsive to the needs and interests of Boston resident," the city had far from a sufficient amount of local origination programs.[28]

As for minority training and employment, Cablevision had made sufficient improvements since 1983. The evaluation document for 1984 wrote, "Cablevision's efforts to provide opportunities for Boston residents, minorities and women on its own workforce has been generally exemplary." Cablevision agreed to provide $250,000 over five years for training. The city expected to receive the training program for the first year by the end of February 1985. The progress, however, was relatively limited to operation-level employees. The evaluation suggested that "more affirmative action is needed at the upper management level."[29]

Getting training did not always mean that trained residents would be able to find a job in Boston's cable industry. Twelve residents of Boston submitted a letter to Mayor Flynn asking the city to investigate Cablevision's hiring practices. They suspected that Cablevision had found its employees elsewhere and not in Boston. They claimed that they had been "trained for future unemployment." They continued to argue, "Cablevision has been breaking its contract with the City; as far as hiring Boston residents is concerned." They were so determined that they were "willing to testify at a public hearing." Interestingly enough, many of these residents who signed the bottom of the letter were from neighborhoods with a high African American population density. Jamaica Plain, Dorchester, Mattapan, and Hyde Park were just some of the neighborhoods in which these representatives resided.[30]

The situation did not go as well in terms of contracting. Cablevision did not have good control over its subcontractors and their affirmative action policy. The positive side of the evaluation was that in May 1984, the city and Cablevision's agreed to set up an official liaison to take responsibility for affirmative action reports. Consequently, the city recognized that it began to see that all contractors were aware of affirmative action requirements.[31]

Cablevision's Failure to Meet the Expectations

Cablevision failed to meet its requirement on procurement for two years in a row. The evaluation analyzed that the minority procurement rate for construction and operation remained very low, only 1 percent. Because Cablevision's goal was 10 percent, it was far from sufficient. The Cable Communications Office recommended three plans to amend the situation. First, contracts should be divided up into smaller segments so that more contractors could be involved and smaller scale firms would be able to bid for contracts. Second, minority firms should get a chance to attend prebid conferences to be better informed about the franchise. Finally, joint ventures between minority and nonminority firms also would encourage minority participation.[32]

At the turn of the year, it had become obvious to residents of Boston that Cablevision could do better with its employment and staffing. On the one hand, Hubert Jessup, general manager of the Access Foundation was making $67,500 annually. The second highest salary was $18,000. Mel King remembers that once Jessup's high salary was disclosed, many residents, including African Americans, voiced their dissatisfaction. Additionally, he replaced Deborah Hill's position with a friend of his who was white. A resident of East Boston, Janet L. Doherty wrote a letter to Mayor Flynn emphasizing that the newly employed officer was white and that Jessup was "incompetent." She closed her letter by saying, "Community producers are pretty powerless. We can only yell and scream, and frankly, I don't enjoy it."[33]

Thomas Cohan gave a general impression about Cablevision's performance in his letter to Mayor Flynn. He wrote:

> I firmly believe that Boston residents are ready to welcome cable television service into their homes. Cablevision's own studies show that people look forward to improved reception, a wider variety of programming such as movies and sports, and more diverse local programming. In this respect, Cablevision has satisfied many of its Boston subscribers. However, the Company has also frustrated and angered many of these same customers by its inability to efficiently respond to service and billing problems. There is a general feeling among many Boston residents that cablevision, because it operates as a de facto monopoly in the city, is perfectly content to treat customers with a "take it or

leave it" attitude. It is this perception, and the willingness of potential subscribers in Boston to, in fact "leave it rather than take it," that has created some of the financial difficulties for the Company.[34]

Cohen believed that 1985 was a critical year for Cablevision to earn the trust of Boston residents back.[35] The future of Cablevision did not look promising. In less than two months, Charles Dolan was quoted in the *Boston Globe* that he wanted to reduce its funding on public access because of the financial difficulty. He called the service "an enormous burden."[36]

By the end of March, it had become unlikely that Cablevision would be able to make the $250,000 payment to support public access programming. Hubert Jessup explained that if the company missed making the payment, it would cause a "meltdown" of access shows, including neighborhood news programs. The situation was serious enough that Cablevision planned to conduct layoffs in May. Without money, the public access programming would be terminated by the end of May. The impact of this possible termination was immense. Paul Phillips from Jamaica Plain was editing a program on teenage suicide and pregnancy with teenagers at the Bromley-Heath housing project. Such positive examples of socially benefiting projects would not see the light of day without public access services.[37]

One way for Cablevision to make up for the financial difficulty was to increase the basic cable service subscription fee. As Tom Cohan explained, the $2 fee was "the cornerstone of Cablevision's original proposal to the City, and one of the chief reasons it was awarded the franchise." In other cities, basic cable service costs between $6 and $10. Cablevision's proposal was therefore particularly attractive to the city. From the perspective of the city and its residents, Cablevision was far from meeting its promise. Not only did Cablevision begin to seek ways to change the basic fee structure that had been frozen for the first five years of the franchise, the studio in Allston-Brighton area had not been built, the wiring of the City Hall had long been overdue, equipment for the municipal studio had not been ordered, and numerous other delays and cancellations abounded.[38]

Many citizens and interested parties sent letters and telegrams to the city regarding their concerns. Charles Spillane and Joseph Quilty from the Boston Building Trades Council asserted that "since Cablevision has come on the scene, they have misled the residents of Boston with rate hikes. Furthermore, they perform shoddy substandard work, refuse to hire union workers and have let go all Boston area residents that worked for them." John Taylor from Dorchester wrote, with frustration, that he did not want the City of Boston [to] cave into the long planned concessions demanded by Cablevision of Boston. Don't make residents pay for Cablevision's mismanagements. By renegotiating that license, the City will create the impression that it doesn't care about the residents who deserve affordable cable TV service and quality community access programming. Cablevision subcontractors should, also, be made to live up to its hiring commitments for Boston residents. The number is currently at an all time low.[39]

A Beginning of an Alternative Media Form for African American Bostonians

Despite the struggles and difficulties that Boston experienced in the early months of 1985, residents' outlook of what cable would bring to them was not solely pessimistic. One of the most significant events was Boston Famine Relief Cablethon 1985. The Boston Neighborhood Network, Roxbury Cable, and other local organizations organized this effort to provide live programs cablecast from the Strand Theater in Dorchester, Harbor Side School in East Boston, and Cablevision in Roxbury. Michael King, son of Mel King, was in charge of reserving studio spaces. Another colleague of his also donated office space next to the old Boston brewery located in Jamaica Plain. This community-based fundraising project aimed to develop a sense of community in Boston while raising money for aid to children in Africa.[40]

The program went on air on Channel 3 from 4 p.m. to 12 a.m. on March 22. Boston's local artists contributed their performances for eight hours. Although an eight-hour program may not have seemed astonishing in the mid-1980s, it was an ambitious plan considering that community members had just started to produce their own content in Boston. In many parts of Boston, the construction had not been completed and many of the equipment and facilities were still on their way. After months of preparation, community members participated in content production that had an objective that was much bigger than most commercial television companies did, while connecting Roxbury to East Boston, to Dorchester, and any other area available to air the content. Michael King and his colleagues raised approximately $20,000 by the end of the program. This effort was one example of how people used the system, the facility, and the equipment to their fullest extent as community members. The project was based on the social capital that efforts around cable television created. Finding production spaces and content, participants used the network of people that they established. When Boston's public access television faced financial difficulty,

Hubert Jessup referred to this project as an example of community efforts. He said, "We got communities working with each other for a common cause. That's one of the great things cable can do. Without it, there's no reason to have cable television." Although King still felt that they could have done even more, it served as proof that a bottom–up social movement mediated by technology was possible.[41]

On April 10, 1985, Cablevision hosted an airing of *The Greatest Place to Be*, a documentary that chronicled a woman's efforts to preserve Jamaica Pond in the midst of social change. The event took place in Curtis Hall in Jamaica Plain.[42] Although some projects were larger in scale or more successful than others, these examples suggest that despite financial and technical difficulties faced by Cablevision and which the city was trying to amend, the residents in Boston had successfully begun to produce their own media content within a few years after the signing of the Final License.

The lack of representation of residents and community producers in the foundation was noted both by city residents and officials. In January, 281 Bostonians signed a petition regarding the lack of fair representation and submitted their signatures to the city on January 27. Two days later, Tom Cohan sent a note to Mayor Flynn stating that many residents had filed grievances regarding the lack of membership representation in the foundation's Board of Trustees.[43]

On February 10, 1986, Boston's cable system showcased another example of resident participation in local politics via cable television. The Access Foundation helped the city produce and cablecast a public hearing session at Faneuil Hall. The live program allowed more than five-hundred people to join the hearing. Without the technology, they would not have been able to do so because of the limited space. These participants were able to watch the proceedings on large television screens set in the City Council chambers. There also was an unknown number of Bostonians who watched the hearing at home via cable television. Mayor Flynn wrote that the program was "truly an effective use of the technology of cable television, as well as a fine example of how institutions, public and non-profit, can share resources to benefit the people of Boston."[44]

The Boston Community Access and Programming Foundation and the Boston Neighborhood Network recorded the School Committee Hearing held on April 29, 1986. The foundation explained that the meeting was put on air on Channels A3 and A8 on the same day. It also aired on May 4. In 1985 and 1986, the city witnessed many programs in which citizens from across Boston partook in municipal politics through local cable technology.[45]

Detroit and Its Post-Agreement Phase

Detroit needed more time than Boston to experience some of the first community-organized examples of cable use in African American communities after signing the franchise agreement. Detroit lacked programs that were comparable to Boston's cablethon during the early construction and implementation period. Detroit's post-agreement phase, however, was rich in efforts by African Americans to maintain local—often Black—ownership of the system, Barden and the city's attempt to deal with litigations, and the city and residents' efforts to make sure that the city would be wired. From this respect, Detroit's development required more time than in Boston. The analysis of Detroit's post-agreement phase involved less African American activism and more legal and administrative strives in the municipal office. Consequently, fewer of the events that took place during this period in Detroit compared with Boston, fall under the scope of this study. A few notable incidents and events, however, deserve our attention.

Although Barden Cablevision won the Final License in December 1985, the firm contemplated the transferring of the franchise to an affiliated entity. It was a strategic move for Barden to facilitate "the financing of the system and the offering of the Local Investment Share to Detroit residents." After the transfer, the partnership company would bear the same responsibility as Barden Cablevision. In order to enable the partnership, Barden Cablevision of Detroit established Barden Cablevision, a co-partnership between Detroit Cable TV and Barden Cablevision of Detroit. Although the partnership agreement was signed on February 14, 1985 between Barden and Detroit Cable TV, different hearings were set for mid to late February to ascertain the legality, feasibility, and effectiveness of the transfer of the franchise.[46]

As described in a previous chapter, non-Detroit ownership continued to be a concern. Despite the voiced disagreement by the local residents and entities, Maclean-Hunter Cable TV, a Canadian firm, was still partnered with Barden Cablevision, under the name Cable Management of Detroit. Don H. Barden and his affiliates owned 51 percent of the stock of the franchisee and Maclean-Hunter owned 49 percent. Although Maclean-Hunter emphasized the benefit that it would bring to the city through ensuring Detroit's funds, citizen concerns were not easily quelled. The application for consent to transfer specifically addressed the issue of securing local ownership.[47]

The process of franchise transfer went smoothly. Once the petition for transfer was submitted to the Detroit Cable Communications Commission on February 14, 1986, the day the partnership agreement was signed, all the necessary paperwork had been submitted by February 21.[48]

Although franchise transfer captured the attention of many Detroit citizens and official, many kept a close eye on Barden's commitment in minority engagement. Citizen Media Alliance, located in Detroit, expressed its concerns for the affirmative action–based training programs that had been promised by Barden. Its letter to Don Barden said:

> Without training in various industry positions an affirmative action plan won't be beneficial to Detroiters or minorities. The aforementioned [Barden Cablevision] [has not] the skills because of prior (and present) denial. Will Barden Cable remedy this? Maybe—maybe not. On-the-job training is vital to ensure that section 16 (and 6. 2-3. public access) are *fully* implemented.[49]

For the city, resident concerns about the partnership between Barden and Maclean-Hunter and Barden's commitment in local engagement appeared relatively minor by mid-March 1986. There was an ongoing threat of lawsuits by cable franchise applicants, particularly City Communications, since the municipal decision to select Barden as the franchisee resurfaced on March 18, 1986. On this day, City Communications, Inc. filed a complaint against the city of Detroit, Barden Cablevision of Detroit, and Maclean-Hunter.

City Communications listed several violations by the defendants. According to the firm, there was an ongoing conspiracy among the defendants. First, City Communications claimed that the Detroit Cable Communications Commission had recommended to Mayor Young that the content of the RFP and Franchise Ordinance be modified mid-June 1983. Barden won the franchise in the summer following the modification. As a result, the signed agreement did not match the original RFP or the ordinance. Despite the unfairness in practice, the City Council approved the franchising of Barden without showing that it had sufficient financing to develop the system in Detroit. What City Communications called a "conspiracy" continued well into 1985. It claimed:

> defendants have unlawfully conspired to permit the further modification and amendment of Barden's proposal and/or award, including: elimination of a promised dual cable system; elimination of approximately $38 million in promised grants to the City; elimination of a promised immediately functioning two way interactive system with a two way addressable cable box; reduction in the number of channels in the system from a promised

112 to 78; increase in construction time from a promised three years to five and one-half years; provision of a 40% or greater equity ownership to defendant MacLean-Hunter; and reduction by approximately 33 per cent of promised construction expenditures for the system, in violation of the RFP, the ordinance, and the rights of plaintiff and the citizens of the City of Detroit.[50]

City Communications continued to argue that the city did not make information on the amendment and modification public.

City Communications claimed that such a "conspiracy" resulted in primarily three major problems for the city. First, it created a monopoly in the cable market. The construction of the system and facilities, as well as the offering of such services had become a monopolized business. Second, City Communications had its right to compete with Barden denied and it had been unfairly treated without being given an equal footing in the bidding process. The last point was most significant for the residents of Detroit. It argued that "public and private institutions and the citizens of the City of Detroit have been deprived of the benefits of competition in the construction of cable television facilities and the provision of cable television service." Not only were Barden and other relevant entities in the city facing serious hardships, residents of Detroit also were undergoing difficult times witnessing the promised advantages that the cable system would bring to their lives.[51]

Despite the castigations by City Communications about Barden's repeated attempts to amend its agreement with the city, Barden continued to make changes to the resolution. The purpose of requesting an amendment was to postpone the offering of the initial local investment share by two months to December 1986. The firm argued that due to the complexity of the corporation and the partnership, Thomson McKinnon Securities, Inc., an investment banking firm, advised that the offering should be delayed to make Barden's shares more attractive. Barden attempted to justify such efforts, citing a part of the resolution that said:

> That the Detroit Cable Communications Commission hereby issues the Authority to Proceed to Barden Cablevision, as provided by and subject to the provisions of the Franchise Agreement, the Franchise Ordinance, and the attached report on Satisfaction of Conditions Subsequent and Issuance of the Authority to Proceed.[52]

This authority document mentioned that the resolution was subject to the provisions of the relevant documents.

Barden's request for an amendment did not end. Although the offering had been extended until December 31, 1986, it had become clear by mid-December that the promise was not going to be met again. The U.S. Securities and Exchange Commission and Barden continued to negotiate the terms on which the offering would take place. The discussion between the two parties continued well into 1987. On January 16, 1987, Barden requested that the offer be extended until June 30, 1987, almost one year after the originally planned date of offering. The request was later approved to make another example of Barden's slow progress in wiring the city.[53]

The lawsuit with City Communications and the delaying of the offering did not mean that the city remained unwired. The wiring process started in late 1986. The city had been divided into grids. The wiring construction took place on a grid basis. By the end of November 1986, four grids had been wired and three more were under construction. The number continued to grow for months to come. By the time the *Detroit Cable Communications Commission Construction Grid Report* came out in April 1987, thirty grids were wired and another thirty were in the process of being wired. The thirty completed grids included more than ninety-five miles of cables with more than one-hundred miles more cable to be installed. Of 96.62 miles of installed cable, 42.55 miles had been activated by April 1987.[54]

These examples of litigation and administrative efforts and the lack of community-based projects show that in Detroit, residents needed more time before they would be able to fully benefit from the arrival of the cable system. It is important to remember that efforts by Black communities should not be evaluated or judged based on how quickly they were able to take advantage of the new technology. It is undeniable, as has been demonstrated in previous chapters, that Detroit's Blacks were heavily involved in the process of cable television system planning and implementation.

Delays During the Post-Agreement Phase

Both in Boston and Detroit, residents hoped that the city would be wired immediately after the signing of the agreement and the greatest number of residents possible would benefit from the new cable system. Bostonians and Detroiters, however, had to wait for a long time until they were able to receive the service they had expected. In both cities, the power politics among relevant parties caused delays. In some cases, financial concerns accounted for the slow process. In some others, the question of ownership made residents anxious. Regardless of the circumstance, African American and other minority residents and organizations made sure that their system would reflect their interests. Their struggles were the embodiment of African American efforts to establish a community medium that would enable affirmative representations in the post-Civil Rights age.

Conclusion

BET Is Not the Answer

Popular culture, by definition, is contested terrain. African American popular culture has been even more controversial. From slave songs and narratives to more contemporary hip-hop culture, the history of Black popular culture is that of African American struggles between being marginalized and being mainstream, and between the journey from three-fifths citizenship to full citizenship. Visual popular culture especially reflects American racial politics, starting during the Civil Rights era as television became increasingly more important to the American lifestyle. This scholarly investigation has expanded the existing scholarship in Media Studies and African American Studies and History that has heavily focused on mainstream broadcast television and produced images to community-based narrowcasting television and processes before the production of televisual images.

Historically, African Americans lacked access to the means of production and consumption of media industries. Radio, newspaper, trade publications, and others excluded African Americans both as producers and consumers. Visual culture was nothing different. African Americans faced severe discrimination, exclusion, and segregation both in the commercialized film and television industries. Their rare appearances in movies or television shows took place at the risk of continuing image distortions that reinforced assumptions about Blacks by white audiences. It contained persisting and underlying racial thoughts. Despite improvement made in the twentieth century, visual popular culture continued to be highly racially contextual. Such negative images had historically hurt African American psychology, lowered their self-esteem, and kept them socially marginalized.

To associate the history of African American visual popular culture only to one of exclusion, however, is to ignore their struggles and agency.

Black history is filled with African Americans' tactful use of media to affect national sentiment and advance their political agenda. Especially after the establishment of African American Studies as an academic discipline in the middle of the twentieth century, scholars have identified numerous examples of African Americans' political use of visual popular culture. These positive exploits nurtured a can-do attitude, brought social change, and functioned as political tools. Popular culture for and by African Americans has been a source of persisting prejudice against them, but also has driven social change, particularly during the second half of the twentieth century. This book has demonstrated how the struggles and efforts within the history of cable television are examples of this.

African American intellectuals' attention to visual culture dates back at least to the early twentieth century. W. E. B. Du Bois' famous concepts of "double consciousness" and the "veil" implicitly emphasize his awareness about the importance of visual images. He argued in 1903 that African Americans had "this sense of always looking at one's self through the eyes of others, of measuring one's soul by the tape of a world that looks on it amused contempt and pity."[1] He was well aware that positive images of Blacks could help improve their social condition, whereas negative ones could easily hurt it. Du Bois, who Shawn Michelle Smith calls "an early *visual* theorist of race and racism," was careful about the kind of African American photographic images that he had prepared for the 1900 Paris Exposition.[2] Du Bois consciously exhibited middle-class images of well-coiffed and well-dressed Black men and women in a photo studio. He was cognizant that such portraits would challenge "the Black image in the eyes of white beholders."[3] An archive of visual images for Du Bois was a tool to reverse the idea of "Negro inferiority."[4]

The strong belief in the power of visual representations that Du Bois, Booker T. Washington, Carter G. Woodson, Marcus Garvey, and other early twentieth-century Black intellectuals and leaders shared continued throughout the century. As visual culture itself developed through the popularization of films and television, such critical discussions on the power of the Black image became more important. Once African American Studies began to be recognized as a discipline on white university campuses, the volume of scholarly investigations on African American life through the lenses of their visual culture increased.[5]

Although reality is more complex, scholars generally have categorized African American images in American visual popular culture into two large groups. The first are those produced by and/or for whites. From minstrelsy on, this group contains images that have long been prevalent and normative in American society. Because black participation in popular culture production progressed slowly and African Americans often lacked the resources required in a highly capitalized American popular cultural industry, whites produced most of the Black images until very recently. Therefore, George M. Frederickson's naming of his seminal work, *The Black Image in the White Mind*, is not only accurate for the period he covers—between 1817 and 1914—but also for most of the twentieth century.[6] Blacks were often portrayed as toms, coons, mulattoes, mammies, and bucks, as Donald Bogle summarizes.[7]

The second type of Black images includes those produced by and/or for Blacks themselves. This group signals African American agency and resistance, unlike the first category of images. Such images are more likely to be positive than those in the first category. These were the images that made African Americans more aware of American racial politics, politicized them during and after the Civil Rights era, helped improve their self-esteem, and encouraged them to cherish their heritage. As Du Bois tried to refute the common prejudice about African Americans through his picture archives and expositions, later visual popular culture proved to have the power for social change. But even these images were not free of complications. Actors and actresses faced financial difficulty when they tried to be faithful to their race in their acting career, even during the era of the Harlem and Chicago renaissances. The same was true with singers, painters, and other artists. "Would I rather make money by subscribing to the dominant but demeaning African Americans or would I rather risk poverty while working for a greater cause?" was a frequent and imminent question for many African Americans. This conundrum is omnipresent in the writing of Langston Hughes, in Katherine Durham's post-performance statement in Louisville, Kentucky, in 1944, and in the work of others. Critics also problematized the perceived lack of their authenticity and questioned if white-dominated society would accept and believe in what those images tried to represent. Cynics would claim that authentic African American images would never be possible in the racist United States.

The study of African American images in film and television shows how whites produced African American images based on what they thought African Americans were like and what African Americans should be; how such white-produced images reinforced African Americans' negative self-image and white's prejudice; how Black-produced images about themselves reflected efforts by African American leaders to change their self-esteem; and how images of African Americans by Blacks actually helped African Americans change their self-image. Mamie Till-Bradley's courageous decision to have an open casket funeral for her son in 1955, and *Jet* and *Look*'s bold step to publish the picture of "Till's savagely disfigured corpse" showed the power of visual images—both positive and negative—was a major social force within Black popular culture.[8] Decades have passed since the Civil Rights era. Many students today only remember this period through audio and visual representations. Martin Luther King Jr.'s recorded speech, a picture of a small Black girl walking to a desegregated school, an image of a police dog and African American protesters, and many others continue to touch us. Violence against African Americans speaks loudly in pictures, even in today's society in which Hollywood movies, computer games, and online fantasies are becoming more violent. The 1960s and 1970s are rich in these examples of African American images working both for and against the good of African Americans.

In order to understand African American public culture during the post-Civil Rights era, studying cable television is an effective method, especially because its public access and local origination reflected many of the fundamental characteristics of public television. Reality-based programs that were popular on narrowcast cable television channels are, according to Barry Dornfeld, a "corpus of images that needs to be analyzed to better understand how practices of producing and consuming media forms contribute to the constitution of American public culture." Case studies on African Americans in Boston and Detroit have revealed not only what public culture for Blacks was in these two cities, but also how such culture had been established by African Americans through their engagement in formulating this particular culture-production technology.[9]

Historical Lessons From Cable Television in Boston and Detroit

The history of the introduction of cable television to Boston and Detroit was a history of African Americans successfully acquiring access to the technology in front of both the television set and the television camera, as well as behind the camera. This book has demonstrated that existing scholarship that has disproportionately focused on broadcasting media and image content has only explored a part of the history of African Americans and their use of media for community justice. The initial discussion of developing a cable television system happened in both cities in the late 1960s and the early 1970s. It took almost a decade before the cities were committed in their feasibility discussion. What was significant from the early phases of the local studies was that African American community leaders and citizens had expressed their interest in the system. They conducted their own research to identify some of the benefits and risks for minority populations associated with cable television, worked with city officials to make sure their voice was reflected, and actively participated in meetings, hearings, and committee activities. As a result, both the cities and cable applicants established RFPs and applications that accommodated many of the needs of local community minority residents.

Although the history of cable television in Boston and Detroit is rich in examples of minority participation, it is important to remember that participatory media was not new when cable television became popular. In the seventeenth and eighteenth centuries, an early form of participatory media existed in magazines. Participatory entertainment continued to thrive in the nineteenth century and into the twentieth when radio and television dramatically changed the dynamics of audience-participation media. As a result, an important question about media came out: "Will they make for a more democratic nation and a more coherent community?" The answer was rarely affirmative. Although audience participation allowed an increase in reflection on the impact of audience voice on media content, the separation between producers and consumers remained clear. This was one of the main reasons racial stereotypes about Blacks continued to exist through the twentieth century.[10]

Cable television was a new period in this history of audience-participation media. A major divergence of the technology compared with its predecessors was that it successfully combined the rise of visual culture and the increased participation of the audience as producers. On the one hand, photography, cinema, and other media had expanded the power of visual images during the first half of the twentieth century. As discussed throughout this book, visual culture affected how Americans saw various ethnic and racial groups and how individual and collective identities were formed. With its appeal to vision, it often was more effective and powerful than other media in identity politics. Cable television was another tool of such televisual image production. On the other hand, it also was a way for audiences to participate in the media. The technology's two-way communication method meant that producers could receive input and feedback from their audience. More than with broadcast television, radio, or films, the voice of the audience was likely to be reflected and heard in cable television.

Cable television further blurred the line between producers and audience. Similar to the rise in awareness that the line between the two does not always have to be clearly marked, as personal websites and user-driven websites became increasing popular in the late 1990s and 2000s, cable television allowed the American public to realize that the new media technology could collapse the bifurcation between the producing minority and the consuming majority. Although it did not take long for major media corporations to tap into the cable industry, the community access and local origination features of the technology enabled production and programming by those who had long remained mere consumers of televisual images. By Mel King's understanding, they were able to produce, program, consume, and share what they thought was important to them. They no longer needed to simply consume what others thought was important to them or thought they hoped to see on their television screen.

This advancement in visual culture was significant in African American communities, as was the case with other minority communities. Shawn Michelle Smith, for example, assessed the impact of photographic images on African American identities. Donald Bogle, James Snead, Ed Guerrero, and others have extensively examined the cause, nature, and consequence of African American images in Hollywood films and other film productions. Similar studies have been conducted with television as well. Beretta E. Smith-Shomade, Donald Bogle, Christine Acham, Steven Classen, Robert Entmann, Herman Gray, and Sasha Torres, just to name a few, have analyzed African American images on television. Despite their methodological or thematic differences, they agreed that visual images of African Americans in the twentieth century had distorted Blackness and had hurt identity and self-image of Blacks in many cases. A small number have resisted using mass media to counter such images, which was effective in many cases, but not enough to fully correct the reality of American visual culture.

Cable television had a potential to change this trend in American popular culture. Not only was it going to enable the public to program what was important to them, it was also going to allow them to produce a more fair representation of themselves. Instead of the stereotypical representations of African Americans that were a reason for Mel King to see his friends at school get in a fight, or that served as the basis for Donald Bogle's book title, more accurate images of African Americans were going to appear on cable television screens. Black images were no longer contingent on how the producing few, who were mostly white, viewed African Americans. With local origination and community access, cable television was going to allow Black images as produced by Blacks to appear on television. Cable television, therefore, was another means to develop the Black image in the Black mind.

Despite the capacity to enable such image production on cable television, it still required effort by African Americans to realize cable television's potential. Both in Boston and Detroit, Black community leaders and residents expressed their interests in cable television before the cities began their full-scale feasibility studies. 1963 marked a significant year, when the power of visual images became obvious in the eyes of many civil rights activists, and the late 1960s and the early 1970s witnessed various discussions about how African Americans and other minorities would best benefit from the narrowcast technology. What they discussed was not simply whether cable systems would be good for their communities, but also how to make them useful for their own purposes. Their involvement with cable commissions and other municipal bodies attest that Black residents were willing to use the technology instead of simply consuming what would be provided to them.

African American involvement during this early part of cable television history in both cities served as the basis for how city offices formulated their RFPs. As much as the documents required an elaboration on the financial and technical capabilities of the candidate corporations and proposed systems, they also mandated a large number of community access and local programming channels, recruitment and hiring of local employees, contributions to local citizens through scholarship, complimentary job training, internship, and other forms of community outreach, a prioritization of local and minority-owned businesses over others for subcontractors, and so on. These policies were included in the RFPs not only because of a requirement by the FCC, but significantly because local residents had been assertive throughout the RFPs' drafting process.

In return, applicants made efforts to meet and surpass requests by the cities. Not only Cablevision in Boston and Barden in Detroit, but also non-franchised candidates promised a significant commitment in local engagement. Specific plans for local origination studios, fellowships for students, the number of minority employees to be hired, channel spaces for African American and other minority viewers, and others were stated in their applications. Similarly, the evaluation of these proposals extensively examined how their plans would affect local residents. The fact that Barden, which was not the most financially stable, won the franchise in Detroit attests to the fact that officials understood the importance of what the company promised to bring to the city, despite its financial weakness.

African American involvement, city officials' commitment, and cable franchise companies' ardent efforts did not guarantee a smooth implementation of the systems, however. In Boston, equipment and facilities were not readily available as was originally planned. Although the construction of the system went relatively smoothly, residents also had to urge Cablevision to meet the original promises made in its application and franchise agreement. Similarly, in Detroit, residents had to keep a close eye on the franchisee to make sure that the system would truly belong to them. During the early development of cable television, a power play between the Detroit City Council and the mayor slowed down the development process. Many residents sent letters and met with city officials to correct the problem. Later, they felt that Barden's plan to involve a Canadian firm was a potential risk, and many were afraid that Barden's partnership would compromise its commitment to the city and its residents. Difficulties and hardships were abundant. These historical facts attest that post-Civil Rights movement efforts by African American community leaders and organizers were not perfect. However, it does not mean that the history of cable television for African Americans does not serve as an instructive historical lesson. How Blacks carried out preventative measures, made plans to achieve a successful implementation of the system, cooperated with city officials to secure a beneficial and fair system for themselves, and made various commitments are all useful components in case studies.

African Americans in Cable Television in a National Context

Academic attention to the history of African Americans and cable television has been insufficient. Many autobiographies, such as those of Robert L. Johnson and Gil Noble, exist. Similarly, numerous popular and trade books discuss the history and significance of televisual images, including some from cable television. Historians, sociologists, media studies scholars, and others also have invested much in studying images of African Americans and television. Chapter 1 of this book extensively recapitulated this intellectual history. However, there has been no scholarly book focusing specifically on African American grassroots movements around cable television with a special emphasis on local communities. Canonical African American historians, including Du Bois, Woodson, and others, as well as activists such as Martin Luther King, Paul Robeson, and a countless number of others have voiced their awareness of the power that images possessed even before the popularization of cable television. This book has attempted to fill this gap. And although far from being comprehensive, this can be the beginning of a wave of new scholarships examining how African Americans obtained and employed their agency for fair representation on television.

What cable television portrayed through public access and local origination programming was similar to what Barry Dornfeld termed "public television documentaries." Even though such programs were not "documentaries" by strict definition and were often fictional, many of them still reflected their "conceptions about their social and biological lives" while "[presenting] viewers with depictions of and assertions about the daily lives, institutions, cultural values, and histories of people like themselves or others." These programs allowed viewers to "grapple with and reproduce understandings of cultural identity and cultural difference." Examining these visual representations and the process of producing them allows us to understand how identities are formed, adjusted, altered, and strengthened.[11]

It is also important to identify the uniqueness of the history of narrowcasting in the context of popularized cable television services. Just one example of the controversy over BET suffices to introduce the conundrum of the production of Black images and issues of ownership and control. As a young student, Mel King became aware of the significance of having ownership and control over the visual culture pertaining to his racial images when he visited a Black theater in North Carolina. One of the major reasons why BET became popular among African Americans was not simply because it focused on African American interests, but also because the company was run by an African American. Robert Johnson allowed many African Americans to feel that they finally owned and controlled a major televisual image production outlet.

Particularly because of the externally imposed value that BET had, Johnson's decision to sell his corporation to Viacom in 2000 became a significant event not just in the media world, but also in African American communities. Although the business deal gave birth to the first African American billionaire in the United States, many branded him a sell-out and questioned why he had not found a Black buyer for the company. Although Johnson's interest was clearly in entrepreneurship and business rather than in fair Black representation, the outcry from Black communities shows their need for a television outlet that shares African American perspectives. Kristal Brent Zook explained that the rise in the number of Black-owned radio and television stations from 30 in 1976 to 260 in the late 2000s was a significant growth. The trend has since been reversed and Blacks have begun to lose ownership of their media. Black ownership of media outlets turned out to not be as successful as expected, both in terms of its size and its impact on media content produced in the United States in general.[12]

The controversy surrounding BET was not a unique case. Many African American businesses depended on white-dominated American society. In some cases, African American businesses underscored Black inferiority. Even Madame C. J. Walker did not escape such an accusation. Despite her well-deserved accolades, her hair care products reinforced the idea that straight hair, a typical characteristic of the white, was more ideal than Black hair. Associations between whiteness and beauty are still prevalent. In other instances, African American businesses failed to maintain African American–centered business models. BET is one such example. Similar cases exist throughout African American entrepreneurial history.

The case of BET shows that unlike initial hopes that many African Americans held in 1980 at the launch of the station or in 1991 when it became the first African American company to be listed on the New York Stock Exchange, Johnson's entrepreneurial project did not provide the solution to the century-long unfair representation of African Americans on the television screen.

Instead, the answer existed at the grassroots level. Local origination and public access channels were small scale but a useful outlet for identity production. African Americans in both Boston and Detroit committed themselves to the introduction and implementation of this identity formation technology. In a broader context, the experiences of these two cities suggest that technology has the potential to bring positive changes to and work as a driver of community justice for African Americans. Their role in the history of cable television demonstrates the importance of participation in decision making, drafting and designing of the cable system, public hearings with city officials and franchise candidates, and other phases, in enabling control and ownership over media technology and the content produced with it.

For future community activists, community leaders, and community members, this framework for African American involvement that found success in Boston and Detroit in the 1970s and 1980s offers different ways to use media technology for community building purposes. Regardless of the medium that a community plans to introduce or use, they are likely to benefit from it by securing a means to reflect the voice of the public from the beginning of any community-building project. It is also important to be aware that instead of relying on external sources to determine the kind of images, messages, or content to offer community residents, media technology as a driver of community justice must invite the participation of residents and reflect their wants. Cable television was successful in addressing the opinions and ideas of local community members because of its narrowcasting capabilities. Because the cable system could offer greater channel capacities than the conventional television system, city residents and officials could expect various types of programming satisfying the specific needs of people whose voice had little been heard.

The history of cable television is a history of African American community empowerment and building, of their access to televisual agency of successful examples of African American infrapolitics, and of the use of media technology as a vehicle of community justice. African American cable activism in Boston and Detroit was a grassroots movement that constituted a part of larger African American efforts to achieve true citizenship during both the Civil Rights and post-Civil Rights periods.

Notes

Introduction

1. The scholarship dealing with the Black image in the Black mind is not new. Foundational works sharing this perspectives include W. E. B. Du Bois, *The Souls of Black Folk* (1903; repri., New York: Vintage Books, 1990); Carter G. Woodson, *The Mis-Education of the Negro* (1933; repri., Drewryville, VA: Khabooks, 2006).

2. J. Fred MacDonald, *Blacks and White TV: African Americans in Television Since 1948* (Chicago, IL: Nelson-Hall Publishers, 1992), xi, xvi–xvii.

3. Donald Bogle, *Prime Time Blues: African Americans on Network Television* (New York: Farrar, Straus and Giroux, 2001), 4. For a more complete history of network television and movies in relation to African American representation and participation, see Donald Bogle, *Blacks in American Films and Television: An Encyclopedia* (New York: A Fireside Books, 1988); Donald Bogle, *Bright Boulevards: The Story of Black Hollywood* (New York: Ballantine Books, 2005); Donald Bogle, *Toms, Coons, Mulattoes, Mammies, & Bucks: An Interpretive History of Blacks in American Films* (New York: Continuum, 2006); Marc Cushman and Linda J. LaRosa, *I Spy: A History and Episode Guide to the Groundbreaking Television Series* (Jefferson, NC: McFarland & Company, Inc., 2007). Bogle not only discussed the lack of Black representation and misrepresentation in visual media, he also argued that such problems of television and movies reflected and reinforced existing race-based thought on African Americans in their own minds, as well as those of non-Black viewers.

4. "Television: Negro Performers Win Better Roles in TV than in Any Other Entertainment Medium," *Ebony* (June 1950), 22.

5. George M. Frederickson, *The Black Image in the White Mind: The Debate on Afro-American Character and Destiny, 1817–1914* (Middletown, CT: Wesleyan University Press, 1971); Mia Bay, *The White Image in the Black Mind: African-American Ideas about White People, 1830–1925* (New York: Oxford University Press, 2000); Robert M. Entman and Andrew Rojecki, *The Black Image in the White Mind: Media and Race in America* (Chicago, IL: University of Chicago Press, 2000), 241.

6. Du Bois, *The Souls of Black Folk*, 8.

7. Ralph Ellison, *Invisible Man* (New York; Vintage International, 1947, 1995), 4.

8. Franz Fanon, *Black Skin White Masks* (New York: Grove Press, 1967), 110.

9. Alice Tait and Todd Burroughs, "Mixed Signals: Race and the Media," in *Race and Resistance: African Americans in the Twenty-First Century*, ed. Herb Boyd (Cambridge, MA: South End Press, 2002), 103.

10. Robin D. G. Kelley, "'We Are Not What We Seem': Rethinking Black Working-Class Opposition in the Jim Crow South," *The Journal of American History* 80, no. 1 (June 1993): 77–78; James C. Scott, *Domination and the Arts of Resistance: Hidden Transcripts* (New Haven: Yale University Press, 1990).

11. See Scott, *Domination and the Arts of Resistance*, 183; Kelly, "'We Are Not What We Seem,'": 77.

12. Tavis Smiley, *The Covenant* (Chicago, IL: Third World Press, 2006); Tavis Smiley, *The Covenant in Action* (Carlsband, CA: Smiley Books, 2006); National Urban League, *The State of Black America 2007: Portrait of the Black Male* (New York: The Beckham Publications Group, 2007); Michael Eric Dyson, *Come Hell or High Water: Hurricane Katrina and the Color of Disaster* (Cambridge, MA: Basic Books, 2006).

13. Entman and Rojecki, *The Black Image in the White Mind*; Darnell M. Hunt, *Channeling Blackness: Studies on Television and Race in America* (New York: Oxford University Press, 2005); John Fiske and John Hartley, *Reading Television* (New York: Routledge, 1978, 2003); John Fiske, *Television Culture* (New York: Routledge, 1987, 2004); John Fiske, *Media Matters: Race and Gender in U.S. Politics* (Minneapolis, MN: University of Minnesota Press, 1996, 1999); Herman S. Gray, *Watching Race: Television and the Struggle for Blackness* (Minneapolis, MN: University of Minnesota Press, 1995, 2004); Herman S. Gray, *Cultural Moves: African Americans and the Politics of Representation* (Berkeley, CA: University of California Press, 2005); Alan Nadel, *Television in Black-and-White America: Race and National Identity* (Lawrence, KS: University Press of Kansas, 2005).

14. U.S. Department of Commerce, Bureau of the Census, *United States Census of Population*, 1950–1990 (Washington, D.C.: U.S. Government Printing Office, various years).

15. The Detroit News, *The Voices of Detroit's Blacks* (Detroit, MI: The Detroit News, 1971), 1.

16. U.S. Department of Commerce, *United States Census of Population*, 1950–1990 (various years).

17. Robert Putman and Lewis M. Feldstein, *Better Together: Restoring the American Community* (New York: Simon & Schuster, 2003), 75–97.

18. Susan E. Eaton, *The Other Boston Busing Story* (New Haven, CT: Yale University Press, 2001), 164.

19. Joe T. Darden, Richard Child Hill, June Thomas, and Richard Thomas, *Detroit: Race and Uneven Development* (Philadelphia, PA: Temple University Press, 1987), 209–20; Heather Ann Thompson, *Whose Detroit?: Politics, Labor, and Race in a Modern American City* (Ithaca, NY: Cornel University Press, 2001), 84–85.

20. Richard W. Thomas, "Black Self-Help in Michigan," in *The State of Black Michigan 1967–2007* ed. Joe T. Darden, Curtin Stokes, and Richard W. Thomas (East Lansing, MI: Michigan State University Press, 2007), 85. Also see, Angela D. Dillard, *Faith in the City: Preaching Radical Social Change in Detroit* (Ann Arbor, MI: The University of Michigan Press, 2007).

21. Sarah-Ann Shaw, "The History of *Say Brother*," *WBGH*, http://main.wgbh.org/saybrother/history.html (accessed March 4, 2009); "Introduction," *WBGH*, http://main.wgbh.org/saybrother/intro.html (accessed March 4, 2009).

22. Juan Williams, *Eyes on the Prize: America's Civil Rights Years 1854–1965* (New York: Penguin Books, 1987), ix–x.

23. Sidney Fine, *Violence in the Model City: The Cavanagh Administration, Race Relations, and the Detroit Riot of 1967* (East Lansing, MI: Michigan University Press, 2007), 184–85.

24. Gordon Castelnero, *TV Land Detroit* (Ann Arbor, MI: The University of Michigan Press, 2006), 8.

25. William A. Lucas and Robert K. Yin, *Serving Local Needs with Telecommunications: Alternative Applications for Public Services* (Santa Monica, CA; The Rand Corporation, 1973), 17.

26. Martin H. Seiden, *Cable Television U.S.A.: An Analysis of Government Policy* (New York: Praeger Publishers, 1972), 193, 196.

27. Linda K. Fuller, *Community Television in the United States: A Sourcebook on Public, Educational, and Governmental Access* (Westport, CT: Greenwood Press, 1994), 156–57.

28. Kelley, "'We Are not What We Seem,'" 78.

29. Lila Abu-Lughod, "The Romance of Resistance: Tracing Transformations of Power through Bedouin Women," *American Ethnologist* 17. no. 1 (February 1990): 42.

30. Michael Foucault, *The History of Sexuality: An Introduction Volume 1* (New York: Vintage Books, 1978), 95–96.

Notes to Chapter 1

1. Donald Bogle, *Bright Boulevards: The Story of Black Hollywood* (New York: Ballantine Books, 2005, 2006), 13.

2. Linda Williams, *Playing the Race Card: Melodramas of Black and White from Uncle Tom to O. J. Simpson* (Princeton, NJ: Princeton University Press, 2001), 98.

3. Anna Everett, *Returning the Gaze: A Genealogy of Black Film Criticism 1909–1949* (Durham, NC: Duke University Press, 2001), 15; Donald Bogle, *Toms, Coons, Mulattoes, Mammies, & Bucks: An Interpretive History of Blacks in American Films* (New York: Continuum, 2006), 7.

4. James Snead, *White Screens Black Images: Hollywood from the Dark Side*. ed. Colin MacCabe and Cornel West (New York: Routledge, 1994), 77.

Notes to Chapter 1

5. Ed Guerrero, *Framing Blackness: The African American Image in Film* (Philadelphia, PA: Temple University Press, 1993), 70.
6. Bogle, *Toms, Coons, Mulattoes, Mammies, & Bucks*, 204.
7. Ibid., 220.
8. Ibid., 222.
9. Ibid.
10. Ibid., 215–19.
11. Thomas Cripps, "'Race Movies' as Voices of the Black Bourgeoisie: *The Scar of Shame*," in *Representing Blackness: Issues in Film and Video* ed. Valerie Smith, (New Brunswick, NJ: Rutgers University Press, 2003), 47; James Gaines, "*The Scar of Shame*: Skin Color and Caste in Black Silent Melodrama," in *Representing Blackness: Issues in Film and Video* ed. Valerie Smith, (New Brunswick, NJ: Rutgers University Press, 2003), 61–81.
12. Valerie Smith, "Introduction," in *Representing Blackness: Issues in Film and Video* ed. Valerie Smith, (New Brunswick, NJ: Rutgers University Press, 2003), 2.
13. Sydney Poitier, *The Measure of a Man: A Spiritual Autobiography* (New York: Harper and Collins, 2000), 112.
14. Bogle, *Toms, Coons, Mulattoes, Mammies, & Bucks*, 219.
15. Guerrero, *Framing Blackness*, 69.
16. Ibid., 93.
17. Ibid., 94.
18. Ibid., 70.
19. Ibid., 72.
20. Ibid., 94.
21. Ibid.
22. Bogle, *Toms, Coons, Mulattoes, Mammies, & Bucks*, 242.
23. Stuart Hall, "What is This 'Black' in Black Popular Culture?" in *Representing Blackness: Issues in Film and Video* ed. Valerie Smith, (New Brunswick, NJ: Rutgers University Press, 2003), 127.
24. Ibid., 129.
25. See Wahneema Lubiano, "But Compared to What?: Reading Realism, Representation, and Essentialism in *School Daze, Do the Right Thing*, and the Spike Lee Discourse," in *Representing Blackness: Issues in Film and Video* ed. Valerie Smith, (New Brunswick, NJ: Rutgers University Press, 2003), 97–122; bell hooks, *Reel to Real* (New York: Routledge, 1996), 227–29.
26. Snead, *White Screens Black Images*, 118–19.
27. Bogle, *Toms, Coons, Mulattoes, Mammies, & Bucks*, 318–23.
28. Snead, *White Screens Black Images*, 118–19.
29. Bogle, *Toms, Coons, Mulattoes, Mammies, & Bucks*, 357.
30. Lubiano, "But Compared to What?" 97–122.
31. Ibid., 99.
32. Guerrero, *Framing Blackness*, 149.
33. Ibid., 154.
34. Cecilia Tichi, *Electronic Heath: Creating an American Television Culture* (New York: Oxford University Press, 1991), 3–10.

35. Donald Bogle, *Prime Time Blues: African Americans on Network Television* (New York: Farrar, Straus and Giroux, 2001), 406.

36. John Fiske and John Hartley, *Reading Television* (New York: Routledge, 1989, 1996), 19.

37. Classen, *Watching Jim Crow*.

38. Herman Gray, "The Politics of Representation in Network Television," in *Channeling Blackness: Studies on Television and Race in America* ed. Darnell M. Hunt (New York: Oxford University Press, 2005), 159.

39. Sasha Torres, *Black White and Color: Television and Black Civil Rights* (Princeton, NJ: Princeton University Press, 2003), 22.

40. Robert M. Entman and Andrew Rojecki, *The Black Image in the White Mind: Media and Race in America* (Chicago, IL: The University of Chicago Press, 2000), 105.

41. Ibid.

42. Michael Eric Dyson, *Is Bill Cosby Right?: Or Has the Black Middle Class Lost Its Mind?* (New York: Basic Books, 2005), 41–43.

43. Bogle, *Prime Time Blues*, 287.

44. About lynching and rape, see Ida B. Wells-Barnett, *On Lynchings* (New York: Humanity Books, 2002); John Fiske, *Media Matters: Race and Gender in U.S. Politics* (Minneapolis, MN: University of Minnesota Press, 1996, 1999), 108.

45. Michael Foucault, *The History of Sexuality: An Introduction* (New York: Vintage Books, 1990).

46. Fiske, *Media Matters*, 110.

47. Sut Jhally and Justin Lewis, "White Responses: The Emergence of 'Enlightened' Racism," in *Channeling Blackness: Studies on Television and Race in America* ed. Darnell M. Hunt (New York: Oxford University Press, 2005), 74–88.

48. Christopher P. Campbell, "A Myth of Assimilation: 'Enlightened' Racism and the News," in *Channeling Blackness: Studies on Television and Race in America* ed. Darnell M. Hunt (New York: Oxford University Press, 2005), 138.

49. Michael Eric Dyson, "Bill Cosby and the Politics of Race," *Z Magazine*, September 1989, 29.

50. Alan Nadel, *Television in Black and White America: Race and National Identity* (Lawrence, Kansas: Kansas University Press, 2005), 183.

51. Gray, "The Politics of Representation in Network Television," 163.

52. Nadel, *Television in Black and White America*, 13.

53. Robert M. Entman, "Modern Racism and the Images of Blacks in Local Television News," *Critical Studies in Mass Communication* 7. no. 4 (1990): 332–45.

54. Dyson, *Is Bill Cosby Right?* 184.

55. Fiske, *Media Matters*, 221–22.

56. Molefi Kete Asante, "Television and Black Consciousness," in *Channeling Blackness: Studies on Television and Race in America* ed. Darnell M. Hunt (New York: Oxford University Press, 2005), 60.

57. William Peters, "The Visible and Invisible Images," in *Race and the News Media* ed. Paul L. Fisher and Ralph L. Lowenstein (Missouri: Anti-Defamation League of B'nai B'rith, 1967), 81–82.

58. J. Fred MacDonald, *Blacks and White TV: African Americans in Television Since 1948* (Chicago, IL: Nelson-Hall Publishers, 1992), 81.

59. Christine Acham, *Revolution Televised: Prime Time and Struggle for Black Power* (Minneapolis, Minnesota: Minnesota University Press, 2004), 28.

60. Asante, "Television and Black Consciousness," 61.

61. Torres, *Black White and Color*, 8.

62. Stokely Carmichael, "Berkeley Speech," in *Stokely Speaks: From Black Power to Pan-Africanism* (Chicago, IL: Lawrence Hill Books, 1965, 2007), 58.

63. Peniel E. Joseph, *Waiting 'Tin the Midnight Hour: A Narrative History of Black Power in America* (New York: An Owl Book, 2006), 151.

64. Torres, *Black White and Color*, 6, 34.

65. MacDonald, *Blacks and White TV*, 105–08, 138.

66. Gray, "The Politics of Representation in Network Television," 160–61.

67. Bogle, *Prime Time Blues*, 240.

68. Gray, "The Politics of Representation in Network Television," 161.

69. Ibid.

70. MacDonald, *Blacks and White TV*, 222–23.

71. Bogle, *Prime Time Blues*, 242–43.

72. Alice Tait and Todd Burroughs, "Mixed Messages: Race and the Media," in *Race and Resistance: African Americans in the 21st Century* ed. Herb Boyd (Cambridge, MA: South End Press, 2002), 101; Brett Pulley, *The Billion Dollar BET: Robert Johnson and the Inside Story of Black Entertainment Television* (New York: John Wiley and Sons, Inc., 2004).

Notes to Chapter 2

1. The expression "minorities" includes but is not limited to racial minorities, or non-whites. Much scholarship on cable television has studied how the technology could affect those who have not been the main participants of television, such as children, non-white populations, low-income family members, physically disadvantaged populations, and so on.

2. Robert L. Hilliard, *Television Station Operation and Management* (Boston, MA: Focal Press, 1989), 8.

3. About the corrective nature of scholarly works, see Manning Marable, "Black Studies and the Black Intellectual Tradition," *Race and Reason* 4 (1997–1998), 3–8.

4. The origin of the first cable television system is rather contestable. Although many academic and business works refer to John Walson's system as the first of its kind, all his records were destroyed in a fire. Other cable television systems, however, do not predate the mid-1948 statistics about the cable television subscription. Some of these examples include Robert J. Tarlton's system established in Lansford, Pennsylvania, approximately sixty-five miles away from Philadelphia but which had trouble receiving signals due to the Allegheny Mountains. Tarlton set

up antennas on a mountain top in 1949 and founded a cable television company, Panther Valley Television. The installation fee was $125 and a subscription fee was $3 per month. Similarly, Ed Parsons in Astoria, Oregon, built experimental antennas to get signals from Seattle for his wife. For more detailed history about these early cable television systems including that of Walson, see Stephen R. Barnett, "Cable Television and Media Concentration, Part I: Control of Cable Systems by Local Broadcasters," *Stanford Law Review* 22, no. 2 (January 1970): 224–25; Franklin M. Fisher, Victor E. Ferral, Jr., David Belsley, Beidger M. Mitchel, "Community Antenna Television Systems and Local Television Station audience," *The Quarterly Journal of Economics* 80, no. 2 (May 1966): 229; Scott L. Gorland and John W. Stoops, *Community Antenna Television (CATV): In Michigan Municipalities* (Ann Arbor, MI: Michigan Municipal League, 1971), 3; Williams S. Comanor and Bridger M. Mitchell, "Cable Television and the Impact of Regulation," *The Bell Journal of Economics and Management Science* 2, no. 1 (Spring 1971): 155; Steiner, Robert L. "Visions of Cablevision; The Prospects for Cable Television in the Greater Cincinnati Area," *A Report to the Stephen H. Wilder Foundation* (Cincinnati, OH: December 1972): 9–10; George R. Townsend and J. Orrin Marlowe, *Cable: A New Spectrum of Communications*. (Spectrum Communications, Inc., 1974), 7–9; Gilbert Gillespie, *Public Access Cable Television in the United States and Canada: With an Annotated Bibliography* (New York: Praeger Publishers, 1975), 19–21; James D. Scott, "Cable Television Strategy for Penetrating Key Urban Markets," *Michigan Business Reports* 58 (1976): 1; Don Schiller, *CATV Program Origination & Production* (Blue Ridge Summit, PA: Tab Books, 1979), 9–10; United Cable Television Corporation, *Cablecasting & Public Access in the Eighties* (Colorado, 1981), 2–4; Jon S. Denny, *Careers in Cable TV: A Complete Guide to Getting a Job—From Receptionist to Producer—in America's Fastest Growing Entertainment Industry* (New York: Barnes and Noble Books, 1983), 9; James W. Roman, *Cable Mania: The Cable Television Sourcebook* (Englewood Cliffs, NJ: Prentice-Hall, Inc., 1983), 1–4; Thomas F. Baldwin and D. Stevens McVoy, *Cable Communication* (Eaglewood Cliffs, NJ: Prentice-Hall, Inc., 1983), 8–9; Jon S. Denny, *Careers in Cable TV: A Complete Guide to Getting a Job—from Receptionist to Producer—in America's Fastest Growing Entertainment Industry* (New York; Barns & Noble Books, 1983), 9–15; Edward V. Dolan, *TV or CATV?: A Struggle for Power* (New York: Associated Faculty Press, 1984), 62; National Cable Television Association, *A Cable Primer* (1984), 4; Joshua Sapan, *Making It in Cable TV: Career Opportunities in Today's Fastest Growing Media Industry* (New York: Perigee Books, 1984), 11–12; Jan Bone, *Opportunities in Cable Television* (Lincolnwood, IL: VGM Career Horizons, 1984), 1; Alison Melnick, "Access to Cable Television: A Critique of the Affirmative Duty Theory of the First Amendment," *California Law Review* 70, no. 6 (December 1982): 1394–95; Thomas Whiteside, "Onward and Upward with the Arts," *New Yorker*, May 20, 1985, 45–87; Thomas Whiteside, "Onward and Upward with the Arts," *New Yorker*, May 27, 1985, 43–73; Thomas Whiteside, "Onward and Upward with the Arts," *New Yorker*, June 3, 1985, 82–105; The National Cable Television Association, *The National Cable Television Association* (Washington, D.C.: The National Cable Television Association, 1990), 6; Robert

W. Crandall and Harold Furchtgott-Roth, *Cable TV: Regulation or Competition?* (Washington, D.C.: The Brookings Institution, 1996), 1–3; Lawrence J. Davis, *The Billionaire Shell Game: How Cable Baron John Malone and Assorted Corporate Titans Invented a Future Nobody Wanted* (New York: Doubleday, 1998), 9–13; Simon Applebaum, "The Great Cable Controversy: Who Launched First?" *Cable Vision*, May 4, 1998, 16–17; Thomas R. Eisenmann, "The U.S. Cable Television Industry, 1948–1995: Managerial Capitalism in Eclipse," *The Business History Review* 74, no. 1 (Spring 2000): 4–5; Brian Lockman and Don Sarvey, *Pioneers of Cable Television* (Jefferson, NC: McFarland & Company, Inc., 2005), 9–24; Joseph N. DiStefano, *COMCASTed: How Ralph and Brian Roberts Took Over America's TV One Deal at a Time* (Philadelphia, PA: Camino Books, Inc. 2005), 28–32.

5. Readers who wish to know more about Walson's invention and business expansion should read sources introduced previously.

6. Martin H. Seiden, *Cable Television U.S.A.: An Analysis of Government Policy* (New York: Praeger Publishers, 1972), vii.

7. Barnett, "Cable Television and Media Concentration," 221.

8. Ibid.

9. Comanor and Mitchell, "Cable Television and the Impact of Regulation," 155–57, Rovert L. Hillard, "Principles and Issues," in *Television Station Operations and Management*, ed. Robert L. Hillard (Boston, MA: Focal Press, 1989), 3–4, 8.

10. Richard Roud, "Cable Television and the Arts," *A Report Prepared for the Sloan Commission on Cable Communications* (March 1971): 2–3.

11. Ibid., 2–3, 17–18.

12. Judy Crichon, "Toward an Immodest Experiment in Cable Television: Modestly Produced," *A Report Prepared for the Sloan Commission on Cable Communications* (March 1971), 1–6.

13. Harald Mendelsohn, "The Neglected Majority: Mass Communications and the Working Person," *A Report Prepared for the Sloan Commission on Cable Communications* (March 1971): 45–46.

14. Ibid., 71–72.

15. Ibid., 74–76.

16. Gilbert Cranberg, "Cable Television and Public Safety," *A Report Prepared for the Sloan Commission on Cable Communications* (May 1971): 1–5, 8–10. Also see John Mansell, "Security and Fire Alarm Services," in *The Community Medium* ed. Nancy Jasuale and Ralph Lee Smith (Arlington, VA: The Cable Television Information Center, 1982), 40–45, 49.

17. Ibid., 18–19.

18. Ithiel de Sola Pool and Hervert E. Alexander, "Politics in a Wired Nation," *A Report Prepared for the Sloan Commission on Cable Communications* (September 1971): 1.

19. Ibid., 25.

20. Herbert S. Dordick and Jane Lyle, *Access by Local Political Candidates to Cable television: A Report of an Experiment* (Santa Monica, CA: The Rand Corporation, 1971).

21. Pool and Alexander, "Politics in a Wired Nation," 25.
22. Ibid., 43.
23. Lionel Kestenbaum, "Common Carrier Access to Cable Communications: Regulatory and Economic Issues," *A Report Prepared for the Sloan Commission on Cable Communications* (August 1971): 1, 4, 7–8.
24. Don R. Le Duc, "The Cable Question: Evolution or Revolution in Electronic Mass Communications," *Annals of the American Academy of Political and Social Science* 400 (March 1972): 129; Richard A. Posner, "The Appropriate Scope of Regulation in the Cable Television Industry," *The Bell Journal of Economics and Management Science*, 3 no. 1 (Spring 1972): 98.
25. Rolla Edward Park, "Prospects for Cable in the 100 Largest Television Markets," *The Bell Journal of Economics and Management Science*, 3. no. 1 (Spring 1972): 130.
26. Le Duc, "The Cable Question," 127, 129–33.
27. Monroe R. Price, "Requiem for the Wired Nation: Cable Rulemaking at the FCC," *Virginia Law Review* 61, no. 3 (April 1975): 545.
28. Ibid., 548–50. Also see Davis, *The Billionaire Shell Game*, 27–28.
29. Richard C. Kletter, "Making Public Access Effective" in *Cable Television: Developing Community Services* ed. Polly Carpenter-Huffman, Richard C. Kletter, and Robert K. Yin (New York: Crane, Russak & Company, Inc., 1974), 8.
30. Walter S. Baer, "Preface" in *Cable Television: Developing Community Services* eds. Polly Carpenter-Huffman, Richard C. Kletter, and Robert K. Yin (New York: Crane, Russak & Company, Inc., 1974), v–vi.
31. Robert K. Yin, *Cable Television: Applications for Municipal Services* (Santa Monica, CA; The Rand Corporation, 1973), ix.
32. Community Research Incorporated, *Cable Communications: The Facts, the Issues and the Possibilities* (Dayton, OH: Community Research Incorporated, 1973).
33. Ibid.
34. Dell Keehn, *A Short Review of the Development of Cable Television to 1973* (Seattle, WA: University of Washington, 1973), 21–25.
35. Leonard Ross, *Economic and Legal Foundations of Cable Television* (Beverly Hills, CA: Sage Publications, 1974), 7–8; Walter S. Baer, *Cable Television: A Handbook for Decisionmaking* (New York: Crane, Russak & Company, Inc., 1974), xv, 1; Robert E. Jacobson, *Municipal Control of Cable Communications* (New York: Praeger Publishers, 1977); 1.
36. Amitani Etizioni, introduction to *Economic and Legal Foundations of Cable Television*, by Leonard Ross, 6.
37. Ross, *Economic and Legal Foundations of Cable Television*, 7–8.
38. Baer, *Cable Television*, 1.
39. Ibid., 7. About the relationship between economy and federal regulations, see Harry M. Trebing, "Broadening the Objectives of Public Utility Regulation," *Land Economics* 53. no. 1 (February 1977): 106–22; Drew Shaffer, "Introduction—Overview," in *Creating Original Programming for Cable TV*, ed. Drew Shaffer and Richard Wheelwright (Washington, D.C.: Communications Press, Inc., 1983), 3.

40. Frank Korman, "Innovations in Telecommunications Technology: A Look Ahead," *Educational Broadcasting Review* 6 (Oct. 1972): 328.

41. Gillespie, *Public Access Cable Television*, 4–5.

42. Charles Tate, "Community Control of Cable Television Systems," in *Talking Back, Citizen Feedback and Cable Technology* ed. Ithiel de Sola Pool (Cambridge, MA: The MIT Press, 1973), 54–55.

43. Ibid., 57–58.

44. Ibid., 60.

45. Marion Hayes Hull, "Minorities," *Cable Handbook 1975–1976: A Guide to Cable and New Communications Technologies* ed. Mary Louise Hollowell (Washington, D.C.: Communications Press, Inc., 1975): 121.

46. Ibid., 126.

47. Ibid., 127–28.

48. Ibid.

49. Ibid.

50. The Cabinet Committee on Cable Communications, *Cable Report to the President*, 1974; Robert H. Finch, *Cable: Report to the President*, 1974, 47–48.

51. Ibid.

52. Rudy Bretz, *Handbook for Producing Educational and Public-Access Programs for Cable Television* (Englewood Cliffs, NJ: Educational Technology Publications, 1976), 25–26.

53. Scott, "Cable Television," 2–5.

54. Townsend, *Cable*, 54, 60–61; Jacobson, *Municipal Control of Cable Communications*, 35.

55. Ibid., iv, 20–21.

56. Ibid., 23–24.

57. Jacobson, *Municipal Control of Cable Communications*, 67.

58. Community Research Incorporated, *Cable Communications*; Baer, *Cable Television*, xv–xvii; Leland L. Johnson and Michael Botein, *Cable Television: The Process of Franchising* (Santa Monica, CA: The Rand Corporation, 1973), iii.

59. Walter S. Baer, *Cable Television: A Summary Overview for Local Decisionmaking* (Santa Monica, CA: The Rand Corporation, 1973), 2.

60. Ibid., 3, 10–13.

61. Johnson and Botein, *Cable Television*, vi.

62. Ibid., 37–38.

63. Richard C. Kletter, *Cable Television: Making Public Access Effective* (Santa Monica, CA: The Rand Corporation, 1973), vii.

64. Robert K. Yin, *Citizen Participation in Planning* (Santa Monica, CA: The Rand Corporation, 1973), 11–12.

65. Ibid., 12.

66. Davis, *The Billionaire Shell Game*, 22–23; United Cable Television Corporation, *Cablecating & Public Access in the Eighties* (Denver, CO: United Cable Television Corporation, 1981), iii.

67. United Cable Television Corporation, *Cablecating & Public Access in the Eighties*, iii.

68. Susan Bednarczyk and others, "Community Programming: Public Access and Local Origination," in *CTIC Cablebooks: Volume 1 The Community Medium*, ed. Nancy Jesuale and Ralph Lee Smith (Arlington, VA: The Cable Television Information Center, 1982), 62–63, 70; United Cable Television Corporation, *Cablecasting & Public Access*, iii, 4–6, 29–30. Susan Wallace and Robert E. Jacobson also shared the same opinion about the importance of cooperation. See Susan Wallace, "Programming Sources," in *Creating Original Programming for Cable TV*, ed. Drew Shaffer and Richard Wheelwright (Washington, D.C.: Communications Press, Inc., 1983), 15; Jacobson, *Municipal Control of Cable Communications*, 2.

69. Stanley M. Basesn and Robert W. Crandall "The Deregulation of Cable Television," *Law and Contemporary Problems*, 44. no. 1 (Winter 1981): 77–124.

70. Jules F. Simon, "The Collapse of Consensus: Effects of the Deregulation of Cable Television," *Columbia Law Review* 81, no. 3 (April 1981): 612–638.

71. Daniel Brenner, "Cable Television and the Freedom of Expression," *Duke Law Journal* 1988. no. 2/3 (April–June, 1988): 387–388. About the First Amendment, also see, Dee Pridgen and Eric Engel, "Advertising and Marketing on Cable Television: Whither the Public Interest?" in *Creating Original Programming for Cable TV* ed. Mary Louise Hollowell (Washington, D.C.: Communications Press, Inc., 1983), 131–39.

72. Melnick, "Access to Cable Television," 1414.

73. Schiller, *CATV Program Origination & Production*, 46–50.

74. United Cable Television Corporation, *Cablecasting & Public Access*, 10. Ann McIntosh and Steve Feldman also emphasized the significance of local study and its positive impact on audience. See Ann McIntosh, and Steve Feldman, "Producing Original Programming for Cable," in *Creating Original Programming for Cable TV*, ed. Drew Shaffer and Richard Wheelwright (Washington, D.C.: Communications Press, Inc., 1983), 23.

75. Ronald R. Frank and Marshall G. Greensberg, *Audience for Public Television* (Beverly Hills, CA: Sage Publications, 1982), 187–90, 193–206.

76. Ibid., 193.

77. Ibid.

78. Gil Noble, *Black is the Color of My TV Tube* (New York: Carol Publishing Group, 1981), 23.

79. Ibid., 34.

80. Ibid., 54.

81. Ibid., 57–58.

82. Ibid., 37.

83. Ibid., 125.

84. Kirsten Beck, *Cultivating the Wasteland: Can Cable Put the Vision Back in TV?* (New York: American Council for the Arts, 1983), 114.

85. Lawrence N. Redd, "The Use of Two-Way Television to Solve Problems of Inequality in Education: A Comment," *The Journal of Negro Education*, 52. no. 4 (Autumn 1983): 450.

86. Dolan, *TV or CATV?* 71–83.

87. Ibid., 77.

88. Susan Wallece, "Programming Sources," in *Creating Original Programming for Cable TV* ed. Mary Louise Hollowell (Washington, D.C.: Communications Press, Inc., 1983), 15.

89. Ibid., 15–17.

90. Sapan, *Making It in Cable TV*, 34–35.

91. Dolan, *TV or CATV?* vii.

92. Ibid., 8.

93. Ibid.

94. Ibid., 8–11.

95. Ibid., 17–19.

96. Ibid., 29–31.

97. Ibid., 67, 88.

98. Bone, *Opportunities in Cable Television*, 105, 132–33.

99. Roman, *Cable Mania*, 57.

100. Ibid., 177–181.

101. James P. Mooney, A Message to *the National Cable Television Association*, by the National Cable Television Association (Washington, D.C.: The National Cable Television Association, 1990), 4; Marc Doyle, *The Future of Television: A Global Overview of Programming, Advertising, Technology, and Growth* (Lincolnwood, IL: NTC Business Books, 1992), 3.

102. Linda K. Fuller, *Community Television in the United States: A Sourcebook on Public, Educational, and Governmental Access* (Westport, CT: Greenwood Press, 1994), 4–5.

103. Donald R. Browne, Charles M. Firestone, and Ellen Mickiewicz, *Television/Radio News and Minorities* (Queenstown, MD: The Aspen Institute, 1994), 27–29.

104. Crandall and Furchtgott-Roth, *Cable TV*, 4–9.

105. David Waterman and Andrew A. Weisse, *Vertical Integration in Cable Television* (Cambridge, MA: The MIT Press, 1997).

Notes to Chapter 3

1. *Social Services and Cable TV*. I1–I3.

2. Kristal Brent Zook, *I See Black People: The Rise and Fall of African American-Owned Television and Radio* (New York: Nation Books, 2008), 2.

3. Ibid., x.

4. Barry Bluestone and Mary Huff Stevenson, *The Boston Renaissance: Race, Space, and Economic Change in an American Metropolis* (New York: Russell Sage Foundation, 2000), 1–3.

5. Katherine L. Bradbury, Anthony Downs, and Kenneth A. Small, *Urban Decline and the Future of American Cities* (Washington, D.C.: Brookings Institution, 1982), 50.

6. Bluestone and Stevenson, *The Boston Renaissance*, 1; Bradbury, Downs, and Small, *Urban Decline and the Future of American Cities*, 50; U.S. Census Bureau, "Census 2000 Demographic Profile Highlights," http://factfinder.census.gov/servlet/SAFFFacts?_event=Search&geo_id=&_geoContext=&_street=&_county=boston&_cityTown=boston&_state=04000US25&_zip=&_lang=en&_sse=on&pctxt=fph&pgsl=010 (accessed March 5, 2009).

7. Susan E. Eaton, *The Other Boston Busing Story: What's Won and Lost Across the Boundary Line* (New Haven, CT: Yale University Press, 2001); Ronald P. Formisano, *Boston Against Busing: Race, Class, and Ethnicity in the 1960s and 1970s* (Chapel Hill, NC: The University of North Carolina Press, 1991); Jon Hillson, *The Battle of Boston* (New York: Pathfinder Press, 1977); Jennifer Hochschild and Nathan Scovronick, *The American Dream and the Public Schools* (New York: Oxford University Press, 2003); Michael J. Ross, *"I Respectfully Disagree with the Judge's Order": The Boston School Desegregation Controversy* (Beverly, MA: Commonwealth Editions, 2006); and Jeanne Theoharis, "They Told Us Our Kids Were Stupid: Ruth Baston and the Educational Movement in Boston," in *Groundwork: Local Black Freedom Movements in America*, ed. Jeanne Theoharis and Komozi Woodard (New York: New York University Press, 2005).

8. Hubie Jones, "Social and Political History of Race in Boston from 1950–2000" (lecture, University of Massachusetts Boston, Boston, MA, December 5, 2005).

9. George Lipsitz, *The Possessive Investment in Whiteness: How White People Profit from Identity Politics* (Philadelphia: Temple University Press, 1998), 1–23.

10. Mel King, *Chain of Change: Struggles for Black Community Development* (Boston: South End Press, 1981), xx; Robert J. Allison, *A Short History of Boston* (Beverly, MA: Commonwealth Editions, 2004): 107; Thomas H. O'Connor, *The Hub: Boston Past and Present* (Boston, MA: Northeastern University Press, 2001), 20–227.

11. Lipsitz, *The Possessive Investment in Whiteness*, 5–11.

12. The Boston Foundation, "Thinking Globally / Acting Locally: A Regional Wake-Up Call," in *A Summary of the Boston Indicators Report 2002–2004*, March 2005, 9.

13. Melvin King, interviewed by the author, Technology Center at Tent City, October 31, 2005; Michael Dukakis, interviewed by the author, Northeastern University, November 7, 2005; Hubie Jones, "Social and Political History of Race in Boston from 1950–2000."

14. Bluestone and Stevenson, *The Boston Renaissance*, 128–30, 209–10.

15. Bluestone and Stevenson, *The Boston Renaissance*, 108–13.

16. O'Connor, *The Hub*, 262; The Boston Foundation, "Racial and Ethnic Diversity," http://www.tbf.org/indicators/civic-health/indicators.asp?id=941&fID=209&fname=Race/Ethnicity (accessed March 5, 2009); U.S. Department of Commerce, Bureau of the Census. "1960 General Population Characteristics,"

(Washington, D.C.,: Government Printing Office, 1962); U.S. Department of Commerce, Bureau of the Census. "1970 General Population Characteristics," (Washington, D.C.,: Government Printing Office, 1972); U.S. Department of Commerce, Bureau of the Census. "1980 General Population Characteristics," (Washington, D.C.,: Government Printing Office, 1982); The Boston Foundation, "Thinking Globally / Acting Locally," 9.

17. City of Boston, *Boston's Economy: Excerpt from the Official Statement of the City of Boston, Massachusetts. $75,000,000 General Obligation Bonds* (Boston: Boston Redevelopment Authority, 1995).

18. Quoted in Bluestone and Stevenson, *The Boston Renaissance*, 348.

19. Ibid.

20. Hochschild and Scovronick, *The American Dream and the Public Schools*, 184; Thomas H. O'Connor, *South Boston: My Home Town* (Boston, MA: Northeastern University Press, 1994, 1998), 216–30; Bluestone and Stevenson, *The Boston Renaissance*, 349.

21. Jacqueline Jones, *American Work: Four Centuries of Black and White Labor* (New York: W. W. Norton and Company, 1998), 372–74; Bluestone and Stevenson, *The Boston Renaissance*, 215. The Boston Foundation, "Thinking Globally / Acting Locally," 21.

22. Edward Murguia, *Political Capital and the Social Reproduction of Inequality in a Mexican Origin Community in Arizona*, ed. Michael Peter Smith and Joe R. Feagin (Minneapolis, MN: University of Minnesota Press, 1995), 310–11.

23. Although Bluestone and Stevenson originally use the term, *industrial change* as the second tenet, I use the term *economic change* to better contextualize this social change in my project.

24. Jacqueline Jones, *American Work*, 301–36; Bluestone and Stevenson, *The Boston Renaissance*, 52; Alan Dawley, *Struggles for Justice: Social Responsibility and the Liberal State* (Cambridge, MA: Harvard University Press, 2000), 256–57.

25. Bluestone and Stevenson, *The Boston Renaissance*, 51.

26. Hard skill and soft skill are the notions often used in the business world. Whereas hard skill brings a tangible end result, soft skill is an interpersonal or analytical skill.

27. U.S. Department of Labor, Bureau of Labor Statistics, *Employment, Hours, and Earnings for States and Local Areas* (Washington, D.C.: Government Printing Office, 1995); Allison, *A Short History of Boston*, 95–96.

28. King, interview; King, *Chain of Change*, 11; Bluestone and Stevenson, *The Boston Renaissance*, 25.

29. Ibid. Although Mel King does not remember seeing any movies with African American actors playing a heroic role, African American cowboys already existed in his childhood. The number of such movies, however, was limited. See Philip Durham and Everett L. Jones, *The Negro Cowboys* (Lincoln, NE: University of Nebraska Press, 1965), 220–31; Julia Leyda, "Black-Audience Westerns and the Politics of Cultural Identification in the 1930s," *Cinema Journal* 42. no. 1 (2002): 46–70.

30. Michael Eric Dyson, *Why I love Black Women* (New York: Basic Books, 2003), 256–57; Franny Nudelman, *John Brown's Body: Slavery, Violence and the Culture of War* (Chapel Hill, NC: The University of North Carolina Press, 2004), 173–76.

31. King, *Chain of Change*, 11; Harvard Sitkoff, *The Struggles for Black Equality 1954–1980* (Toronto: Harper and Collins, 1981), 76.

32. King, *Chain of Change*, 12–13.

33. Ibid., 14.

34. King, *Chain of Change*, xx–xxi.

35. Sasha Torres, *Black White and in Color: Television and Black Civil Rights* (Princeton, NJ: Princeton University Press, 2003), 91–93.

36. King, interview; Dukakis, interview.

37. King, interview.

38. City of Boston Finance Commission, *Statement* (Boston, MA: 1971), 1–2. Government Documents Department, Boston Public Library.

39. Whitewood Stamps, *Cable in Boston: A Basic Viability Report* (Newton, MA: 1974), 4–6. Government Documents Department, Boston Public Library; Boston Consumers' Council, *Report of the Boston Consumers' Council to the Honorable Kevin H. White, Mayor on the Development of a Cable Television System* (November 1973), I. Government Documents Department, Boston Public Library.

40. City of Boston Finance Commission, *Statement*, 1–3, 11.

41. Konrad K. Kalba, "Cable Television and the Development of the Boston Region," in *The Boston Development Strategy Research Project: Volume 1* (Boston, MA: 1972). Government Documents Department, Boston Public Library.

42. Ibid., 677.

43. Ibid., 678–79.

44. Ibid., 679.

45. Ibid., 681–82.

46. Ibid., 682.

47. Ibid., 682–83.

48. Ibid., 694–95.

49. Ibid., 694.

50. Ibid., 695.

51. Ibid.

52. Ibid., 712–18.

53. Boston Consumers' Council, *Report of the Boston Consumers' Council*, I.

54. Ibid., 1.

55. Ibid., 4, 24.

56. Ibid., 66, 69.

57. Ibid., 68.

58. Ibid., 66.

59. Ibid., 70.

60. Ibid., 23, 74.

61. Ibid., 23.

62. Ibid., 73.
63. Ibid., 21–22.
64. Ibid., 69.
65. Ibid.
66. Ibid.
67. Ibid., 85.
68. Ibid., 25–29.
69. Ibid., 88.
70. Ibid., 22.
71. Ibid., 88.
72. Ibid., 68.
73. Whitewood Stamps, *Cable in Boston*, 3.
74. Cable Television Review Commission, *Report of the Mayor's Cable Television Review Commission* (Boston, MA: October 1979), 5. Government Documents Department, Boston Public Library.
75. Whitewood Stamps, *Cable in Boston*, 53.
76. Ibid., 7–9; Rob McCausland, "Cable TV for Boston?: A Lesson in Politics," *Community Television Review* (May 1980): 18.
77. McCausland, "Cable TV for Boston?" 18.
78. Cable Television Review Commission, *Report of the Mayor's Cable Television Review Commission*, 1.
79. Ibid.
80. Ibid.
81. Ibid., 7–8.
82. Ibid., 16–17.
83. Ibid., 17, 25.
84. Ibid., 21.
85. Ibid., 25–26.
86. Ibid., 39–40.
87. Ibid., 33–35.
88. Allison Kuhlein, *Michigan Cable Television*, VHS, Ann Arbor, MI: Michigan Municipal League.
89. For the general overview of the history of Detroit in the twentieth century, see Reynolds Farley, Sheldon Danziger, and Harry J. Holzer, *Detroit Divided* (New York: Russell Sage Foundation, 2000).
90. For detailed discussion about post-war Detroit, see Thomas J. Sugrue, *The Origins of the Urban Crisis: Race and Inequality in Postwar Detroit* (Princeton, New Jersey: Princeton University Press, 1996, 2005). On preceding problems from the pre-war period, see Richard W. Thomas, *Life for Us is What We Make It: Building Black Community in Detroit, 1915–1945* (Indianapolis, IN: Indiana University Press, 1992).
91. U.S. Department of Commerce, Bureau of the Census, *United States Census of Population*, 1940–1950 (Washington, D.C.: U.S. Government Printing Office, various years).

Notes to Chapter 3

92. Sugrue, *The Origins of the Urban Crisis*, 23.

93. U.S. Department of Commerce, Bureau of the Census, *United States Census of Population*, 1970 (Washington, D.C.: U.S. Government Printing Office, 1970).

94. The rapid change in the racial composition of the city of Detroit in terms of African American and non-African American ratio was also affected by whites moving away from the city to suburbs.

95. Sugrue argues that in June 1948, 65 percent of job ads discriminated against Black job seekers. Sugrue, *The Origins of the Urban Crisis*, 94.

96. Ibid., 130–31, 138, 139.

97. U.S. Department of Commerce, Bureau of the Census, *City and County Data Books* (Washington D.C.: U.S. Government Printing Office, 1947, 1977); Sugrue, *The Origins of the Urban Crisis*, 144.

98. Sugrue, *The Origins of the Urban Crisis*, 105.

99. Ibid., 151, 269–70.

100. For more detailed discussion about deindustrialization in other U.S. urban cities concurrent with Detroit, see William Julius Wilson, *The Truly Disadvantaged: The Inner City, the Underclass, and Public Policy* (Chicago, IL: The University of Chicago Press, 1987). I have made and will make reference to the deindustrialization case in the city of Boston. My understanding about Boston's history, particularly in relation to deindustrialization and its impact on the African American community has been deepened by Bluestone and Stevenson, *The Boston Renaissance*; Thomas H. O'Connor, *Building a New Boston: Politics and Urban Renewal 1950 to 1970* (Boston, MA: Northeastern University Press, 1993); O'Connor, *South Boston*; O'Connor, *The Hub*.

101. Elliot Liebow, *Tally's Corner: A Study of Negro Streetcorner Men* (Maryland: Rowman & Littlefield Publishers, Inc., 1967, 2003), 145.

102. The Detroit News, "A Searching Look—the City, the Future" in *The Voice of Detroit's Blacks* (Detroit, MI: The Detroit News, 1971), 2–6.

103. Richard W. Thomas, "Black Self-Help in Michigan" in *The State of Black Michigan, 1967–2007* ed. Joe T. Darden, Curtis Stokes, and Richard W. Thomas (East Lansing, MI: Michigan State University Press, 2007), 81–95; Richard W. Thomas, "The Black Self-Help Tradition in Michigan, 1967–2007" in *The State of Black Michigan, 1967–2007* ed. Joe T. Darden, Curtis Stokes, and Richard W. Thomas (East Lansing, MI: Michigan State University Press, 2007), 97–110.

104. WJBK-TV2 Storer Broadcasting Company, *Southeast Michigan Community Needs '70* (Detroit, January 1970), 1.

105. Ibid., 5, 10.

106. Ibid., 6.

107. The Cable TV Study Committee, *Cable Television in Detroit: A Study in Urban Communications* (Detroit, May 1972), v.

108. The Cable TV Study Committee, "Letter of Transmittal," *Cable Television in Detroit: A Study in Urban Communications* (Detroit, May 1972).

109. Ibid.

110. Ibid., 20.
111. Ibid.
112. The Cable TV Study Committee, *Cable Television in Detroit*, 5.
113. Ibid., 12.
114. Ibid., 20.
115. Ibid., 37.
116. Marshall McLuhan, *Understanding the Media: The Extensions of Man* (Cambridge, MA: MIT Press, 1964, 1994), 68.
117. Carter G. Woodson, *The Mis-Education of the Negro* (1933; repri., Drewryville, VA: Khabooks, 2006), xiii.
118. Ibid., 192.
119. The Cable TV Study Committee, *Cable Television in Detroit*, 10, 63.
120. Ibid., 66–67.
121. Ibid.
122. Ibid., 27.
123. The Detroit News, "Job Bias Fades; Still It's a Factor" *The Voice of Detroit's Blacks*, 46–50.
124. Ibid., 111.
125. Ibid., 159–60.
126. Ibid., 111–12.
127. Ibid., 111–12, 160.
128. Ibid., 8–9.
129. Ibid.
130. Ibid., 76.
131. N. E. Feldman, *Cable Television: Opportunities and Problems in Local Program Origination* (Rand Corporation. A report prepared for the Ford Foundation in 1970).
132. The Cable TV Study Committee, *Cable Television in Detroit*, 74–75.
133. Ibid., 80–81.
134. Cable Television Advisory Committee to Coleman Young, March 1981, Mahaffey Collection, Burton Historical Collection, Detroit Public Library.
135. James H. Bradley to Coleman Young, 9 February 1979, Mahaffey Collection, 17:1, Burton Historical Collection, Detroit Public Library; Coleman Young to City Council, 15 February 1979, Mahaffey Collection, 17:1, Burton Historical Collection, Detroit Public Library.
136. City Planning Commission to City Council Members, 7 August 1979, Mahaffey Collection, 17:1, Burton Historical Collection, Detroit Public Library.
137. William F. Rushton, "Turn on the Tube: Plug Your Community into Cable TV," *Planner* (August 1979): 17.
138. James H. Bradley to Lois Pincus, 2 November 1979, Mahaffey Collection, 17:1, Burton Historical Collection, Detroit Public Library.
139. The Detroit Cable Television Advisory Committee, *The Detroit Cable Television Advisory Committee Report* (Detroit, MI: November 13, 1979), Mahaffey Collection, 17:1, Burton Historical Collection, Detroit Public Library.

140. The Detroit Cable Television Advisory Committee, *The Detroit Cable Television Advisory Committee Report* (Detroit, MI: May 5, 1980), Mahaffey Collection, 17:2, Burton Historical Collection, Detroit Public Library.

141. Lois Pincus to City Council Member, 17 October 1980, Mahaffey Collection, 17:2, Burton Historical Collection, Detroit Public Library; Maryann Mahaffey to Lois Pincus, 3 November 1980, Mahaffey Collection, 17:2, Burton Historical Collection, Detroit Public Library.

142. Eugene Strobel, 10 February 1981. Mahaffey Collection, 17:3, Burton Historical Collection, Detroit Public Library.

143. Charlie J. Williams to Coleman Young, 6 March 1981. Mahaffey Collection, 17:3, Burton Historical Collection, Detroit Public Library; Coleman Young to City Council, 9 March 1981. Mahaffey Collection, 17:3, Burton Historical Collection, Detroit Public Library.

144. Kay D. Schloff to Charles Williams, 6 March 1981, Mahaffey Collection, 17:3, Burton Historical Collection, Detroit Public Library.

145. Michigan Library Association Telecommunications Committee, *Cable Telecommunications Resource Guide* (Lansing, MI: Michigan Library Association, 1982).

146. The Detroit Cable Television Advisory Committee, "Request for Proposals," *The Detroit Cable Television Advisory Committee: Recommendations to Coleman A. Young* (Detroit, MI: March 1981), Municipal Reference Library of the City of Detroit.

147. Ibid.

148. Ibid.

149. Sharlan M. Douglas to City Council, 20 March 1981. Mahaffey Collection, 17:3, Burton Historical Collection, Detroit Public Library; Maryann Mahaffey to Sharlan M. Douglas, 30 March 1981. Mahaffey Collection, 17:3, Burton Historical Collection, Detroit Public Library.

150. Michael E. Turner, *Memorandum*, April 15, 1981, Mahaffey Collection, 17:4, Burton Historical Collection, Detroit Public Library.

151. E. Barbara Wilson, *Recommendations*, April 22, 1981, Mahaffey Collection, 17:4, Burton Historical Collection, Detroit Public Library.

152. Ibid.

153. Michael E. Turner, *Memorandum*, April 24, 1981, Mahaffey Collection, 17:4, Burton Historical Collection, Detroit Public Library.

154. Tim Kiska, "Council Makes Cable TV Power Play," *Detroit Free Press*, April 30, 1981.

155. Nolan Finley, "Power Struggle May Stall Detroit Cable TV," *Detroit News*, April 30, 1981.

156. Michael E. Turner, *Memorandum*, May 13, 1981, Mahaffey Collection, 17:4, Burton Historical Collection, Detroit Public Library.

157. Monroe Walker, "Young May Get His Way Yet on Cable TV Panel," *Detroit News*, May 16, 1981.

158. Nolan Finley, "Detroit Debates Cable TV 'Need,'" *Detroit News*, June 2, 1981.

159. Luther Jackson, "Be Careful on Cable TV, Council Advised," *Detroit Free Press*, June 2, 1981.

160. *Cable Television: In the Public Interest*, 24 April 1981, Mahaffey Collection, 17:4, Burton Historical Collection, Detroit Public Library; *The Cable Licensing Process: A Practical Guide for Municipal Officials and Cable Advisory Committees*, Mahaffey Collection, 17:4, Burton Historical Collection, Detroit Public Library.

161. "Cable Seminar Program," *Cable Television: Franchising, Administration and Local Use*, 28 November, 1981, Mahaffey Collection, 17:5, Burton Historical Collection, Detroit Public Library.

162. Kay D. Schloff to City Council, 10 February 1982, Mahaffey Collection, 17:7, Burton Historical Collection, Detroit Public Library; Patricia Chargot, "Candidates for City Cable TV Panel Down to 9," *Detroit Free Press*, March 30, 1982.

163. Morris H. Goodman to Maryann Mahaffey, 20 March 1981, Mahaffey Collection, 17:5, Burton Historical Collection, Detroit Public Library.

164. Donald Snider to Maryann Mahaffey, 14 January 1982, Mahaffey Collection, 17:7, Burton Historical Collection, Detroit Public Library.

165. Erma Henderson to City Council Members, 10 February 1982, Mahaffey Collection, 17:7, Burton Historical Collection, Detroit Public Library.

166. Richard B. Anderson to Alta Harrison, 8 March 1982, Mahaffey Collection, 17:8, Burton Historical Collection, Detroit Public Library; Alta Harrison, *Resume*, Mahaffey Collection, 17:8, Burton Historical Collection, Detroit Public Library.

167. Richard B. Anderson to Joseph R. Jordan, 8 March, 1982, Mahaffey Collection, 17:8, Burton Historical Collection, Detroit Public Library; Joseph R. Jordan, *Resume*, Mahaffey Collection, 17:8, Burton Historical Collection, Detroit Public Library; Maryanne Mahaffey, Personal Memo, Mahaffey Collection, 17:8, Burton Historical Collection, Detroit Public Library.

168. Maryanne Mahaffey, Personal Memo, Mahaffey Collection, 17:9, Burton Historical Collection, Detroit Public Library.

169. *CATV Interview Questions*, March 1982, Mahaffey Collection, 17.9, Burton Historical Collection, Detroit Public Library; *Suggested Interview Questions: Cable Television Commissioners*, March 1982, Mahaffey Collection, 17.9, Burton Historical Collection, Detroit Public Library.

170. Barbara-Rose Collins to Council Members, 23 April 1982, Mahaffey Collection, Burton Historical Collection, Detroit Public Library.

171. Ken Coleman, "Black Cable TV: 'Economically and Politically It's Out Last Option!'" *Sepia* (April 1982): 37.

172. Ibid., 38.

173. Ibid., 41.

174. Coleman A. Young, *Mayor's Press Release*, May 20, 1982, Mahaffey Collection, 17.10, Burton Historical Collection, Detroit Public Library.

175. Geraldine L. Daniels to Maryann Mahaffey, 9 June 1982, Mahaffey Collection, 17.10, Burton Historical Collection, Detroit Public Library.

176. Federal Communications Commission, *Cable Television Employment Statistics*, July 20, 1981. Mahaffey Collection, 17.10, Burton Historical Collection, Detroit Public Library.

177. ELRA Group, *Cablemark Survey: Public Interest in Cable Television, Detroit, Michigan* (Detroit, MI: 1982).

178. Ibid.

179. Ibid.

180. Michael E. Turner to City Council, 30 July 1982, Mahaffey Collection, 17.10, Burton Historical Collection, Detroit Public Library; Detroit Cable Communications Commission, *Cable Communications Code of Conduct*, July 20, 1981. Mahaffey Collection, 17.10, Burton Historical Collection, Detroit Public Library; Detroit Cable Communications Commission, *Request for Proposals*, July 20, 1981. Mahaffey Collection, 17.10, Burton Historical Collection, Detroit Public Library.

181. Michael E. Turner and Kathryn A Bryant to Detroit Cable Communications Commission, 2 August 1982, Mahaffey Collection, 17.10, Burton Historical Collection, Detroit Public Library.

182. Michael E. Turner to City Council, 4 August 1982, Mahaffey Collection, 17.10, Burton Historical Collection, Detroit Public Library.

Notes to Chapter 4

1. U.S. Department of Commerce, Bureau of the Census, *United States Census of Population*, 1980 (Washington, D.C.: U.S. Government Printing Office, 1980).

2. Peter D. Hart Research Associates, Inc., *A Survey of Boston Residents' Attitudes toward Cable Television* (April 1981): 5.

3. Cable Television Review Commission, *Report of the Mayor's Cable Television Review Commission* (Boston, MA: October 1979). Government Documents Department, Boston Public Library.

4. Robert McCausland, "Cable TV for Boston?: A Lesson in Politics," *Community Television Review* (May, 1980): 18.

5. Ibid; Martin Kessel, "Boston Cable Franchise Pits Public Interest vs. Private Profit," *The Citizen Advocate*, September, 1980.

6. Dan Jones, and Martin Kessel, "Cable TV Picture if Clear: Tough Advisory Board Needed," *The Boston Globe*, August 20, 1980; Kessel, "Boston Cable Franchise Pits Public Interest vs. Private Profit."

7. Mayor's Office, *A Cable Television "Primer"* (Boston, 1980).

8. Ibid., 16.

9. Coordinating Council on Drug Abuse, "Telecommunications Position Paper," (September 11, 1980), 2–3.

10. Ibid., 3–4, 6.

11. Ibid., 10.

12. Eugenie Beal to Rick Borten, "Cable TV," 15 September, 1980. Boston City Archive.

13. Doug Herberich to Paul D. Horn, "Memorandum," 16 September 1980. Boston City Archive.

14. Boston Police to Rick Borten, October 15, 1980, Boston City Archive.

15. Kevin White, "Boston TV Enters the Cable Era and It's Not All Entertainment," *Boston Herald American*, December 22, 1980; Robert McCausland, "Boston—The Development of a Public Access Plan" *Community Television Review* (July, 1981): 25.

16. Boston Cablevision Services, *Cable Television Proposal of Boston Cablevision Services, Inc.,* 3/D/271 Boston City Archive; From Herbert P. Wilkins to Charles W, 31 October 1980, 3/D/271, Boston City Archive.

17. Boston Cablevision Services, *Cable Television Proposal*, 4–8.

18. Ibid., 37.

19. New York Times Corporation, "Exhibit 13," *Application for a Cable Television License for the City of Boston, Massachusetts*, 3/J/148, Boston City Archive.

20. New York Times Corporation, "Exhibit 15," *Application for a Cable Television License for the City of Boston, Massachusetts*, 3/J/148, Boston City Archive.

21. New York Times Corporation, "Affirmative Action Plan for Minorities," *Application for a Cable Television License for the City of Boston, Massachusetts*, 3/J/148, Boston City Archive.

22. Rollins Cablevision, "Thomas R. Bird," 3/J/146, Boston City Archive.

23. Rollins Cable Vision, *Technical Supplement and Operational Manual*, 3/J/146, Boston City Archive.

24. From Ralph J. Swett to Kevin White, 3 November 1980, 1–4, 3/J/144, Boston City Archive.

25. Ibid., 1–1.

26. Ibid., 1–1 – 1–2.

27. Ibid., 1–2.

28. Ibid., 1–10.

29. Ibid., 1–10 – 1–11.

30. Ibid., 1–12 – 1–13, 5–36 – 5–37.

31. Ibid., 1–13 – 1–14.

32. Ibid., 5–11.

33. Ibid.

34. Triune Cable of Boston, *Cable Television Proposal: Application of Tribune Cable of Boston for a Licence* Boston, MA, November 3, 1980, 3/J/144, Boston City Archive.

35. Triune Cable of Boston, *Cable Television Proposal*, Exhibit E.

36. Ibid.

37. Warner Amex Cable Communications Company, "Executive Summary," *A Warner Amex Proposal to Construct and Operate a Cable Television in Boston*, November 3, 1980, S5, 1–1 – 1–3, Government Documents Department, Boston Public Library.

38. Gustave M. Hauser to Kevin White, November 1, 1980, Government Documents Department, Boston Public Library.

39. Urban Research Association, *Warner Amex Cable Television: Cable Television Penetration Study of Boston, Massachusetts*, 3/J/145, Boston City Archive.

40. Warner Amex, "Executive Summary," S2.

41. Ibid., S6.

42. Ibid., 2–1.

43. Ibid., 2–21.

44. Charles F. Dolan to Kevin White, October 31, 1980 3/D/271 Boston City Archive.

45. Ibid.

46. Ibid.

47. Ibid.; Cablevision, "Form B100," 3/D/271, Boston City Archive.

48. Dolan to White, 31 October 1980; Cablevision, "Exhibit E" *Form 100*, 3/D/271, Boston City Archive.

49. Cablevision, "Exhibit E."

50. "Editorial," *The Boston Globe* December 30, 1980; "Boston Cable Study Recommends Public Corp. Birddog Local Access," *Variety*, December 30, 1980; White, "Boston TV"; City of Boston, *Facilities Planning for Boston Cable Communications System/A Proposal*, January 19, 1981, 29, 34, 37–40; McCausland, "Boston," 26.

51. White, "Boston TV."

52. Boston Police Department. January 21, 1981. Boston City Archive.

53. Boston Water and Sewer Commission, "Interoffice Communication," February 13, 1981. Boston City Archive.

54. Public Facilities Department, "Interoffice Communication," March 17, 1981. Boston City Archive.

55. Boston Housing Authority, "Interoffice Communication," April 8, 1981. Boston City Archive.

56. City of Boston, *Facilities Planning*, 1, 41–44.

57. Ibid., 14–15, 25.

58. Ibid., 16–18.

59. Ibid., 41.

60. Ibid.

61. Ibid., 19–20.

62. Ibid., 21–22.

63. Ibid., 22–23.

64. Ibid., 36.

65. About the summary of position papers, see Ibid., 49–63.

66. Ibid., 64.

67. Ibid., 64–65.

68. Henry P. Becton, Jr. to Joanne Anderson, 19 January 1981, Governmental Document Department, Boston Public Library.

69. Liam M. Kelly to Joanne Anderson, 19 January 1981, Governmental Document Department, Boston Public Library.

70. Research Program on Communications Policy, *Report on Municipal Uses of Cable Television for the City of Boston* 3/D/271, Boston City Archive.

71. City of Boston, *Issuing Authority Report*, i, xi, 3/D/271 Boston City Archive.

72. Ibid.

73. Ibid., i–ii.

74. Ibid., ii.

75. Ibid., ii–iv.

76. Ibid., vi–viii.

77. Ibid., xix–xxi.

78. Ibid., 45.

79. Ibid., 47.

80. Ibid., 73.

81. Ibid.

82. Ibid., 73–74.

83. Martin Kessel, "Let the Public Help in Decision on Cable Television," *The Boston Globe* February 25, 1981, 15.

84. McCausland, "Boston," 26–27.

85. Ibid., 27.

86. Gustave M. Hauser to Kevin White, 22 April 1981, 3/J/144, Folder 71, Boston City Archive; Warner Amex Cable Communications Company, "Overall Summary," *Application for Cable Communications License for Boston, Massachusetts* 3/J/144, Folder 71, Boston City Archive; Warner Amex Cable Communications Company, "Ownership Structure," *Application for Cable Communications License for Boston, Massachusetts*, 3/J/145, Boston City Archive; Warner Amex Cable Communications Company, "System Design and Construction," *Application for Cable Communications License for Boston, Massachusetts*, 3/J/145 Boston City Archive.

87. Warner Amex, "Overall Summary."

88. Ibid.

89. Warner Amex Cable Communications Company, "Community Communications: Boston's Own Programming," *Application for Cable Communications License for Boston, Massachusetts*, 78–78b, 3/J/144, Folder 76, Boston City Archive.

90. Warner Amex Cable Communications Company, "Community Communications: Boston's Own Programming," *Application for Cable Communications License for Boston, Massachusetts* 78–78b, 3/J/144, Folder 76 Boston City Archive; Warner Amex Cable Communications Company, "Community Communications: Boston's Own Programming" *Application for Cable Communications License for Boston, Massachusetts* Exhibit E. 2. 3., 3/J/145, Folder 81, Boston City Archive.

91. Warner Amex Cable Communications Company, *Application for Cable Communications License for Boston, Massachusetts*, 45–45b, 3/J/144, Folder 74, Boston City Archive; Thelma Cromwell Moss to Peter G. Meade, 16 April 1981, 3/J/144, Folder 74, Boston City Archive.

92. Jorge N. Hernandez to Peter Meade, 16 April 1981, 3/J/144, Folder 76, Boston City Archive.

93. Warner Amex Cable Communications Company, *Application for Cable Communications License for Boston, Massachusetts*, 64d–64e, 3/J/144, Folder 75, Boston City Archive.

94. Warner Amex Cable Communications Company, *Application for Cable Communications License for Boston, Massachusetts*, 71–71a, 75a, 3/J/144, Folder 75, Boston City Archive.

95. Ibid., 16–17; Allison Sloan, "The Man Who Hated Commercials," *Forbes* October 27, 1980.

96. Cablevision Systems Boston Corporation, "Executive Summary," *The Boston Plan: Application for Cable Communications License for Boston, Massachusetts*, 1, 3/J/142, Folder 23, Boston City Archive.

97. Ibid., 1–2, 5, 6.

98. Peter D. Hart Research Associates, Inc., *A Survey of Boston Residents' Attitudes toward Cable Television*, 2.

99. Ibid., 3, 6

100. Cablevision Systems Boston Corporation, "Executive Summary," 3.

101. Ibid., 14; Cablevision Systems Boston Corporation, "Administration and Regulation," *The Boston Plan: Application for Cable Communications License for Boston, Massachusetts*, 67, 3/J/142, Folder 25, Boston City Archive.

102. Cablevision Systems Boston Corporation, "Administration and Regulation," *The Boston Plan: Application for Cable Communications License for Boston, Massachusetts*, 71, 3/J/142, Folder 25, Boston City Archive.

103. Ibid.

104. Cablevision Systems Boston Corporation, "Exhibit E. 3. 1a" *The Boston Plan: Application for Cable Communications License for Boston, Massachusetts*, 71, 3/J/142, Folder 28, Boston City Archive; Cablevision Systems Boston Corporation, "Exhibit F. 1," *The Boston Plan: Application for Cable Communications License for Boston, Massachusetts*, 71, 3/J/142, Folder 28, Boston City Archive.

105. Kalba Bowen Associates, Inc., *Institutional Network Design for the City of Boston: Prepared for Cablevision Systems Boston Corporation* (April, 1981): 45–47, 3/J/143, Folder 33, Boston City Archive.

106. Kenneth D. Wade to Donna Garifano, 6 April 1981, 3/J/142, Folder 28, Boston City Archive.

107. Cablevision Systems Boston Corporation, "Equal Employment Opportunity Program," *The Boston Plan: Application for Cable Communications License for Boston, Massachusetts*, 3/J/142, Folder 28, Boston City Archive.

108. McCausland, "Boston—The Development of a Public Access Plan," 27.

109. Detroit Cable Communications Commission, "General Instructions," *Request for Proposals for the Provision of Cable Communications Services in the City of Detroit, Michigan* (Detroit, MI: City of Detroit, 1982).

110. James P. Robinson to Coleman Young and the City Council, 16 August 1982; Detroit Cable Communications Commission, *Request for Proposals*.

111. Detroit Cable Communications Commission, "Cable Communications Franchise Schedule," *Request for Proposals*; Detroit Cable Communications Com-

mission, "Request for Proposals," *Request for Proposals*, 1–2; Michael E. Turner and Kathryn A. Bryant to Coleman Young, 19 August 1982. Mahaffey Collection, 17:10, Burton Historical Collection, Detroit Public Library.

112. Detroit Cable Communications Commission, "Request for Proposals," *Request for Proposals* (Detroit, MI: City of Detroit, 1982), 7.

113. Ibid,

114. Michael E. Turner and Kathryn A. Bryant to Coleman Young, 19 August 1982.

115. Augustin K. Fosu, "Michigan Black–White Unemployment Patterns, 1971–1986: Differences by Gender," in *The State of Black Michigan 1988* ed. Frances S. Thomas (East Lansing, MI: Urban Affairs Program at Michigan State University, 1988), 3–7.

116. Detroit Cable Communications Commission, "Request for Proposals," 10–11.

117. Ibid., 12.

118. Michael E. Turner and Kathryn A. Bryant to Coleman Young, 19 August 1982.

119. Detroit Cable Communications Commission, "Request for Proposals," *Request for Proposals*, 15–16.

120. Detroit Cable Communications Commission, "Certificate of Non Discrimination," *Official Application Forms for Proving Cable Communications Services*.

121. Detroit Cable Communications Commission, "Employment Practices," *Official Application Forms for Proving Cable Communications Services*, Form M, 1.

122. Ibid, 1–4.

123. Center for Creative Communications, "Addendum," *The Request for Proposal*.

124. Ibid.

125. Ibid.

126. The Detroit Public Schools Cable Television Advisory Committee, "Educational Cable Communications Recommendations: Part 1," *The Request for Proposal*; "Form D," *The Request for Proposal*.

127. Barden Cablevision, "Introductory Overview and Summary of Proposal," *Cable Television Application*, Detroit, MI: December 10, 1982.

128. Barden Cablevision, "Don H. Barden," *Cable Television Application*.

129. Barden, "Introductory Overview and Summary of Proposal."

130. Ibid.

131. Ibid.

132. Ibid.

133. Gregory J. Reed to Don H. Barden, 1 December 1982, Barden Cablevision, *Cable Television Application*.

134. Clarence J. Hall to Don H. Barden, 6 December 1982, Barden Cablevision, *Cable Television Application*.

135. Wardell C. Croft to Don H. Barden, 3 December 1982, Barden Cablevision, *Cable Television Application*.

136. R. A. Stribling to Larry Baskerville, 29 November 1982, Barden Cablevision, *Cable Television Application*; George Riley to Larry Baskerville, 1 December 1982, Barden Cablevision, *Cable Television Application*; Booker Sabbath to Don H. Barden, 24 November 1982, Barden Cablevision, *Cable Television Application*; William L. Richardson to Don H. Barden, 29 November 1982, Barden Cablevision, *Cable Television Application*; John Chiodo to Don H. Barden, 1 December 1982, Barden Cablevision, *Cable Television Application*.

137. Kenneth A. Booker to Ray A. Leporati, 3 December 1982, Barden Cablevision, *Cable Television Application*; William C. Eubanks to Don H. Barden, 3 December 1982, Barden Cablevision, *Cable Television Application*; Simon J. Jones to Ray A. Leporati, 3 December 1982, Barden Cablevision, *Cable Television Application*; Beatrice Campbell to Ray A. Leporati, 3 December 1982, Barden Cablevision, *Cable Television Application*; Wally Triplett to Don H. Barden, 3 December 1982, Barden Cablevision, *Cable Television Application*; Willie L. Welch to Don H. Barden, 7 December 1982, Barden Cablevision, *Cable Television Application*; Kedra Huddleston to Don H. Barden, 7 December 1982, Barden Cablevision, *Cable Television Application*; Alvin S. Dukes to Ray A Leporati, 6 December 1982, Barden Cablevision, *Cable Television Application*; Roderick I. Parks to Ray A. Leporati, 7 December 1982, Barden Cablevision, *Cable Television Application*; Bryan Shillingford to Ray A. Leporati, 6 December 1982, Barden Cablevision, *Cable Television Application*; Raurice Mopkins to Larry Bafkerbille, 2 December 1982, Barden Cablevision, *Cable Television Application*; Mopkins International Services to Don H. Barden, Barden Cablevision, *Cable Television Application*; Armand Hall to Don H. Barden, 29 November 1982, Barden Cablevision, *Cable Television Application*; Glenn E. Wash to Don H. Barden, 23 November 1982, Barden Cablevision, *Cable Television Application*; N. G. Conyers to Don H. Barden, 6 December 1982, Barden Cablevision, *Cable Television Application*; Max Mobley to Don H. Barden, 6 December 1982, Barden Cablevision, *Cable Television Application*; Alton Lewis to Don H. Barden, 8 December 1982, Barden Cablevision, *Cable Television Application*; Agnes L. Scott to Don H. Barden, 3 December 1982, Barden Cablevision, *Cable Television Application*; Michael Jarard to Don H. Barden, 30 November 1982, Barden Cablevision, *Cable Television Application*; LeRoy W. Haynes to Don H. Barden, 2 December 1982, Barden Cablevision, *Cable Television Application*; Peter L. Blackburn to Don. H. Barden, 30 November 1982, Barden Cablevision, *Cable Television Application*.

138. George Busigo to Don H. Barden, 7 December 1982, Barden Cablevision, *Cable Television Application*; Johnny O. Elias to Don H. Barden, 8 December 1982, Barden Cablevision, *Cable Television Application*; Ernest L. Dixon to Don H. Barden, 1 December 1982, Barden Cablevision, *Cable Television Application*; William E. Andrews to Don H. Barden, 2 December 1982, Barden Cablevision, *Cable Television Application*; J. Ronald White to Don H. Barden, 3 December 1982, Barden Cablevision, *Cable Television Application*; William D. Murphy to Larry Baskerville, 2 December 1982, Barden Cablevision, *Cable Television Application*; Wiley Wilson to Don H. Barden, 3 December 1982, Barden Cablevision, *Cable Television Application*; Herman Malone to Don H. Barden, 1 December 1982, Barden Cablevision, *Cable

Television Application; A. Gus Warren to Don H. Barden, 2 December 1982, Barden Cablevision, *Cable Television Application*; Rosemary Alford-Whitehurst to Don H. Barden, 29 November 1982, Barden Cablevision, *Cable Television Application*; Shinzo Saiki to Don H. Barden, 1 December 1982, Barden Cablevision, *Cable Television Application*; Steve Reece to Don H. Barden, 3 December 1982, Barden Cablevision, *Cable Television Application*; Shakespeare L. Brissett to Larry Baskerville, 1 December 1982, Barden Cablevision, *Cable Television Application*; Alvin Taylor to Don H. Barden, Barden Cablevision, *Cable Television Application*; John M. Worrell to Don H. Barden, 1 December 1982, Barden Cablevision, *Cable Television Application*; Robert L. Pitts to Larry Baskerville, 3 December 1982, Barden Cablevision, *Cable Television Application*; Antoinette Bianchi to Don H. Barden, 2 December 1982, Barden Cablevision, *Cable Television Application*; Julius L. Johnson, Jr. to Larry Baskerville, 2 December 1982, Barden Cablevision, *Cable Television Application*; Alice B. Soltysiak to Ray A. Leporati, 23 November 1982, Barden Cablevision, *Cable Television Application*; William E. Maben to Don H. Barden, 4 December 1982, Barden Cablevision, *Cable Television Application*.

139. Joseph E. Madison to Don H. Barden, 7 December 1982, Barden Cablevision, *Cable Television Application*.
140. Barden Cablevision, "Ray Leporati," *Cable Television Application*.
141. Barden Cablevision, "Lawrence R. Baskerville," *Cable Television Application*.
142. Barden Cablevision, "Dr. David M. Miller," *Cable Television Application*.
143. Barden Cablevision, "Wade Briggs," *Cable Television Application*.
144. Barden Cablevision, *Cable Television Application*, Form N.
145. Ibid.
146. Barden Cablevision, *Cable Television Application*, Form I.
147. Barden Cablevision, "Introductory Overview and Summary of Proposal."
148. Barden Cablevision, *Cable Television Application*, Form J, J-3.
149. Ibid.
150. Ibid.
151. Ibid.
152. Ibid., Form K.
153. Ibid.
154. Ibid.
155. Ibid.
156. Ibid; Barden Cablevision, "Barden Job Opportunity and Retraining Communication," *Cable Television Application*.
157. Ibid.
158. Barden Cablevision, *Cable Television Application*, Form K.
159. Ibid., Form K.
160. Ibid.
161. Ibid.
162. Ibid.

Notes to Chapter 4

163. Ibid., Form M.
164. Ibid.; John P to Parry to Ray A. Leporati, 3 December 1982. Barden Cablevision, *Cable Television Application*.
165. Barden Cablevision, *Cable Television Application*, Form M.
166. Ibid.
167. Robert L. Green to Michael Turner, 3 December1982, *Application for Cable Communication System Franchise for the City of Detroit, Michigan* (Detroit, MI: City Communications, 1982); City Communications, *Application for Cable Communication System Franchise for the City of Detroit, Michigan* (Detroit, MI: City Communications, 1982).
168. Robert L. Green to Michael Turner, 3 December 1982, *Application for Cable Communication System Franchise*.
169. City Communications, *Application for Cable Communication System Franchise*, Form A.
170. Ibid.
171. Ibid., Form A, Form D 2–2c.
172. Ibid.
173. Ibid., Form I 22-23b.
174. Ibid., Form J, L la.
175. Ibid., Form J.
176. Ibid., Ibid, Form L lb.
177. Ibid., Form J (Supplement).
178. Ibid.
179. Ibid., Form K 1-1a, 3b.
180. Ibid., Form K 3-3a.
181. Ibid.
182. Ibid., Form K 3e-3f, 3k-3l.
183. Ibid., Form K 3m-3o.
184. Ibid., Form K 3o-3t.
185. Ibid., Form K 3y-3ee.
186. Ibid., Form K, 3ff-3ii.
187. Ibid., Form L, 1-3h.
188. Ibid., Form M, 1c4-1d.
189. Ibid., Form M, 1p.
190. Ibid., Form M, 1i-1n.
191. Ibid., Form M, 1v-1w.
192. Ibid., Form N, 1a-1g, Exhibit 1.
193. Detroit Inner-Unity Bell Cable System, *Highlights of the Proposed Detroit cable System*, 1982, 3-4, Mahaffey Collection, Burton Historical Collection, 18:3, Detroit Public Library; Detroit Inner-Unity Bell Cable Systems, *A Cable Communications Proposal for the City of Detroit*, December 1982, Form A.
194. Wendell Cox to Coleman A. Young, 10 December 1982, *A Cable Communications Proposal for the City of Detroit*.

195. Detroit Inner-Unity Bell Cable Systems, *A Cable Communications Proposal for the City of Detroit*, December 1982, Form D; Cox Broadcasting Corporation, *1981 Annual Report Cox Broadcasting Corporation*, 3.

196. Detroit Inner-Unity Bell Cable System, *Highlights of the Proposed Detroit cable System*, 3–4.

197. Wendell Cox to Coleman A. Young, 10 December 1982, *A Cable Communications Proposal for the City of Detroit*; Bruce N. Burnham to Sydney Small, 9 December 1982, *A Cable Communications Proposal for the City of Detroit*.

198. Detroit Inner-Unity Bell Cable Systems, *A Cable Communications Proposal for the City of Detroit*, December 1982, Form M.

199. Ibid.

200. Ibid.

201. Detroit Inner-Unity Bell Cable Systems, *A Cable Communications Proposal for the City of Detroit*, December 1982, Form M; Detroit Inner-Unity Bell Cable System, *Highlights of the Proposed Detroit Cable System*, 5.

202. Ibid.

203. Jack Martin to Oscar King, 30 November 1982, *A Cable Communications Proposal for the City of Detroit*, December 1982; Gregory J. Reed to Oscar King, 1 December 1982, *A Cable Communications Proposal for the City of Detroit*, December 1982; Charles R. Scales, Jr. to Oscar King, 30 November 1982, *A Cable Communications Proposal for the City of Detroit*, December 1982.

204. Detroit Inner-Unity Bell Cable Systems, *A Cable Communications Proposal for the City of Detroit*, December 1982, Form K.

205. Ibid.

206. Ibid.

207. Ibid., 5–7.

208. Detroit Inner-Unity Bell Cable Systems, *A Cable Communications Proposal for the City of Detroit*, December 1982, Form K.

209. Detroit Inner-Unity Bell Cable System, *Highlights of the Proposed Detroit cable System*, 11–12; Detroit Inner-Unity Bell Cable Systems, *A Cable Communications Proposal for the City of Detroit*, December 1982, iv.

210. Detroit Inner-Unity Bell Cable Systems, *A Cable Communications Proposal for the City of Detroit*, December 1982, Exhibit B.

Notes to Chapter 5

1. John E. Ward, "Warner Amex Proposal," May 22, 1981, Boston City Archive; John E. Ward, "Cablevision Proposal," May 22, 1981, Boston City Archive; John E. Ward, *Matrix Comparison of Technical Features: Warner and Cablevision Proposals to the City of Boston*, May 29, 1981; John E. Ward, "Warner Amex Proposal," May 22, 1981, Boston City Archive.

2. Richard Borten, *Site Visit Report*, June 9, 1981, 3/J/145, Folder 105, Boston City Archive; Dan Jones, *Site Visit Report*, June 9, 1981, 3/J/145, Folder 105, Boston City Archive.

3. Richard A. Borten, *Site Visit Report*, June 17, 1981, 3/J/145, Folder 105, Boston City Archive; Dan Jones, *Site Visit Report*, June 17, 1981, 3/J/145, Folder 105, Boston City Archive.

4. Jay Cowlrs, *Notes of Clarification Hearing for Boston CATV License*, June 12, 1981, 3/J/145, Folder 86, Boston City Archive; *Second Clarification Hearing*, June 16, 1981, 3/J/145, Folder 86, Boston City Archive.

5. Commonwealth of Massachusetts, Boston Public Library, *Public Hearing*, June 23, 1980, 3–4.

6. Ibid., 4–5.
7. Ibid., 7–8.
8. Ibid., 9–11.
9. Ibid., 12.
10. Ibid., 18–19.
11. Ibid., 20.
12. Ibid,. 22–23.
13. Ibid.
14. Ibid., 23–24.
15. Ibid., 27–29.
16. Ibid., 39.
17. Ibid., 57–58.
18. Ibid., 58–60.
19. Ibid.
20. Ibid., 65–73.
21. Ibid., 113.

22. Ibid., 65–73; David O. Wicks, Jr. to Kevin White, 23 June 23, 1981, 3/J/146, Boston City Archive.

23. Commonwealth of Massachusetts, *Public Hearing*, June 23, 1980, 74.
24. Ibid., 76.
25. Ibid., 105.
26. Ibid., 103–05.
27. Ibid., 116–17.

28. Commonwealth of Massachusetts, Boston Public Library, *Public Hearing*, June 25, 1980, 3–5.

29. Ibid., 6–7.
30. Ibid., 7–10.
31. Ibid., 10–12.

32. For detailed conversation between Perkins and two cable applicants, see Ibid., 13–17.

33. Ibid., 17–19.
34. Ibid., 31–35.
35. Ibid., 51–57.
36. Ibid., 61–68.
37. Ibid.
38. Ibid., 39–41; 59–61; 103–06.
39. Ibid., 69–73; 81–85.

40. Ibid., 114–19.
41. Ibid., 19–25.
42. Malarkey, Taylor & Associates, Inc., *Analysis of Cable Television Proposals: Final Report Prepared for City of Boston, Massachusetts* (June 30, 1981), II12–15; II62.
43. Ibid., I164.
44. Ibid., II66–67.
45. Ibid., II67–68; II76.
46. Ibid., III1.
47. Ibid., III52.
48. Ibid., III53–III54.
49. Ibid., III58.
50. Ibid., III56, III63. III72.
51. Sheila Mahony to Richard A. Borten, 15 July, 1981, 3/J/146, Boston City Archive.
52. Charles F. Dolan to John MacIsaac, 16 July, 1981, 3/J/146, Boston City Archive.
53. Warner Amex Cable Communications, "Comments or Corrections," July 14, 1981, 3/J/145, Folder 85, Boston City Archive.
54. Ibid.
55. Kevin White, *Issuing Authority Final Decision and Statement of Reasons*, August 12, 1981.
56. Ibid., 3.
57. Ibid., 3.
58. Herbert Swartz, "The Big Change is Cable," *The Boston Business Journal* (January 18, 1982), 1.
59. White, *Issuing Authority Final Decision*, 4–6.
60. Ibid., 7.
61. "Access Governance Still Undecided," *Cable Access News* 5 (November 1981): 1–3.
62. Ibid.
63. Fietcher Roberts, "Four Are Weighed for Cable TV Board," *The Boston Globe*, December 24, 1981; "Mayor Leaning to Non-Elected Access Board: First Four Members Named," *Cable Access News* 7 (January 1982): 1–2.
64. Robert McCausland, "Boston Community Access and Programming Foundation: What Lies Ahead?" *Uplink/Downlink* (Spring 1982): 20–22.
65. City of Boston, *Provisional Cable Television License: Granted to Cablevision Systems Boston Corporation*, 1.
66. Ibid., 10.
67. Ibid., 1.
68. Ibid., 46.
69. Ibid., 45.
70. Ibid., 49.
71. Ibid., 83.
72. Ibid., 86.
73. *A Plan for Boston's Public Institutional Network*, May 7, 1982, 1.

74. *Cable Access News*, (April 1982), 1–3.
75. Ibid.,2.
76. *Cable Access News*, (May 1982), 3.
77. *A Plan for Boston's Public Institutional Network*, 1.
78. Ibid.
79. Ibid., 3.
80. *Cable Access News*, (June 1982), 7.
81. From Kathleen M. Crred to Margie Cohen, 4 November, 1982, 3/J/143, Boston City Archive.
82. "Employment Policy & Training Program," November 22, 1982, 1, 3/J/146, Folder 125, Boston City Archive.
83. Ibid.
84. Ibid., 2.
85. Ibid., 2–3.
86. Ibid., 4.
87. From Laurence M. Bloom to Richard A. Borten, 7 December, 1982, 3/J/146, Folder 119, Boston City Archive.
88. From Laurence M. Bloom to Richard A. Borten, 1 December, 1982, 3/J/146, Folder 119, Boston City Archive.
89. From Laurence M. Bloom to Richard A. Borten, 7 December, 1982, 3/J/146, Folder 124, Boston City Archive.
90. Ibid.
91. From Kathleen O'Keeffe to East Boston Social Center, 23 November, 1982, 3/J/146, Folder 124, Boston City Archive.
92. Ibid.
93. *Assignments of Provisional Cable Television License* (December 10, 1982), 3/J/145, Folder 108, Boston City Archive.
94. *Final Cable Television License*, December 15, 1982. 3/J/145, Folder 109, Boston City Archive.
95. Ibid., 5, 7, 10.
96. Ibid., 23.
97. Ibid., 44–45.
98. Ibid., 46.
99. From Coleman Young to City Council, 14 December, 1982, Mahaffey Collection, 17:10, Burton Historical Collection, Detroit Public Library.
100. Tim Kiska, "He's Hoping to Wire Detroit for Cable TV," *Detroit Free Press*, January 31, 1983; Mireille Grangenois, "Minority Firms Focus on Cable Plans in Detroit," *USA Today*, April 20, 1983.
101. Joseph E. Madison to Coleman Young, March 9, 1983, Mahaffey Collection, 17:12, Burton Historical Collection, Detroit Public Library.
102. Maryanne Mahaffey, 1983, Mahaffey Collection, 17:12, Burton Historical Collection, Detroit Public Library.
103. Detroit Cable Communications Commission, *Evaluation of Proposals for the Provision of Cable Communications Services in the City of Detroit, Michigan: Preliminary Report*, March 1983, VI19–VI20.

104. Ibid., VI21–VI23.
105. Ibid., VI27–VI29.
106. Ibid., VI32–VI44.
107. Ibid., VIII1–VIII7.
108. George J. Dunmore, "Providing Careers for Minorities in Cable TV," *Michigan Chronicle*, May 21, 1983.
109. Clarence Welcome to Maryann Mahaffey, 2 May, 1983, Mahaffey Collection, 17:13, Burton Historical Collection, Detroit Public Library. Emphasis in original.
110. Robert Green to Michael Turner, 13 April, 1983, Mahaffey Collection, 17:16, Burton Historical Collection, Detroit Public Library.
111. Tim Kiska, "Shooting Stars in City Cable War," *Detroit News*, April 15, 1983.
112. Michael Turner to Maryann Mahaffey, 28 April, 1983, Mahaffey Collection, 17:13, Burton Historical Collection, Detroit Public Library.
113. Michael Turner to Maryann Mahaffey, 13 May, 1983, Mahaffey Collection, 17:13, Burton Historical Collection, Detroit Public Library.
114. Larry Baskerville to Detroit Board of Education, 7 June, 1983, Mahaffey Collection, 17:14, Burton Historical Collection, Detroit Public Library.
115. City Communications, *Presentation: Before the Board of Education Detroit Public Schools*, June 7, 1983, Mahaffey Collection, 18:2, Burton Historical Collection, Detroit Public Library.
116. Oscar W. King, III to Barbara Wilson, 7 June, 1983, Mahaffey Collection, 18:3, Burton Historical Collection, Detroit Public Library; Detroit Inner-Unity Bell, *Detroit Inner-Unity Bell cable Summary of Educational Recommendations*, Mahaffey Collection, 18:3, Burton Historical Collection, Detroit Public Library.
117. James P. Robinson to Coleman Young, 13 June, 1983, Mahaffey Collection, 17:14, Burton Historical Collection, Detroit Public Library.
118. George J. Dunmore, "Providing Careers for Minorities in Cable TV," *Michigan Chronicle*, May 21, 1983; "Wiring Detroit," *Cablevision*, May 9, 1983: 183.
119. Coleman Young to City Council, March 25, 1983, Mahaffey Collection, 17:13, Burton Historical Collection, Detroit Public Library.
120. James P. Robinson to Coleman A. Young, 17 June, 1983, in *Evaluation of Proposals for the Provision of Cable Communications Services in the City of Detroit, Michigan: Final Report*, June 1983.
121. Detroit Cable Communications Commission, *Evaluation of Proposals for the Provision of Cable Communications Services in the City of Detroit, Michigan: Final Report*, June 1983, I.
122. Ibid., II1–II2; III1–III4.
123. Ibid., II3.
124. Ibid., II4.
125. Ibid., VI1–VI14.

126. Ibid., VI14–VI18
127. Ibid., VI27–VI28.
128. Ibid., VII1–VII6.
129. Ibid., III11.
130. Ibid., VI1–VI3, VI14–42.
131. James P. Robinson to Coleman Young, 13 June, 1983, 2–3, Mahaffey Collection, 17:14, Burton Historical Collection, Detroit Public Library.
132. Detroit Cable Communications Commission, *Resolution Recommending Award of Detroit Cable Franchise* June 20, 1983.
133. Coleman A. Young to the City Council, July 12, 1983, Mahaffey Collection, 17:15, Burton Historical Collection, Detroit Public Library.
134. Susan Paul, "Mayor, Commission Back Barden in Detroit," *Multichannel News*, July 18, 1983; Bruce Alpert, "Cable TV Firm Owner Flirts with Big Time," *Detroit News*, July 18, 1983.
135. Josephine A. Powell to the City Council, 14 July, 1983. Mahaffey Collection, 17:15, Burton Historical Collection, Detroit Public Library.
136. Erma Henderson to the City Council, 14 July, 1983, Mahaffey Collection, 17:15, Burton Historical Collection, Detroit Public Library.
137. Don H. Barden to Michael Turner, 23 July, 1983, Mahaffey Collection, 17:16, Burton Historical Collection, Detroit Public Library.
138. Gilbert A. Maddox to the City Council, 25 July, 1983, Mahaffey Collection, 17:16, Burton Historical Collection, Detroit Public Library.
139. Ibid.
140. Ibid.
141. Michael Turner, News Release, July 27, 1983, Mahaffey Collection, 17:16, Burton Historical Collection, Detroit Public Library.; Tim Kiska, "2 Courts Bar Decision on Detroit Cable," *Detroit Free Press*, July 27, 1983; Bruce Alpert, "Legal Challenged Mar Cable TV Pact," *Detroit News*, July 31, 1983.
142. William Richardson to the City Council, 25 July, 1983, Mahaffey Collection, 17:16, Burton Historical Collection, Detroit Public Library.
143. Dorothy J. Brown, Maryann Mahaffey, 25 July, 1983, Mahaffey Collection, 17:16, Burton Historical Collection, Detroit Public Library.
144. Marc Stepp to Erma Henderson, July 25, 1983, Mahaffey Collection, 17:16, Burton Historical Collection, Detroit Public Library.
145. Coleman A. Young, *Mayor's Press Release*, July 28, 1983, Mahaffey Collection, 17:16, Burton Historical Collection, Detroit Public Library.
146. Mel Ravitz, "Statement of Mel Ravitz on Cable TV Vote," July 27, 1983, Mahaffey Collection, 17:16, Burton Historical Collection, Detroit Public Library.
147. Honingman, Miller, Schwartz and Cohn to Detroit Cable Communications Commission, 22 February, 1984, Mahaffey Collection, 17:19, Burton Historical Collection, Detroit Public Library.
148. Michael Turner to City Council, 3 February, 1984, Mahaffey Collection, 17:19, Burton Historical Collection, Detroit Public Library; "Canadians to

Wire City for Cable TV," *The Detroit News*, January 6, 1984; David Kushma and Tim Kiska, "Canadians to Build, Run Detroit Cable Franchise," *Detroit Free Press*, January 6, 1984.

149. Ibid.

150. Mel Ravitz to Michael Turner, 1 February, 1984, Mahaffey Collection, 17:16, Burton Historical Collection, Detroit Public Library.

151. M. D. Farrell-Donaldson to the City Council, 5 March, 1984, Mahaffey Collection, 17:19, Burton Historical Collection, Detroit Public Library.

152. Michael Turner to the City Council, 6 July, 1984, Mahaffey Collection, 17:19, Burton Historical Collection, Detroit Public Library.

153. M. D. Farrell-Donaldson to City Council, 19 June, 1984, Mahaffey Collection, 17:19, Burton Historical Collection, Detroit Public Library.

154. Susan Paul, "Cable TV Firm, City Will Miss Deadline," *The Detroit News*, July 31, 1984.

155. Michael Turner to City Council, Memorandum, August 1, 1984, Mahaffey Collection, 17:20, Burton Historical Collection, Detroit Public Library; Stephen Advokat, "Cable Firm Seeks OK for Smaller City System," *Detroit Free Press*, August 1, 1984.

156. Mel Ravitz to Council members, 13 August, 1984. Mahaffey Collection, 17:20, Burton Historical Collection, Detroit Public Library.

157. Don Barden to Michael Turner, 21 August, 1984, Mahaffey Collection, 17:20, Burton Historical Collection, Detroit Public Library.

158. Percy Sutton to Coleman Young and Erma Henderson, 31 August, 1984, Mahaffey Collection, Burton Historical Collection, Detroit Public Library; Bruce Alpert, "Downsized Cable System Defended," *The Detroit News*, September 7, 1984; "Wireless: Detroit Should Reopen the Bidding Process for Cable TV," *Detroit Free Press*, September 7, 1984, Danton Wilson, "Barden Cablevision Chief Answer Critics," *Michigan Chronicle*, September 15, 1984; Arlena Sawyers, "Barden gets 2nd Chance: Cable TV Pact Extended," *The Detroit News*, September 19, 1984.

159. Horace L. Sheffield to Maryann Mahaffey, 21 September, 1984, Mahaffey Collection, Burton Historical Collection, Detroit Public Library.

160. James Robinson to Coleman Young, 31 August, 1984, Mahaffey Collection, 17:20, Burton Historical Collection, Detroit Public Library; Coleman Young to City Council, 4 September, 1984, Mahaffey Collection, 17:20, Burton Historical Collection, Detroit Public Library.; Mel Ravitz to Council members, 11 September, 1984, Mahaffey Collection, 17:20, Burton Historical Collection, Detroit Public Library; Mel Ravitz to Council members, 18 September, 1984, Mahaffey Collection, 17:20, Burton Historical Collection, Detroit Public Library; Arlena Sawyers, "Cable TV Pact Extended," *The Detroit News*, September 19, 1984.

161. M. D. Farrell-Donaldson to James Robinson, 1 October, 1984, Mahaffey Collection, 17:20, Burton Historical Collection, Detroit Public Library James Robinson to M. D. Farrell-Donaldson, 8 October, 1984, Mahaffey Collection, 17:20, Burton Historical Collection, Detroit Public Library.

162. Josephine A. Powell and Phillip S. Brown to City Council, 19 October, 1984, Mahaffey Collection, 17:20, Burton Historical Collection, Detroit Public Library.

163. Bruce Alpert, "Barden May Retain TV Pact," *The Detroit News*, December 6, 1984.

164. John E. Boddy to City Council, 7 December, 1984, Mahaffey Collection, 17:20, Burton Historical Collection, Detroit Public Library.

165. Mel Ravitz to City Council, 10 January, 1985, Mahaffey Collection, 17:22, Burton Historical Collection, Detroit Public Library; Michael E. Turner to City Council, 21 January, 1985, Mahaffey Collection, 17:22, Burton Historical Collection, Detroit Public Library.

166. City Council to Michael E. Turner, 18 February, 1985, Mahaffey Collection, 17:22, Burton Historical Collection, Detroit Public Library.

167. Law Department to city Council, 27 February, 1985, Mahaffey Collection, 17:22, Burton Historical Collection, Detroit Public Library.

168. Josephine A. Powell and Phillip S. Brown to City Council, 2 April, 1985, Mahaffey Collection, 17:22, Burton Historical Collection, Detroit Public Library; Josephine A. Powell and Phillip S. Brown to City Council, April 30, 1985, Mahaffey Collection, 17:22, Burton Historical Collection, Detroit Public Library.

169. Carlton T. Stanton to Phillip Brown, 2 April, 1985, Mahaffey Collection, 17:22, Burton Historical Collection, Detroit Public Library.

170. Peter J. Christiano to Josephine Powell, 4 April, 1985, Mahaffey Collection, 17:22, Burton Historical Collection, Detroit Public Library.

171. Josephine A. Powell and Phillip S. Brown to City Council, 29 April, 1985, Mahaffey Collection, 17:22, Burton Historical Collection, Detroit Public Library.

172. Mel Ravitz to Council members, 10 January, 1985, Mahaffey Collection, Burton Historical Collection, Detroit Public Library.

173. Michael Turner to City Council, 21 January, 1985, Mahaffey Collection, Burton Historical Collection, Detroit Public Library; City Council to Michael Turner, 18 February, 1985, Mahaffey Collection, Burton Historical Collection, Detroit Public Library; Josephine A. Powell and Phillip S. Brown to City Council, 21 February, 1985, Mahaffey Collection, Burton Historical Collection, Detroit Public Library.

174. Donald Pailen to City Council, 27 February, 1985, Mahaffey Collection, Burton Historical Collection, Detroit Public Library.

175. Josephine A. Powell and Phillip S. Brown to City Council, 29 April, 1985, Mahaffey Collection, Burton Historical Collection, Detroit Public Library; Mel Ravitz to Council Members, 6 May, 1985, Mahaffey Collection, Burton Historical Collection, Detroit Public Library.

176. Detroit Cable Communications Commission, *Cable Communications Franchise Agreement between the City of Detroit and Barden Cablevision of Detroit, Inc. as Amended and Restated* December, 1985.

177. Ibid., 23, 57.
178. Ibid., 30.
179. Ibid., 38.
180. Ibid., 39.
181. Ibid., 40–41.
182. Ibid., 41.
183. Ibid., 42.
184. Ibid., 41.
185. Ibid., 88.
186. Ibid., Form M.
187. Ibid.
188. Ibid., 90.
189. Ibid., 91.
190. Ibid., "Schedule 16.8."

Notes to Chapter 6

1. *Final Cable Television License*, December 15, 1982, 3/J/145, Folder 109, Boston City Archive.

2. Detroit Cable Communications Commission, *Cable Communications Franchise Agreement between the City of Detroit and Barden Cablevision of Detroit, Inc. as Amended and Restated* December, 1985, 61.

3. Kevin H. White to Charles Dolan, 30 December, 1983, Boston City Archive.

4. Arthur W. Thompson to Scott Stevens, 12 October, 1983, Boston City Archive.

5. *Monthly EEO and Compliance Report*, December 9, 1983. Detroit City Archive; Kevin H. White to Charles Dolan, 30 December, 1983.

6. Kevin H. White to Charles Dolan, 30 December, 1983.

7. Arthur W. Thompson to Margie Cohen Stanzler, 20 December, 1983, Boston City Archive.

8. Mayor's Office of Cable Communications, "Rules of procedure for the Cable Performance Evaluation Session," Detroit City Archive.

9. Martin Kessel, "Comments of Martin Kessel to the City of Boston Regarding Cablevision's Compliance with Basic Service Requirements," December 13, 1983, 5–6, Detroit City Archive.

10. Ibid.

11. Kevin H. White to Charles Dolan, 30 December, 1983, Boston City Archive.

12. Jacob Bernstein, *Comments on Cable Television Developments in Boston: Summary Evaluation End Year One*, December 20, 1983.

13. Ibid.

14. The Global Village Associates, Inc, "Statement on Cablevision Performance 1982–1983," December 20, 1983, Boston City Archive.

15. Ibid.

16. Thomas P. Cohan to Art Thompson, 2 April, 1984, 1984, Boston City Archive; Diane Modica, "Future Mechanism for Handling Cable Consumer Complaints," 1984, Boston City Archive; Dwight Golann to Charles Dolan, 17 April, 1984, 1984, Boston City Archive.

17. Ibid.

18. Raymond L. Flynn to Charles Dolan, 12 April, 1984, 1984, Boston City Archive.

19. Ibid.

20. Raymond L. Flynn. "Mayoral Column," April 30, 1984, 1984, Boston City Archive.

21. Tom Cohan to Raymond L. Flynn, 19 April, 1984, 1984, Boston City Archive

22. For more detailed information about Attorney General's investigation, see Attorney General, "Issues to be Aware of before Meeting with Cablevision," 1984, Boston City Archive.

23. Ibid.

24. Tom Cohan to Raymond L. Flynn, "Memo," May 8, 1984, 1984, Boston City Archive; Arthur W. Thompson to Raymond L. Flynn, May 14, 1984, 1984, Boston City Archive; Tom Cohan to Raymond L. Flynn, "Memo," May 17, 1984, 1984, Boston City Archive.

25. Tom Cohan to Raymond L. Flynn, 24 October, 1984, 1984, Boston City Archive.

26. Raymond L. Flynn to James Kofalt, 1 November, 1984, 1984, Boston City Archive, Raymond L. Flynn, "Press Release," November 5, 1984, 1984, Boston City Archive.

27. George A. Russell, Jr. to Raymond L. Flynn, 4 December, 1984, 1984, Boston City Archive; Tom Cohan to Raymond L. Flynn, "Update on Analysis of Cablevision's Financial Situation," December 12, 1984, 1984, Boston City Archive.

28. City of Boston's Office of Cable Communications, *1984 Annual Performance Evaluation Report on Cablevision of Boston*, 14, 1984, Boston City Archive.

29. Ibid., 29.

30. Letter to Raymond L. Flynn, 1984, Boston City Archive.

31. City of Boston's Office of Cable Communications, *1984 Annual Performance Evaluation Report on Cablevision of Boston*, 29–30, 1984, Boston City Archive.

32. Ibid., 30.

33. Janet L. Doherty to Raymond L. Flynn, 6 January, 1985, 1985, Boston City Archive; Melvin King, interviewed by the author, Technology Center at Tent City, October 31, 2005.

34. "Financial Problems Threaten Public Programs on Hub Cable," *The Boston Globe*, March 16, 1985. For Cablevision's financial difficulty, see "Report to mayor on Cablevision's Finances," 1985, Boston City Archive.

35. Thomas P. Cohan to Raymond L. Flynn, 26 February, 1985, 1985, Boston City Archive.

36. Ibid.

37. Sarah Snyder, "Finances Threaten Public Programs" *Boston Globe*, March 16, 1985; Sarah Snyder, "Public Access TV May Not Get Paid," *Boston Globe*, March 26, 1985.

38. Tom Cohan to Raymond L. Flynn, "Memo," March 17, 1985, 1985, Boston Public Library; "Cablevision Briefing," March 17, 1985, 1985, Boston Public Library.

39. Charles F. Spillane and Joseph M. Qulity, June 3, 1985, 1985, Boston City Archive; John E. Taylor to Raymond L. Flynn, May 31, 1985, 1985, Boston City Archive.

40. "Boston for Africa: Feed the Children of Eritrea and Tigre," 1985, Boston City Archive; Melvin King, interview.

41. Ibid; Snyder, "Public Access TV May Not Get Paid," March 26, 1985.

42. Cablevision, "The Greatest Place to Be," April 3, 1985, 1985, Boston City Archive.

43. Victoria Hull to Raymond L. Flynn, 27 January, 1986, 1985, Boston City Archive; Tom Cohan to Raymond L. Flynn, 29 January, 1986, 1985, Boston City Archive.

44. Raymond L. Flynn to Hubert Jessup, 20 February, 1986, 1985, Boston City Archive.

45. Hubert Jessup to the School Committee, 1 May, 1986, 1985, Boston City Archive.

46. Michael E. Turner to City Council, 14 February, 1985, Mahaffey Collection, 17:28, Burton Historical Collection, Detroit Public Library; *Certificate of Partnership*, Mahaffey Collection, 17:28, Burton Historical Collection, Detroit Public Library.

47. Barden Cablevision of Detroit, Inc. *Application for Consent to Transfer Franchise and Detroit CATV System*, February 14, 1986, Mahaffey Collection, 17:28, Burton Historical Collection, Detroit Public Library; *Partnership Agreement*, Mahaffey Collection, 17:28, Burton Historical Collection, Detroit Public Library; J B. Gage to Barden Cablevision, 20 February, 1986, Mahaffey Collection, 17:28, Burton Historical Collection, Detroit Public Library.

48. Michael E. Turner to City Council, 24 February, 1986, Mahaffey Collection, 17:28, Burton Historical Collection, Detroit Public Library.

49. Alvin Xex to Don Barden, 13 March, 1986, Mahaffey Collection, 17:28, Burton Historical Collection, Detroit Public Library. Emphasis in original.

50. *City Communications, Inc., v. City of Detroit, Barden Cablevision of Detroit, Inc., and MacLean-Hunter*, Mahaffey Collection, 17:28, Burton Historical Collection, Detroit Public Library; Abigail Elias to City Council, 25 March, 1986, Mahaffey Collection, 17:28, Burton Historical Collection, Detroit Public Library.

51. Ibid.

52. Don H. Barden to Detroit Cable Communications Commission, 18 September, 1986, Mahaffey Collection, 17:28, Burton Historical Collection, Detroit Public Library; Kathryn A. Bryant to City Council, 13 October, 1986, Mahaffey Collection, 17:28, Burton Historical Collection, Detroit Public Library.

53. Don H. Barden to Detroit Cable Communications Commission, 16 January, 1987, Mahaffey Collection, 17:28, Burton Historical Collection, Detroit Public Library; Charles E. Colding, Sr. to City Council, 30 January, 1987, Mahaffey Collection, 17:28, Burton Historical Collection, Detroit Public Library.

54. *Detroit Cable Communications Commission Construction Grid Report*, April 30, 1987, Mahaffey Collection, 17:28, Burton Historical Collection, Detroit Public Library.

Notes to Conclusion

1. W. E. B. Du Bois, *The Souls of Black Folk* (New York: Vintage Books, 1990), 8.

2. Shawn Michelle Smith, *The Photography on the Color Line: W. E. B. Du Bois, Race, and Visual Culture* (Durham, NC: Duke University Press, 2004, 2005), 2.

3. Ibid., 90.

4. Ibid., 7.

5. Carter G. Woodson, "Negro Life and History in Our Schools," *Journal of Negro History* 4 (July, 1919): 278; Lawrence P. Crouchett, "Early Black Studies Movements," in *The African American Studies Reader* ed. Nathaniel Norment, Jr. (Durham, NC: Carolina Academic Press, 2001), 192–198; Vivian V. Gordon, "The Coming of Age of Black Studies," in *The African American Studies Reader* ed. Nathaniel Norment, Jr. (Durham, NC: Carolina Academic Press, 2001), 212–220.

6. George M. Frederickson, *The Black Image in the White Mind: The Debate on Afro-American Character and Destiny, 1817–1914* (Middletown, CT: Wesleyan University Press, 1971).

7. Donald Bogle, *Blacks in American Films and Television: An Encyclopedia* (New York: A Fireside Books, 1988); Donald Bogle, *Prime Time Blues: African Americans on Network Television* (New York: Farrar, Straus and Giroux, 2001); Donald Bogle, *Bright Boulevards: The Story of Black Hollywood* (New York: Ballantine Books, 2005); Donald Bogle, *Toms, Coons, Mulattoes, Mammies, & Bucks: An Interpretive History of Blacks in American Films* (New York: Continuum, 2006).

8. Jeffrey O. G. Ogbar, *Black Power: Radical Politics and African American Identity* (Baltimore, MD: Johns Hopkins University Press, 2004), 38–39; Steven Classen, *Watching Jim Crow: The Struggles over Mississippi TV, 1955–1969* (Durham, North Carolina: Duke University Press, 2004), 33.

9. Barry Dornfeld, *Producing Public Television: Producing Public Culture* (Princeton, NJ: Princeton University Press, 1998), 5.

10. Jason Loviglio, *Radio's Intimate Public: Network Broadcasting and Mass-Mediated Democracy* (Minneapolis, MN: The University of Minnesota Press, 2005), 125–126.

11. Dornfeld, *Producing Public Television*, 5.

12. Brett Pulley, *The Billion Dollar BET: Robert Johnson and the Inside Story of Black Entertainment Television* (Hoboken, NJ: John Wiley & Sons, Inc., 2004),

186–205; John T. Barber, *The Black Digital Elite: African American Leaders of the Information Revolution* (Westport, CT: Praeger, 2006), 172–77; Kristal Brent Zook, *I See Black People: The Rise and Fall of African American-Owned Television and Radio* (New York: Nation Books, 2008), xiv.

Bibliography

Archives

Boston Public Library, Government Documents Department.
Detroit Public Library, Mahaffey Collection, Burton Historical Collection.
Municipal Reference Library of the City of Detroit.

Newspapers and Magazines

Boston Business Journal
Boston Globe
Boston Herald American
Cable Access News
Cablevision
Chronicle of Higher Education
Detroit Free Press
Detroit News
Michigan Chronicle
Multichannel News
New York Times
Toronto Star
USA Today

Sources

Abu-Lughod, Lila. "The Romance of Resistance: Tracing Transformations of Power through Bedouin Women." *American Ethnologist* 17. no. 1 (February 1990): 41–55.

Allison, Robert J. *A Short History of Boston*. Beverly, MA: Commonwealth Editions, 2004.

Asante, Molefi Kete. *The Afrocentric Idea*. Philadelphia, PA: Temple University Press, 1998.
———. *Afrocentricity: The Theory of Social Change*. Chicago, IL: African American Images, 2003.
Barber, John T. *The Black Digital Elite: African American Leaders of the Information Revolution*. Westport, CT: Praeger, 2006.
Bay, Mia. *The White Image in the Black Mind: African-American Ideas about White People, 1830–1925*. New York: Oxford University Press, 2000.
Bluestone, Barry, and Mary Huff Stevenson. *The Boston Renaissance: Race, Space, and Economic Change in an American Metropolis*. New York: Russell Sage Foundation, 2000.
Bogle, Donald. *Blacks in American Films and Television: An Encyclopedia*. New York: A Fireside Books, 1988.
———. *Prime Time Blues: African Americans on Network Television*. New York: Farrar, Straus and Giroux, 2001.
———. *Bright Boulevards: The Story of Black Hollywood*. New York: Ballantine Books, 2005.
———. *Toms, Coons, Mulattoes, Mammies, & Bucks: An Interpretive History of Blacks in American Films*. New York: Continuum, 2006.
"Boston Cable Study Recommends Public Corp. Birddog Local Access." *Variety* (December 30, 1980).
The Boston Foundation. "Racial and Ethnic Diversity." http://www.tbf.org/indicators/civic-health/indicators.asp?id=941&fID=209&fname=Race/Ethnicity.
———. "Thinking Globally / Acting Locally: A Regional Wake-Up Call." In *A Summary of the Boston Indicators Report 2002–2004*. March 2005.
Bradbury, Katherine L., Anthony Downs, and Kenneth A. Small, *Urban Decline and the Future of American Cities*. Washington, D.C.: Brookings Institution, 1982.
The Cable TV Study Committee. *Cable Television in Detroit: A Study in Urban Communications*. Detroit, May 1972.
Castelnero, Gordon. *TV Land Detroit*. Ann Arbor, MI: The University of Michigan Press, 2006.
City of Boston. *Boston's Economy: Excerpt from the Official Statement of the City of Boston, Massachusetts. $75,000,000 General Obligation Bonds*. Boston: Boston Redevelopment Authority, 1995.
Classen, Steven. *Watching Jim Crow: The Struggles over Mississippi TV, 1955–1969*. Durham, NC: Duke University Press, 2004.
Coleman, Ken. "Black Cable TV: 'Economically and Politically It's Out Last Option!'" *Sepia* (April 1982).
Coordinating Council on Drug Abuse. *Telecommunications Position Paper*. September 11, 1980.

Crouchett, Lawrence P. "Early Black Studies Movements." In *The African American Studies Reader* edited by Nathaniel Norment, Jr, 192–98, Durham, NC: Carolina Academic Press, 2001.

Cushman, Marc, and Linda J. LaRosa. *I Spy: A History and Episode Guide to the Groundbreaking Television Series*. Jefferson, NC: McFarland & Company, Inc., 2007.

Dagbovie, Pero Gaglo. *Black History: "Old School" Black Historians and the Hip Hop Generation*. Troy, MI: Bedford Publishers, Inc., 2006.

Darden, Joe T., Richard Child Hill, June Thomas, and Richard Thomas. *Detroit: Race and Uneven Development*. Philadelphia, PA: Temple University Press, 1987.

Dawley, Alan. *Struggles for Justice: Social Responsibility and the Liberal State*. Cambridge, MA: Harvard University Press, 2000.

Delgado, Richard, and Jean Stepancic, *Critical Whiteness Studies: Looking behind the Mirror*. Princeton, NJ: Princeton University Press, 1997.

Detroit News. *The Voices of Detroit's Blacks*. Detroit, MI: The Detroit News, 1971.

Dibbell, Julian. "A Rape in Cyberspace." *Village Voice* 38, no. 51 (December 21, 1993): 36–42.

Dillard, Angela D. *Faith in the City: Preaching Radical Social Change in Detroit*. Ann Arbor, MI: The University of Michigan Press, 2007.

Dornfeld, Barry. *Producing Public Television: Producing Public Culture*. Princeton, NJ: Princeton University Press, 1998.

Durham, Philip, and Everett L. Jones. *The Negro Cowboys*. Lincoln, NE: University of Nebraska Press, 1965.

Dyson, Michael Eric. *Why I Love Black Women*. New York: Basic Books, 2003.

———. *Come Hell or High Water: Hurricane Katrina and the Color of Disaster*. Cambridge, MA: Basic Books, 2006.

Du Bois, W. E. B. *The Philadelphia Negro: A Social Study*. 1899. Reprint, Philadelphia, PA: University of Pennsylvania Press, 1996.

Eaton, Susan E. *The Other Boston Busing Story: What's Won and Lost across the Boundary Line*. New Haven, CT: Yale University Press, 2001.

Ellison, Ralph, *Invisible Man*. 1947. Reprint, New York; Vintage International, 1995.

ELRA Group. *Cablemark Survey: Public Interest in Cable Television, Detroit, Michigan*. Detroit, MI: 1982.

Entman, Robert, and Andrew Rojecki. *The Black Image in the White Mind: Media and Race in America*. Chicago, IL: The University of Chicago Press, 2000.

Fanon, Franz. *Black Skin White Masks*. New York: Grove Press, 1967.

Farley, Reynolds, Sheldon Danziger, and Harry J. Holzer. *Detroit Divided*. New York: Russell Sage Foundation, 2000.

Feldman, N. E. *Cable Television: Opportunities and Problems in Local Program Origination* Rand Corporation. A report prepared for the Ford Foundation in 1970.

Fine, Sidney. *Violence in the Model City: The Cavanagh Administration, Race Relations, and the Detroit Riot of 1967.* East Lansing, MI: Michigan University Press, 2007.

Fiske, John. *Television Culture.* New York: Routledge, 1987, 2004

——. *Media Matters: Race and Gender in U.S. Politics.* Minneapolis, MN: The University of Minnesota Press, 1996.

Fiske, John, and John Hartley. *Reading Television.* New York: Routledge, 1978, 2003.

Formisano, Ronald P. *Boston Against Busing: Race, Class, and Ethnicity in the 1960s and 1970s.* Chapel Hill, NC: The University of North Carolina Press, 1991.

Fosu, Augustin K. "Michigan Black–White Unemployment Patterns, 1971–1986: Differences by Gender." In *The State of Black Michigan 1988*, edited by Joe T. Darden, Curtin Stokes, and Richard W. Thomas, 3–8, East Lansing, MI: Michigan State University, 1988.

Foucault, Michael. *The History of Sexuality: An Introduction Volume 1.* New York: Vintage Books, 1978.

Frederickson, George M. *The Black Image in the White Mind: The Debate on Afro-American Character and Destiny, 1817*–1914. Middletown, CT: Wesleyan University Press, 1971.

Fuller, Linda K. *Community Television in the United States: A Sourcebook on Public, Educational, and Governmental Access.* Westport, CT: Greenwood Press, 1994.

Gordon, Vivian V. "The Coming of Age of Black Studies." In *The African American Studies Reader* edited by Nathaniel Norment, Jr, 212–20, Durham, NC: Carolina Academic Press, 2001.

Grant, Donald L., and Mildred Bricker Grant. "Some Notes on the Capital 'N.'" *Phylon* 36 no. 4 (1975): 435–43.

Gray, Herman S. *Watching Race: Television and the Struggle for Blackness.* Minneapolis, MN: University of Minnesota Press, 1995, 2004.

——. *Cultural Moves: African Americans and the Politics of Representation.* Berkeley, CA: University of California Press, 2005.

Hillson, Jon. *The Battle of Boston.* New York: Pathfinder Press, 1977.

Hochschild, Jennifer, and Nathan Scovronick. *The American Dream and the Public Schools.* New York: Oxford University Press, 2003.

hooks, bell. *Reel to Real.* New York: Routledge, 1996.

Hunt, Darnell M. *Channeling Blackness: Studies on Television and Race in America.* New York: Oxford University Press, 2005.

Jones, Hubie. "Social and Political History of Race in Boston from 1950–2000." Lecture, University of Massachusetts Boston, Boston, MA, December 5, 2005.

Jones, Jacqueline. *American Work: Four Centuries of Black and White Labor.* New York: W. W. Norton and Company, 1998.

Kelley, Robin D. G. " 'We Are Not What We Seem': Rethinking Black Working-Class Opposition in the Jim Crow South." *The Journal of American History* 80, no. 1 (June 1993): 75–112.

Kessel, Martin. "Boston Cable Franchise Pits Public Interest vs. Private Profit." *The Citizen Advocate*. (September, 1980).

King, Mel. *Chain of Change: Struggles for Black Community Development*. Boston: South End Press, 1981.

Kuhlein, Allison. *Michigan Cable Television*. VHS, Ann Arbor, MI: Michigan Municipal League.

Landsberg, Alison. *Prosthetic Memory: The Transformation of American Remembrance in the Age of Mass Culture*. New York: Columbus University Press, 2004.

Leuda, Julia. "Black-Audience Westerns and the Politics of Cultural Identification in the 1930s." *Cinema Journal* 42. no. 1 (2002): 46–70.

Liebow, Elliot. *Tally's Corner: A Study of Negro Streetcorner Men*. Maryland: Rowman & Littlefield Publishers, Inc., 1967, 2003.

Lipsitz, George. *The Possessive Investment in Whiteness: How White People Profit from Identity Politics*. Philadelphia: Temple University Press, 1998.

Loviglo, Jason. *Radio's Intimate Public: Network Broadcasting and Mass-Mediated Democracy*. Minneapolis, MN: The University of Minnesota Press, 2005.

Lucas, William A. and Robert K. Yin. *Serving Local Needs with Telecommunications: Alternative Applications for Public Services*. Santa Monica, CA; The Rand Corporation, 1973.

MacDonald, J. Fred. *Blacks and White TV: African Americans in Television Since 1948*. Chicago, IL: Nelson-Hall Publishers, 1992.

Marable, Manning. "Black Studies and the Black Intellectual Tradition." *Race & Reason* (1997–1998): 3–4.

McCausland, Robert. "Boston—The Development of a Public Access Plan." *Community Television Review* (July 1981): 25–29.

———. "Boston Community Access and Programming Foundation: What Lies Ahead?" *Uplink/Downlink* (Spring 1982): 20–22.

———. "Cable TV for Boston?: A Lesson in Politics." *Community Television Review* (May 1980): 18.

McLuhan, Marshall. *Understanding the Media: The Extensions of Man*. Cambridge, MA: MIT Press, 1964, 1994.

Michigan Library Association Telecommunications Committee. *Cable Telecommunications Resource Guide*. Lansing, MI: Michigan Library Association, 1982.

Murguia, Edward. "Political Capital and the Social Reproduction of Inequality in a Mexican Origin Community in Arizona." In *The Bubbling Cauldron: Race, Ethnicity, and the Urban Crisis*, edited by Michael Peter Smith, and Joe R. Feagin, 304–22, Minneapolis, MN: University of Minnesota Press, 1995.

Nadel, Alanl. *Television in Black-and-White America: Race and National Identity*. Lawrence, KS: University Press of Kansas, 2005.

Nakamura, Lisa. *Cybertypes: Race, Ethnicity, and Identity on the Internet.* New York: Routledge, 2002.
National Urban League. *The State of Black America 2007: Portrait of the Black Male.* New York: The Beckham Publications Group, 2007.
Novick, Peter. *That Noble Dream: The "Objectivity Question" and the American Historical Profession.* New York: Cambridge University Press, 1988.
Nudelman, Franny. *John Brown's Body: Slavery, Violence, and the Culture of War.* Chapel Hill, NC: The University of North Carolina Press, 2004.
O'Connor, Thomas H. *Building a New Boston: Politics and Urban Renewal 1950 to 1970.* Boston, MA: Northeastern University Press, 1993.
———. *South Boston South Boston: My Home Town.* Boston, MA: Northeastern University Press, 1994, 1998.
———. *The Hub: Boston Past and Present.* Boston, MA: Northeastern University Press, 2001.
Ogbar, Jeffrey O. G. *Black Power: Radical Politics and African American Identity.* Baltimore, MD: Johns Hopkins University Press, 2004.
Peter D. Hart Research Associates, Inc. *A Survey of Boston Residents' Attitudes toward Cable Television.* April 1981.
Pulley, Brett. *The Billion Dollar BET: Robert Johnson and the Inside Story of Black Entertainment Television.* Hoboken, NJ: John Wiley & Sons, Inc., 2004.
Putman, Robert, and Lewis M. Feldstein. *Better Together: Restoring the American Community.* New York: Simon & Schuster, 2003.
Robins, Kevin. "Cyberspace and the World We Live in." In *The Cybercultures Reader edited by* David Bell, and Barbara M. Kennedy, 77–95, New York: Routledge, 2000.
Ross, Michael J. *"I Respectfully Disagree with the Judge's Order": The Boston School Desegregation Controversy.* Beverly, MA: Commonwealth Editions, 2006.
Rushton, William F. "Turn on the Tube: Plug Your Community into Cable TV." *Planner* (August 1979).
Scott, James C. *Domination and the Arts of Resistance: Hidden Transcripts.* New Haven: Yale University Press, 1990.
Seiden, Martin H. *Cable Television U.S.A.: An Analysis of Government Policy.* New York: Praeger Publishers, 1972.
Shaw, Sarah-Ann. "Introduction." *WBGH.* http://main.wgbh.org/saybrother/intro.
———. "The History of *Say Brother.*" *WBGH.* http://main.wgbh.org/saybrother/history.html.
Sitkoff, Harvard. *The Struggles for Black Equality, 1954–1980.* Toronto: Harper and Collins, 1981.
Sloan, Allison. "The Man Who Hated Commercials." *Forbes* (October 27, 1980).
Smiley, Tavis. *The Covenant.* Chicago, IL: Third World Press, 2006.
———. *The Covenant in Action.* Carlsband, CA: Smiley Books, 2006

Smith, Shawn Michelle. *The Photography on the Color Line: W. E. B. Du Bois, Race, and Visual Culture*. Durham, NC: Duke University Press, 2004, 2005.

Sugrue, Thomas J. *The Origins of the Urban Crisis: Race and Inequality in Postwar Detroit*. Princeton, New Jersey: Princeton University Press, 1996, 2005.

Susman, Warren I. *Culture as History: The Transformation of American Society in the Twentieth Century*. Washington, D.C.: Smithsonian Institution, 2003.

Tait, Alice, and Todd Burroughs. "Mixed Signals: Race and the Media." In *Race and Resistance: African Americans in the Twenty-First Century*, edited by Herb Boyd, 101–08, Cambridge, MA: South End Press, 2002.

"Television: Negro Performers Win Better Roles in TV than in Any Other Entertainment Medium." *Ebony* (June 1950): 22–24.

Terborg-Penn, Rosalyn. "Naming Ourselves: The Politics and Meaning of Self-Designation." In *The Columbia Guide to African American History since 1939*, edited by Robert L. Harris Jr., and Rosalyn Terborg-Penn, 91–100, New York: Columbia University Press, 2006.

Theoharis, Jeanne. "They Told Us Our Kids Were Stupid: Ruth Baston and the Educational Movement in Boston." In *Groundwork: Local black Freedom Movements in America*, edited by Jeanne Theoharis and Komozi Woodard, 17–44, New York: New York University Press, 2005.

Thomas, Richard W. *Life for Us is What We Make It: Building black Community in Detroit, 1915–1945*. Indianapolis, IN: Indiana University Press, 1992.

———. "Black Self-Help in Michigan." In *The State of Black Michigan, 1967–2007*, edited by Joe T. Darden, Curtin Stokes, and Richard W. Thomas, 81–96, East Lansing, MI: Michigan State University Press, 2007.

———. "The Black Self-Help Tradition in Michigan, 1967–2007." In *The State of Black Michigan, 1967–2007* edited by Joe T. Darden, Curtin Stokes, and Richard W. Thomas, 97–110, East Lansing, MI: Michigan State University Press, 2007.

Thompson, Heather Ann. *Whose Detroit?: Politics, Labor, and Race in a Modern American City*. Ithaca, NY: Cornel University Press, 2001.

Torres, Sasha. *Black White and in Color: Television and Black Civil Rights*. Princeton, NJ: Princeton University Press, 2003.

Tosseto, Louis. "Editorial." *Wired* 1, no. 1 (March/April 1993): 10Turkle, Sherry. *Life on the Screen*. New York: Touchstone, 1995.

Turkle, Sherry. "Ghosts in the Machine." *Sciences* 35. no. 6 (November/December 1995): 36–39.

U.S. Bureau of the Census. "1960 General Population Characteristics." Washington, DC., 1962.

———. "1970 General Population Characteristics." Washington, DC, 1972

———. "1980 General Population Characteristics." Washington, DC, 1982

———. "Census 2000 Demographic Profile Highlights." http://factfinder.census.gov/servlet/SAFFFacts?_event=Search&geo_id=&_geoContext=&_street=&_county=boston&_cityTown=boston&_state=04000US25&_zip=&_lang=en&_sse=on&pctxt=fph&pgsl=010.

———. *City and County Data Books.* Washington, DC, 1947, 1977.

———. *United States Census of Population, 1950.* Washington, DC, 1951.

———. *United States Census of Population, 1960.* Washington, DC, 1961.

———. *United States Census of Population, 1970.* Washington, DC, 1971.

———. *United States Census of Population, 1980.* Washington, DC, 1981.

———. *United States Census of Population, 1990.* Washington, DC, 1991.

U.S. Department of Labor. Bureau of Labor Statistics. *Employment, Hours, and Earnings for States and Local Areas.* Washington, DC, 1995.

Williams, Juan. *Eyes on the Prize: America's Civil Rights Years, 1854–1965.* New York: Penguin Books, 1987.

Wilson, William Julius. *The Truly Disadvantaged: The Inner City, the Underclass, and Public Policy.* Chicago, IL: The University of Chicago Press, 1987.

WJBK-TV2 Storer Broadcasting Company. *Southeast Michigan Community Needs '70.* Detroit, January, 1970.

Woodson, Carter G. "Negro Life and History in Our Schools." *Journal of Negro History* 4 (July 1919): 273–80.

———. *The Mis-Education of the Negro.* 1933. Reprint, Drewryville, VA: Khabooks, 2006.

Zook, Kristal Brent. *I See Black People: The Rise and Fall of African-American Owned Television and Radio.* New York: Nation Books, 2008.

Index

Abetta Corporation, 115
Access; *See* Access channel; Access program
Access and Programming Corporation, 130, 132, 135, 139–40, 142, 182, 186–87
Access channel, 9, 36, 38, 40–41, 44, 49–61 passim, 109, 255, 258; in Boston, 75, 115, 118–21, 123–24, 131, 138, 142, 182–85 passim, 190–91, 197–98, 204; in Detroit, 91–92, 99–100, 145, 150, 154, 157, 159, 164, 166, 174–75, 207–8, 212–13, 226
Access Hollywood, 14
Access programming, 44, 51, 53, 54; in Boston, 82, 115, 123, 135, 183, 190, 235, 237, 240–41; in Detroit, 98, 156, 159–60, 168, 173, 175, 226
American Civil Liberties Union (ACLU), 44, 176
Affirmative action 17, 44; in Boston, 115–16, 197–98, 200, 204, 233, 238; in Detroit, 145–46, 172, 173, 208, 219, 244
African American disposable income, 2, 55
African American history, xi, xv, xvi, 6, 12, 13, 30–32, 165, 176, 250
African American psychology; *See* psychology

Afrocentric media, 4
Agency, 2, 6, 7, 17, 20, 249, 251, 256, 258
Albee, Bob, 179
Allston, 118, 236, 240
American Cablevision of Boston, 115, 152
Amos 'n' Andy Show, 62
Analysis of Cable Television License, 189–93
Application;*See* Cable application
Asante, Molefi Kete, 4, 29
Assignment of Provisional Cable Television License, 203–5
Atkins, Thomas, 72
Aureolio, Richard, 179

Back Bay, 66, 118
Baer, Walter, 43, 45–46, 50
Barden Cablevision; franchise agreement and, 9, 177, 212–15, 225–28, 243–46; as a Black-owned cable company, 87, 149, 158; application by, 148–61, 205–6; as a minority-owned company, 148–50, 158, 160, 217; existing systems of, 149, 152–53; financial investment of, 149, 160, 206, 208, 210, 219, 211, 221, 225; minority involvement and, 150, 153–54, 158–62, 211, 214–15, 244; local origination and, 150–51, 156,

312 Index

Barden Cablevision *(continued)* 158, 159, 206, 212–13, 226, 255; community access and, 150–51, 155, 156–58, 206, 212–13, 226, 255; local investors and, 153, 243; service structure of, 154, 212, 214, 221; employment opportunities of; 157–58, 160, 161, 207–9, 215, 226–28; franchise agreement with Detroit, 177, 212–15, 225–28, 243–46; facilities and, 206, 208, 213, 225–26; equipment and, 208, 213, 225–26; affirmative action and, 208–9; presentation to city officers, 209; and educational institutions, 209–10; ownership of, 212, 243; financial plan and, 214; hearings, 215–16; delay caused by, 216–25; 228; and MacLean-Hunter, 218–20, 230, 243–45, 256
Barden, Don H., 149, 152, 206, 212, 215, 218, 222, 244
Baskerville, Lawrence R., 152, 209
Bay Village, 188
Beard, Charles, 179, 181, 184, 186
Bell, Bill, 180, 182
Bell, Edward F., 161, 162
Bernstein, Jacob, 232–33
Best of Boston, 140
Birth of a Nation, the, 18, 20, 166
Black Corner, The, 48
Black Entertainment Television, 1, 4, 32, 37, 60, 27–28; and Boston, 119, 122, 137, 140–41, 180, 190, 192; and Detroit, 154–55, 164–65, 166
Black history; *See* African American history
Black Power, 64, 88
Blackside Film & Video Production, 13
Blaxploitation, 2, 14, 21–22, 28
Bloom, Laurence M., 201–2
Bogle, Donald, 2, 18–20, 24, 32, 251, 254

Bone, Jan, 59
Borten, Richard, 77, 178–80, 183, 186, 192, 193, 195, 201
Boston Access and Information Corporation, 135
Boston Alive, 135
Boston Almanac, 141
Boston Cable Access Advisory Committee, 125
Boston Cable Television Access Coalition, 111, 186, 187, 196, 199
Boston Cable Television Review Commission, 81–84 passim, 111, 112
Boston Cablevision Systems, Inc., 115–16
Boston City Council, 72, 73, 179, 242
Boston Community Access and Programming Foundation, 196, 198–99, 200, 205, 232, 236, 239, 242
Boston Community Cable Association, 125, 187
Boston Community Television, 133
Boston Community Television Corporation (BCTC), 133
Boston Consumers' Council, 77
"Boston Elders Service," 121, 134
Boston Famine Relief Cablethon 1985, 241, 243
Boston Housing Authority, 124–25
Boston Police Department, 114, 122, 124
Boston Production Center, 190
Boston Public Library, 80, 124, 128, 179, 186, 199
Bradley, James H., 96–97
Brighton, 118, 240
Briggs, Wade, 153, 212
Broadcasting: marginalization of African Americans in, 1, 7; white domination of, 6, 146; difference from narrowcasting, 8, 36, 39, 41, 95; radio and, 14; failure to address

African American interests, 32, 43, 58, 148, 168; government protection and, 58; as privilege, 58; ownership, 63; cost of, 76; Times Mirror and, 118; studies on, 252; and industry; *See* Television industry
Bryant, Kathryn A., 211, 225
Busing, 11–12, 56, 65, 71
Butterball, Jr.; *See* Briggs, Wade

Cable in Boston: A Basic Viability Report, 80
Cablemark Survey: Public Interest in Cable Television in Detroit, 107–8
Cable television; for community building, xvi, 4, 11, 15, 38, 50, 53, 64, 258; history of, 1, 2, 3, 4–6, 8, 14, 15, 18, 35–62 passim, 64, 252–56; equal participation and, 33, 52, 100; equal employment and, 48, 51, 115–16, 141, 143, 144, 146, 160–61, 169, 172, 198, 204, 208, 226–27; equal access to, 76, 84–85, 131, 154; for social and community justice, 1, 41, 64, 70, 252, 258, 259; scholarship on 12, 17, 35–62 passim; in Massachusetts, 15; for political campaign, 41; as a democratic medium, 43, 109; as a reflection of public culture, 252; as an empowerment tool, xvi, 14, 36, 64; as a master's tool, 3; as a job creation tool, 38, 147; as a community tool, 38, 49, 50, 59, 62, 83, 110, as a business tool, 61; as a political tool, 3, 8, 24, 66, 117, 157; as a visual culture tool 12, 38, 76, 110, 250, 253; as a tool for social change, 3, 17, 28, 66, 219; urban areas and, 15, 2, 36, 39, 40–52 passim, 57–70 passim, 75, 85, 92, 96, 152, 159, 162–63, 169, 193; and working class population, 40, 66, 92; surveillance and, 40–41; and foreseen benefits for African Americans, 3–14 passim, 36, 39–44, 46, 49, 52, 55, 60, 63–65, 69, 246, 353–58 passim; foreseen benefits for Boston and, 72–84, 112–14, 117–21, 125, 129–37 passim, 180–82, 188, 195, 199, 206, 233–34, 240, 242; foreseen benefits for Detroit and, 90–99, 148, 150, 153–54, 157, 160, 163, 165, 169, 171, 175–76, 214, 218, 223, 227, 243–46; origin of, 37, 266–68 (n. 4); commercialism and, 39–40; as a two-way technology, 113, 124, 129–30, 133, 199, 253; *See also*, Empowerment; Federal Communications Commission; Employment; Women; Narrowcasting
Cable Television Information Center Associates, 53, 77, 98, 207, 211, 214
Cable Television in Detroit: A Study in Urban Communications, 87
Cable Television "Primer," A, 112, 115
Cable TV Study Committee, 87–88, 89, 93, 96
Cablevision Systems Boston Corporation; franchise agreement and, 9, 177, 193–94, 194, 196–97, 203–5; preliminary application and, 115, 121–23; service structure of, 122, 138, 180–84, 194, 201–2, 204, 234, 240; local origination and, 122–23, 140–41, 182, 189, 191, 200, 202, 232, 240; amended application and, 133, 137–42, 177, 189; existing systems of, 137, 191; corporate vision of, 137–38, 182; financial investment of, 139, 142, 180–84, 189, 192–93, 201, 204, 235; facilities and, 129, 142, 182, 186, 187, 232, 237; equipment and, 139, 142, 182; employment

Cablevision Systems Boston Corporation *(continued)*
  opportunities of, 115–16, 141, 196, 197, 201, 204, 230–31, 234–36, 238, 239; community access and, 140–41, 180–84, 185–86, 189, 200, 202, 237, 240; minority involvement and, 140, 180–84, 189, 193, 205, 231, 235, 238; study conducted by Kalba Bowen Associates, 141; site visit by city officials and, 178; public hearings attended by, 179–84, 186–88, 203–4, 231–32; financial difficulties of, 184, 237–38, 240; ownership of, 186, 187; local organization partnership with, 187, 202, 235; construction schedule of, 188, 192, 201, 229, 230, 232, 234–36; affirmative action and, 200–1, 238; delay in 1983 and, 232–33; lack of accountability and, 234; failure to meet the agreement by, 234–37, 239–41, 255; boycott of, 237; programs aired with, 242
Campus Boston, 123, 140
Carroll, Harold, 179, 186
Cass Institute of Technology, 160, 210
Center for Creative Communications, 147, 174
Channel; availability of, 37, 42; availability in Boston, 74, 76, 114, 120–22, 134–35, 137–38, 142, 180–97 passim, 232; availability in Detroit, 95, 107–8, 145, 148, 154, 164–65, 175, 207–8, 212–13, 221–26 passim 244; control of, 48, 54; for commercial purposes, 37–38, 60, 90, 92, 97; for education, 92, 136, 210; allocation in Boston, 78, 117, 132, 190; allocation in Detroit, 90, 92, 96, 208; for lease, 101, 145, 155, 164–65; for special interests in Detroit, 108, 147, 158, 164–65; for special interests in Boston, 122, 136, 140–41, 194, 200–4; downstreaming and, 131, 136, 147, 225; upstreaming and, 131, 147; pay-per-view amd, 121, 134, 166, 212, 213; Public Institutional Network and, 139, 199; *See also* access channel, Channel abundance, Local origination channel, Minorities, Narrowcasting, Satellite
Channel abundance, 41–42, 50, 56, 61, 114, 117, 137, 167
Charlestown, 67, 81, 127
Chinatown, 118
Church; *See* religious institutions
Chuzmir, Stuart F., 137
Citizens Advisory Board, 84–85, 112, 233
City Communications Inc.; 161–69, 205–10, 212–15, 216–17, 219, 244–46
City Council; *See* Boston City Council; Detroit City Council
City of Detroit Cable Communications Service Franchise Agreement as Amended and Restated, 225–28
City Planning Commission, 97
Civil Rights era; *See* Civil Rights movement
Civil Rights movement, 8, 11, 25, 29–30, 33, 71, 86
Clark, John, 29, 178
Cleveland, Clyde, 142, 211, 225
Cognitive map, 12, 67, 69, 93, 127
Cohan, Thomas P., 234–42 passim
Cohen, Margie, 179, 231, 240
Coleman, Ken, 106
Collins, Barbara-Rose, 106, 142, 206, 211, 225
Collins, James P., 81
Colorblind, xvii–xviii, 28, 60, 62
Commercial programming, 136, 175

Commercial television, 44, 55, 147, 163, 241, 249
Community Access Board, 118
Community access channel; *See* access channel
Community Research Incorporated (CRI), 43–44, 50
Connell, Eileen, 183, 187–88
Contractor's Association of Boston, 135–36
Control; of media and African Americans, 10, 25, 38, 41, 47, 62, 257–58; of cable television by the federal government, 36, 39, 42, 54, 58; of cable television by state, 38; of cable television by public, 39, 43–46, 48–49; of cable television in Boston, 72, 75–79 passim, 113, 123–27, 134, 182, 185–89, 195, 233, 238; of cable television in Detroit, 91–93, 104–6, 110, 208, 219, 223, 230
Coordination Council on Drug Abuse (CCDA), 113–14
Cosby Show, The, 2, 25–28
Cox, Wendell, 170, 173

Davidson, David, 178, 183
Dearborn Heights, 149
Deindustrialization, 85, 86
Demography; change in Boston 11, 15, 65–66, 68–69, 71, 74, 126–27; change in Detroit, 11–12, 85–86
Deregulation, 61, 76
Detroit Cable Communications Commission, 99–100, 102, 105–8, 142, 162, 207, 209, 211, 214–27, 243–46
Detroit Cable Television Advisory Committee, 96–104, 107
Detroit City Council, 96, 97–98, 99, 101–3, 105, 107–8, 142–43, 149, 205–6, 209, 211, 215–17, 220, 221–25, 244, 256
Detroit Inner-Unity Bell Cable System (DIUB), 170–76, 205–10, 212–13, 215, 217, 219, 222
Detroit Medical Center Corporation, 147
Detroit Public Access Corporation, 206, 212, 215
Detroit Public Benefit Corporation, 171, 175
Detroit Public Schools, 147, 148, 162, 174
Detroit Public Television, 174
Detroit Symphony, 147
Detroit Urban League, 159, 163, 176
"Development without Displacement," 12
Dialogue, 140
Discrimination; media and, xviii, 16, 35, 249; FCC and, 51, 63; in Boston, 65–67, 71, Cablevision policy on, 204–5; in Detroit, 85–86, 92–93, 94, the RFP in Boston and, 100, 144, 145–46
Disposable income; *see* African American disposable income
Distant signal, 39, 42, 49, 92; *See also* FCC restrictions on the top 100 markets
Diversity, xvi; cable television and, 41, 42; in Detroit, 95, 102, 110, 155, 168, 180, 207; in Boston, 118, 124, 132, 138, 141, 198
Dolan, Charles F., 121–23, 137, 178, 179, 180, 181–83, 192, 198, 203, 234–36, 240
Dolan, Edward V., 57–59
"Don't Dump on Us," 12
Dorchester, 80, 118, 186, 233, 238, 240–41
Dornfeld, Barry, 252, 256
Douglas, Sharlan M., 101

Du Bois, W.E.B., 3, 5, 250–51, 256
Dudley Street Neighborhood Project, 12
Dunmore, George J., 209, 211

East Boston, 81, 118, 134, 185, 188, 202, 239, 241
Eberhard, David, 142, 211, 225
Economy; in the U.S., xvii, 53; white domination of, 2; African Americans and, 5, 15, 21, 23, 27, 30–31, 86; media industry and, 7, 21, 53; cable television studies on, 12, 15, 36, 40–47 passim, 50–51, 54, 57; in Boston, 65–66, 68, 126, 201; cable television in Boson and, 75–78, 80–81, 83, 112, 119, 130, 136, 189, 195, 231–34; in Detroit, 85, 93–94, 149, 219; cable television in Detroit and, 90, 84, 106, 154, 156–59, 162, 169–70, 221–22
Education; African Americans and, 2; cable television and, 36, 43, 46–48, 52, 55, 56–57, 60; media and, 60; in Boston, 65, 67–68; Mel King and, 71; cable television in Boston and, 72–74, 76, 79–80, 85, 112–13, 125–27, 128, 130–31, 139; in Detroit, 89–91, 92–93, 105, 106, 143, 206, 208, 209–10, 215; the RFP in Detroit and, 98, 145–46, 148; Rollins Cablevision and, 117; Times Mirror and, 118; Warner Amex and, 121, 136, 187, 192; Cablevision and, 123, 198, 199; Barden Cablevision and, 149–50, 152–53, 157, 159–60, 210, 214, 226–27; City Communications and, 163–64, 165, 167, 210; DIUB and, 174–75 210, 213; *See also* Programming; Training
Educational channel, 92, 131, 136
Educational institution, 36; in Boston, 72, 112–13, 117, 118, 121, 123, 126, 196, 187, 188–89; in Detroit, 106, 148, 157, 163, 174–75, 210, 213–14
Educational programming, 55–57 passim, 60; in Boston, 80, 130, 136; in Detroit, 145, 159, 164–65, 210
Educational service, 43, 85, 90, 143, 148, 208, 210, 226
Edwards, Carl, 92–93
Edwards, Esther G., 89, 212
Elderly population, 55, 83, 112, 115, 120–21, 127, 134, 184–85, 190–91, 193
ELRA Group, Inc., 107
Employment; African Americans and, 24, 26, 30, 109; cable television and, 44, 48–49, 59, 84, 110; in Boston, 67, 68, 114; cable television in Boston and, 76, 130, 131–32, 188, 200; in Detroit, 85, 90; cable television in Detroit and, 93, 94, 96, 101, 105, 143, 144, 145–46, 208–9, 214; Cablevision and, 115–16, 141, 196, 197, 204, 230–31, 234–36, 238, 239; Warner Amex and, 135, 186, 187; Barden Cablevision and, 150, 153, 157–58, 160–61, 215, 226–27; City Communications and, 164, 167, 169, 210; DIUB and, 172, 175
Empowerment, xvi, 2, 5–6, 12, 14, 36, 64, 70, 96, 105–6, 258
Entry barrier, 10, 42, 45
Epstein, Peter, 179, 183
Equipment, 2, 9, 10, 53, 54, 57, 61; Detroit and, 94; New York Times Corporation and, 116; Times Mirror and, 118–19; Cablevision and, 123, 139, 141, 190, 198, 231, 232, 237, 240, 241; Boston and, 125–26, 255; Warner Amex and, 135, 192; Barden Cablevision and, 150, 152, 155, 157, 159, 178, 206–8, 213, 225–28; City Communications and,

165, 168, 208, 210, 213; DIUB and, 174, 208, 213
Erasure, xvi, xviii, 1, 17, 26–27, 29, 35, 46, 52, 61
Ethnic Arts Channel, 147
Evaluation of Proposals for the Provision of Cable Communications Services in the City of Detroit, Michigan, 207–11
Evaluation of Proposals for the Provision of Cable Communications Services in the City of Detroit, Michigan: FINAL REPORT, June, 1983, 211–16
Eyes on the Prize, 13

Facility, 9, 10, 53; New York Times Corporation and, 116; Times Mirror and, 18–19; Tribune Company and, 119; in Boston, 83, 124, 125–26, 128–29, 132, 133, 178, 189, 255; Warner Amex and, 121, 134, 136; in Detroit, 144, 147, 152, 176, 226; Barden Cablevision and, 155–57, 159, 206–8, 210, 212, 227; DIUB and, 175–213; Cablevision and, 139–40, 178, 181, 182, 186, 195, 198–200, 222, 232, 237, 241; City Communications and, 165, 210, 213, 245; DIUB and, 174, 176, 210
Fanon, Franz, 4
Farrell-Donaldson, M. D., 223
FCC v. Midwest Video Corporation, 53
Federal Communications Commission (FCC), 9, 36, 38–39, 42, 45–46, 49, 50–51, 52–54, 58, 63, 82, 84, 91, 104, 107, 116, 170, 180, 221, 255; restriction on the top 100 markets, 39, 42, 45, 49; See also *FCC v. Midwest Video Corporation*, Control, Deregulation
Fee; minimization and affordability, 10, 96, 122; for Walson's system, 37; studies on, 38; Cablevision and, 122, 142, 182, 202, 233, 240; Warner Amex and, 134, 142, 191, 194; in Boston, 139, 189; City Communications and, 164; Barden Cablevision and, 178, 212, 215, 225; *See also* Installation fee
Final Cable Television License to Cablevision of Boston, 203–5
Final license; See *Final Cable Television License to Cablevision of Boston*; *Evaluation of Proposals for the Provision of Cable Communications Services in the City of Detroit, Michigan: FINAL REPORT, June, 1983*; *City of Detroit Cable Communications Service Franchise Agreement as Amended and Restated*
Finn, Daniel J., 196, 198
Film industry, 2, 18–24, 55, 249
Flynn, Raymond L., 234–39, 242
Foundation 70, 72
Frank, Roland E., 54–55
Frustrations, 10, 43; among Detroit residents, 98, 101, 221–23; among Boston residents, 240
Fuller, Linda K., 60–61

Gallery, 123
Garrity, Jr., Arthur, 65
Gay community, 84, 188
Gone with the Wind, 19, 166
Gospel Time, 48
Greater Roxbury Community Development Corporation, 141
Greatest Place to Be, the, 242
Green, Robert L., 162, 209
Greensberg, Marshall G., 54–55
Grigsby, Charles, 78–79
Guess Who's Coming to Dinner, 20

Hall, Stuart, 22–23
Hampton, Henry, 13
Harrison, Alta, 105–7, 143, 211, 225
Hauser, Gus, 183–85

318 Index

Hearing, 253, 258; by Foundation 70, 72; in May 1971, 72–73; in October 1972, 78–80; in June 1981, 142–43, 177, 179–91, 192, 193–94, 204; in December 1983, 231–32; in April 1986, 242; in Boston, 102, 107, 114, 124, 238, 242; in Detroit, 107, 206, 211, 215, 223, 243
Henderson, Erma, 89, 103, 104, 142, 206, 211, 215, 222, 225
Hetzel, Otto, 89, 127, 143, 211, 225
Hiring; *See* Employment
Hollywood, 2, 18–22
Hood, Nicholas, 142, 211, 225
Housing, 26, 30, 43; in Boston, 76, 87, 124–25, 127, 240; *See also* Boston Housing Authority
Hudson, Lester D., 161–62
Hull, Marrion H., 47–49
Humphrey, Hubert, 167
Humphrey Occupational Resource Center, 140
Hyde Park, 65, 118, 238

Income; *See* Disposable income; Low income residents
Infrapolitics, 4–5, 9, 259
Inkster, 148–49, 151, 224
Installation, 46; in Boston, 111, 196; Warner Amex and, 120, 134, 191; Cablevision and, 122, 201–2, 203; in Detroit, 146; Barden Cablevision and, 154, 227; City Communications and, 164
Internship, 176, 255; in Boston, 136, 235; in Detroit, 150, 169, 174, 208, 210
Interview; by WJBK-TV2, 88; of Candidates for the Detroit Cable Communications Commission, 105–7; by ELRA Group, 107, by Warner Amex, 120; by Peter D. Hart Research Associates, 138–39, by City Communications, 169
Inquilinos Boricuas en Accion, 136
Issuing Authority Final Decision and Statement of Reasons, 193–94
Issuing Authority Report, 129

Jackson, Jesse, 13, 162
Jamaica Plain, 80, 118, 189, 200, 238, 240–42
Jean, Martha, 14
Jessup, Hubert, 239–42
Johnson, Dawson V., 186
Johnson, Leland, L., 51
Johnson, Robert, 1, 32, 256–58
Johnson, William T., 163
Jones, Dan, 111–12, 178–79, 183
Jones, Hubie, 65
Jordan, Joseph R., 105–7, 143, 211, 225

Kalba, Konrad K., 73–77
Kelley, Jack, 142, 211, 225
Kessel, Martin, 111–12, 132–33, 196, 232
King, Coretta Scott, 162, 209
King Jr., Martin Luther, 3, 25, 29, 33, 252–56
King, Melvin, 13, 69–72, 75–77, 138, 182–83, 239, 241–42, 254, 257
King, Michael, 241
King III, Oscar W., 170, 173
Kiska, Tim, 206
Kletter, Richard C., 43, 51

Lamont, Ned, 180–82
Latino, 110; in Boston, 66, 68, 127, 136, 188, 200, 204; in Detroit, 146
Leatherman-Massey, Jean, 107, 143, 211, 25
Lee, Spike, 22–24, 32–33, 71
Legal studies, 36, 38–39, 52–60 passim

Leporati, Raymond A., 152, 212
Like It Is, 55
Local origination channel, 60, 255, 258; in Boston, 121, 138, 140–42, 188, 192, 200; in Detroit, 155–56, 159, 175, 207–8, 212–13
Local origination programming, 11–12, 14, 26, 47, 51, 60, 69, 109–10, 252, 255–58 passim; in Boston, 74, 76, 82–83, 128, 131, 132, 135–37 passim, 139–42 passim, 179; in Detroit, 90–91, 95, 145, 210; Boston Cablevision Systems and, 115; New York Times Corporation and, 116; Times Mirror ad, 118; the Tribune Company and, 119–20; Warner Amex and, 121, 134, 135, 192; Cablevision and, 123, 128, 180, 187, 190, 197, 199–200, 203, 204, 232, 238; facilities for, 121, 134, 156, 187, 192, 208, 212, 213; Barden Cablevision and, 150, 155–56, 158–59, 206, 208, 212, 213, 215–16, 228; City Communications and, 163, 165–66, 207–8; DIUB and, 173–76, 208, 213
Low income residents; in Boston, 12, 76–77; cable television's benefit for, 39; in Detroit, 96, 106

MacDonald, J. Fred, 2, 31
MacLean-Hunter Cable TV, 218–19, 243–45
Maddox, Gilbert A., 163, 215
Madison, Joseph E., 152, 206
Mahaffey, Maryann, 98, 101, 105, 107, 206, 209, 217; and Alta Harrison, 105; and RFP, 142; and the review process, 206–7; and final report, 211; and final agreement, 225
Mahony, Sheila, 178, 180–82, 188, 192

Maintenance; in Detroit, 93–94, 146; Barden Cablevision and, 151, 227; City Communications and, 166; Cablevision and, 196, 198, 203, 236, 237
Malarkey, Taylor & Associates, Inc., 189–93
Manufacturing industry, 11; in Boston, 65, 68, 126; in Detroit, 85–86, 93, 158
Massachusetts Institute of Technology (MIT), 13, 69, 73, 77, 81, 129, 178, 183
Mattapan, 67, 81, 118, 233, 238
Mayor's Office of Cable Communications, 125, 129, 200, 234
McCausland, Robert, 187–88
McGhee, Reginald D., 93–94
Meade, Peter, 178–79, 183, 187
Media affirmative action, 17
Media Council, 96–97
Miller, David M., 152–53
Minorities, 16, 39, 41; negative impact of media on, 3; cable television's foreseen benefits for, 8, 36, 44, 59, 229; the radio industry in Detroit and, 14; fair representation through cable television for, 15, 246–47, 253; cable content produced by 35, 257; lack of attention to in studies, 38, 39, 62; emphasis on in studies, 44, 110; institutions for, 44, 47–48, 161, 169; public access for, 50–51, 56, 61; differences within, 55; educational discriminations against, 67; access to Boston cable television, 69; Mel King and, 69; concerns in Boston, 74; cable television in Detroit and, 95, 106; 146–48, 154; lease channels for, 101, 155, 164; Times Mirror and, 118; Warner Amex and, 120–21, 135–37, 185–

320 Index

Minorities *(continued)*
    86, 192; Cablevision and, 140–41,
    180, 190, 202, 204, 231–39 passim;
    Barden Cablevision and, 149–50,
    153–55, 158–61 passim, 208–9,
    215, 217, 227, 228, 231, 244;
    City Communications and, 164,
    167, 168–69, 208–9; DIUB and,
    172–73, 175, 206, 208–9
Minority audience, 45, 50, 58–59, 60,
    255; in Detroit, 95, 110; in Boston,
    114
Minority contractor, 135, 150,
    153–54, 217, 233, 236
Minority employment, 44–45, 48, 49,
    255; in Boston, 84, 130–32, 198; in
    Detroit, 92–94, 98, 105, 107, 110,
    144, 146–47, 207, 208, 214; Boston
    Cablevision Systems and, 116;
    New York Times Corporation and,
    116; Cablevision and, 141, 231,
    233, 235, 236–38, 239; Barden
    Cablevision and, 149–50, 153–54,
    158, 161, 208–9, 215, 217, 227,
    231; City Communications and,
    167, 169, 208–9; DIUB and, 172,
    208–9; Warner Amex and, 185–86
Minority investor, 212–13
Minority majority, 11
Minority ownership; 14, 48–49, 59,
    78, 106; of business in Boston,
    132, 198, 231, 235; of business in
    Detroit, 144, 153–54, 158, 168–69,
    173, 228, 255; *See also* Barden
    Cablevision
Minority participation in cable
    television, 47, 48–49, 110–11,
    253; in Boston, 78; in Detroit,
    93, 100–1, 106–7, 111, 144–46;
    Warner Amex and, 120, 135, 185;
    Cablevision and, 140–41, 239;
    Barden Cablevision and, 150, 154,
    158, 244; City Communications
    and, 168; DIUB and, 172, 206

Minority population; in Boston, 8, 36,
    44, 59, 229; in Detroit, 90, 106,
    110, 188, 189
Minstrelsy, 18, 250
Morgan, Lester, 107, 143, 211
Movie industry; *See* Film industry

Narrowcasting, 2, 8, 41, 49, 64, 69,
    74, 87, 199, 249, 257–58
National Association for the
    Advancement of the Colored People,
    5, 44, 70; in Detroit, 149, 152,
    156, 161, 162–63, 168–69, 172,
    176, 206, 209, 217; in Boston, 200
National Black Media Coalition, 47
National Science Foundation, 50, 83
National Telecommunications and
    Information Administration (NTIA),
    125, 126, 152
National Urban League, xii, 5, 44, 47,
    161, 169, 172
New Detroit Urban League, 168
New York Times Corporation, 115–16
Noble, Gil, 55–56, 256
North End, 65
Northeastern University, 13, 80, 196

Occupational opportunity. *See*
    Employment
Omnibus service, 192, 202
Ordinance; in Boston, 72–73; in
    Detroit, 97–98, 99–104, 143, 206,
    215, 216, 219, 220–21, 223–25,
    244–45
Ownership; African American
    media production and, 32, 257;
    cable television and, 48–49, 59,
    63; in Boston, 66, 75–77, 83,
    246; Roxbury Cablevision and,
    78–79, in Detroit, 90, 93–94,
    100, 104, 106, 143–44, 145,
    246; Barden Cablevision and,
    150, 153, 212, 219, 243, 245;
    City Communications and, 163,

212, 245; DIUB and, 170, 212, Cablevision and, 186, 189

Participatory democracy, 43
Participatory media, 253
Peoples, John, 142–43, 211, 225
Perspective, 123, 140
Pincus, Lois P., 89, 97–98
Pluralism *See* Diversity
Poitier, Sydney, 20–31, 26, 32
Policymaking, 47, 84, 107, 181, 233
Post-racial society, xvii, 32; *see also* Colorblind
Poverty, 22, 26, 31, 85–86, 96, 150, 156, 167, 251
Powell, Josephine A., 215, 223
Prejudice; *See* Stereotype
Price, Monroe E., 42–44
Procurement; in Boston, 131–32; DIUB and, 173; Warner Amex and, 185–88; Cablevision and, 185–86, 197–98, 231, 236, 239; Barden Cablevision and, 219, 226–27
Production; of content, 9, 49, 112, 114, 156, 241; visual culture and, 13, 17, 21, 32, 50, 249, 251–58 passim, in the television industry, 14, 30, 31–32; in the film industry, 8, 21–23, 166; training for, 9, 123, 136, 159, 200, 226, 237; facility for, 10, 119, 136, 156, 165, 181, 186; minorities and, 35, 39, 40, 55, 96, 106, 140; quality of, 54, 60; Mel King and, 69–72; African Americans and, 75, 96; exclusion of African Americans from, 84, 96; community access programs and, 105, 115, 135, 137; equipment for, 121, 135, 141, 159, 210, 212, 225; of local origination programs, 137, 140, 155, 163, 165, 181, 187; employment opportunities in, 146; institutions and, 163; City Communications and, 166–69

passim; DIUB and, 174, 210; Cablevision and, 181, 186, 187, 190, 198, 200, 202, 237; Warner Amex and, 185; Barden Cablevision and, 212, 221, 225
Profitability, 10, 38, 39, 58, 199
Programming; for education, 55–57 passim, 60; in Spanish, 55, 84, 136, 140, 172, 190; in a non-English language, 82, 115, 120, 140, 190, 192, 237; *See also* Access programming; Commercial Programming; Educational programming, Local programming, Sports
Procurement, 131–32, 173, 185–86, 197–98, 219, 226–27, 231, 236, 239
Provisional Cable Television License: Granted to Cablevision Systems Boston Corporation, 196–200, 200–1, 203–4
Psychology, 3, 5–6, 12, 29, 31, 40, 66, 167–69, 249
Public access; *See* Access channel

QUBE; *See* Warner Amex Cable Communications Company

Race film, 18
Racism, xvi, xvii–xviii, 2, 6, 16, 21, 24, 30–31, 32, 65, 71, 86, 148, 250; *See also* Discrimination
Rand Corporation, 15, 44, 50, 95
Rate; *See* Subscription; Fee
Ravitz, Mel, 143, 211, 218, 219, 221, 222–24, 225
Readville, 118
Reed, Gregory J., 151, 173
Regulation; by FCC, 38, 39, 42–54 passim, 60, 61, 73, 76, 79, 91; by state, 73, 79, 98; in Boston, 116, 197–98, 205; in Detroit, 156; *Also see* Federation Communications Commission

Religion, 44, 70; cable television in Boston and, 116, 126, 141, 191, 193, 205; cable television in Detroit and, 165, 172, 226–27
Religious institutions, 12, 50, 57–58, 81; in Detroit, 12, 88–89, 106–7, 150, 153, 161–62, 169, 175; in Boston, 205
Representation; See Stereotype
Request for Proposals; See *Issuing Authority Report*; *Request for Proposals for the Provision of Cable Communications Services in the City of Detroit, Michigan*
Request for Proposals for the Provision of Cable Communications Services in the City of Detroit, Michigan, 142–48
Resident's Attitude Study, 81
Residential zoning; See segregation
Resistance, 6, 16, 27, 35, 53, 251
Reregulation, 39, 42, 45, 49–50, 61
Riot; in 1943, 12, 85; in 1967, 12, 14, 85–88
Rivera, Thomas, 78–79
Robinson, James P., 107, 143, 211, 214, 223, 225
Rollins Cablevision of Boston, Inc., 115, 117
Roots, 2, 25, 30–33
Roslindale, 65, 118
Roxbury, 67, 69, 71, 74, 78–80, 127, 203, 233, 237, 241
Roxbury Cablevision, 78, 241
Roxbury Community College, 118, 200
Roxbury Task Force on Cable TV, 233

Salary, 59, 239
Sapan, Joshua, 57–58
Satellite, 56, 82; Times Mirror and, 119; Cablevision and, 122, 139, 141, 202; Warner Amex and, 134, 137; Barden Cablevision and, 154–55, 159, 213; City Communications and, 164–65; DIUB and, 175
Say Brother, 13

Scar of the Shame, The, 20
Schaffner, Maureen F., 77–78
Segregation 3, 249, in Boston, 65–66, 68, 70; in Detroit, 85
Self-esteem, 6–7, 31, 33, 71, 72, 76, 86–87, 105, 214, 219, 249, 251
Self-help, 12–13, 57, 87, 192
Self-representation, 1, 64, 76
Self-sufficiency, 13, 70
Service structure; See Tier structure
Sights and Sounds of the City, 166
Signal; for off-the-air television, 37, 42, 43, 74, 90, 92, 131, 138, 164, 197, 213; from outside of a community, 42; See also Distant signal
Smith, David, 78
Somerville, 15
Sounds, 123
South Boston, 67, 74, 118, 186, 200
South End, 69, 71, 80, 118, 141, 200
Southeast Michigan Community Needs '70, 87, 92
Spanish News Network, 140, 190; See also Programming
Sports; in Detroit, 108, 150, 155–56, 164, 166–67; in Boston, 118–19, 120, 130, 135, 140, 194, 200, 239
Spring, Micho, 179, 182–83, 186, 196
Stereotype, 2, 18, 25, 28, 32, 50, 62, 75, 95, 127, 253; See also Minstrelsy; Racism
Strictly Speaking, 14
Subscription, 10, 61, 96, 108; Warner Amex and, 121, 191–92; Cablevision and, 122, 138, 182, 194, 201–2, 240; City Communications and, 164; DIUB and, 175; Barden Cablevision and, 178
Sugrue, Thomas J., 85–86
Surveillance, 40–41

Tatta, John, 121, 137, 178, 180, 203

Index 323

Television; broadcasting and, 7, 36, 41, 47, 56, 59, 95, 97, 117, 130, 139, 146, 155–56, 249, 253; network television, 2–3, 5, 29, 38, 41–42, 59, 95; African American viewing habit and, 25, 111; *See also* Satellite Television industry, 24–32, 73, 144, 249

Thompson, Arthur, 231, 236

Tier structure 189, 207; Warner Amex and, 120–21, 134, 191–92; Barden Cablevision and, 154–55, 207, 212, 213; City Communications and, 164, 212; Cablevision and, 189–90; 192–93; DIUB and, 208, 212–13

Till, Emmett, 70, 251–52

Times Mirror Cable Television of Boston, Inc., 115, 117–19, 152

Tokenism, 31, 51, 55, 90, 118

Training 2, 9, 10, 36, 51, 53, 59, 92–93; little need in the cable industry, 49; reason for unemployment, 49; in Detroit, 92–93, 101, 105, 144–46, 147, 176, 214–15, 219; in Boston, 114, 125, 129; Boston Cablevision and, 116; Times Mirror and, 118; Warner Amex and, 136, 185; Cablevision and, 141, 196–200, 200–1, 231, 134–35, 238; Barden Cablevision and, 150, 156–61, 210, 227, 244; City Communications and, 164, 167–69; DIUB and, 172–47; *See also* Vocational program

Tribune Cable of Boston, 115, 119–20

Turner, Michael E., 101–2, 108, 211, 215, 217, 220–21, 223, 224, 225

TV 9 case, 63

UHF, 42, 49

Unemployment, 43, 85–86, 89, 94–95, 108, 144, 157, 164, 211, 219

United Cable Television Corporation, 53, 54

United Community Services of Metropolitan Detroit, 147

University Cultural Center Association, 147, 174

Urban Communications Institute of City Communications, 169

Visual culture, 2, 12, 13, 17, 32–33, 35, 61, 70, 249–50, 253–54, 257 *See also* Discrimination

Vocational program; 157, 198, 201, 210

Wage, 66, 68

Wayne County Community College, 160, 174, 210

Wayne State University, 89, 105–7, 147, 163, 174, 210

Walson, John, 37

Ward, John E., 178, 183

Warner Amex Cable Communications Company; preliminary application and, 115, 120; financial condition of, 117, 184, 192–93; perceived indifference to local minority interests, 118–21; difference from Tribune Company, 120; QUBE, 120–21, 133–35, 192; elderly audience and, 120–21, 134, 184–85, 190–93; service structure of, 120–24, 134, 142, 191–92, 195; local origination and, 121, 135, 142; identified interest groups, 121; amended application and, 133, 178, 189, existing systems of, 133, 178, 190–91; facilities and, 134–35, 141–42, 187, financial investment of, 135, 142, 187, 195; access programs and, 135, 141–42, 185–86; minority involvement and, 135–37; education and, 136; similarity with Cablevision, 141; site visit by city officials and, 178–79; public hearings attended by, 179–81, 183–87, 189; construction schedule of, 188, 238; loss to Cablevision, 194–95

Webb, Myrna, 106–7, 143, 211, 225
West Roxbury, 65, 118, 203
WGBH, 13, 81, 128, 188
WGPR, 14
White, Kevin; documents submitted to before the issuing of CFP, 77, 115, 123; disinterests in cable television, 80, 81–82, 111–12; appointing Cable Television Review Commission, 81–82; correspondence with Times Mirror, 117; disinterest in minorities 120; meeting with Cablevision, 121; reviewing proposals, 124; statements on cable television, 124, 130, 131–32; Cable Access Advisory Committee and, 128; issuing of the CFP, 129; Access Corporation and, 130, 133, 195–96; correspondence with a banker, 184; franchise award, 193–94, 196, 203; Warner Amex and, 195
White flight, 11, 13, 85, 127
Wicks, Jr., David O., 180–81, 184
WJBK-TV2, 87–88
WJLB, 14
Women; degraded visual images of, 33; programs for, 55; occupations of, 66; employment in Boston and, 82, 84, 131, 135, 141, 185, 198, 200–1, 204–5, 231, 234–38 passim; employment in Detroit and, 101, 144, 150, 158, 160–61, 162, 167–68, 172–73, 206, 208, 228, employment opportunities for, 110; programs in Boston for, 115, 117, 121, 131; as the head of a household, 127; channels for 136, 190–91, unemployment of, 144; programs in Detroit for, 155, 168
Woodson, Carter G., 5, 90–91, 250, 256

Young, Coleman, 13, 85, 96, 108; RFP and, 99, 142; causing delay in decision-making, 103–4; selection of Cable Communication Commissioners and, 106–7; review of proposals and, 205–9, 214; awarding franchise to Barden and, 215, 217–20, 222, 244
Youth, 105, 115, 140, 141, 153, 162, 167–68

Zeff, A. Robert, 161